Ergebnisse der Mathematik und ihrer Grenzgebiete

Band 63

Herausgegeben von P. R. Halmos · P. J. Hilton
R. Remmert · B. Szőkefalvi-Nagy

Unter Mitwirkung von L. V. Ahlfors · R. Baer
F. L. Bauer · A. Dold · J. L. Doob
S. Eilenberg · M. Kneser · G. H. Müller
M. M. Postnikov · B. Segre · E. Sperner

Geschäftsführender Herausgeber: P. J. Hilton

Derek J. S. Robinson

Finiteness Conditions and Generalized Soluble Groups

Part 2

Springer-Verlag Berlin Heidelberg New York
1972

Derek J. S. Robinson
University of Illinois
Urbana, Ill.

AMS Subject Classifications (1970):

Primary 20E15, 20E25, 20E99
Secondary 20F30, 20F35, 20F45, 20F50, 20F99

ISBN 978-3-642-05712-0

Preface

This book is a study of group theoretical properties of two disparate kinds, firstly finiteness conditions or generalizations of finiteness and secondly generalizations of solubility or nilpotence. It will be particularly interesting to discuss groups which possess properties of both types.

The origins of the subject may be traced back to the nineteen twenties and thirties and are associated with the names of R. Baer, S. N. Černikov, K. A. Hirsch, A. G. Kuroš, O. J. Schmidt and H. Wielandt. Since this early period, the body of theory has expanded at an increasingly rapid rate through the efforts of many group theorists, particularly in Germany, Great Britain and the Soviet Union. Some of the highest points attained can, perhaps, be found in the work of P. Hall and A. I. Mal'cev on infinite soluble groups.

Kuroš's well-known book "The theory of groups" has exercised a strong influence on the development of the theory of infinite groups: this is particularly true of the second edition in its English translation of 1955. To cope with the enormous increase in knowledge since that date, a third volume, containing a survey of the contents of a very large number of papers but without proofs, was added to the book in 1967. Despite this useful addition and the books of M. Hall, E. Schenkman and W. R. Scott, which deal with finite as well as infinite groups, there is a clear need for a detailed account of the theory of finiteness conditions and of generalized soluble and nilpotent groups. The present work represents an attempt to meet this need.

I have sought to collect the most important results in the theory, which are scattered throughout the literature, and to present them in a compact and accessible form with improved and shortened proofs wherever possible. The original aim was to supply full proofs of all the theorems mentioned. However, strict adherence to this rule would probably have doubled the length of the book—it is now possible to write a complete book about locally finite groups for example. Accordingly a compromise has been reached by which certain results are, for one reason or another, stated without proof or, in a few cases, with abbreviated proofs. Of course the choice of what detail to present is in the end a personal one with which the reader may take issue.

Concerning the contents of the book, I might mention that Chapter 1 is of a rather more general nature than is strictly necessary: this has the twofold advantage of supplying a unified basis and of collecting many definitions and elementary results which might otherwise have impeded the development of later chapters. Part of Chapter 2 is also elementary: thereafter the chapters speak for themselves; their admittedly ephemeral aim is to bring the reader abreast of current developments.

Generally speaking, standard notation has been employed. I have followed, albeit reluctantly, the alphabetic notation of Kuroš and Černikov for classes of generalized soluble and nilpotent groups (SN, SI, Z etc.). It would seem a formidable task to find appropriate names for so many classes, although most of these can be rendered symbolically by means of Philip Hall's class operations (which are frequently employed here). However, where suitable terms exist they have been adopted. Thus, following Baer, I prefer "hyperabelian" to "SI^*" and "hypercentral" to "ZA". I might add that the word "series" is used here in contexts where some authors would employ "normal series" or "normal system". Also it seems preferable to speak of a "residual class" rather than a "semi-simple class".

Needless to say, the reader is expected to have a good basic knowledge of the theory of groups, including abelian groups, and of some of the more familiar parts of commutative algebra and ring theory. Otherwise the presentation is, with very few exceptions, self-contained. As a general rule a chapter will depend upon previous chapters in that results established therein may be used.

It is a pleasure to record my gratitude to the many friends and colleagues who have contributed their advice and criticism or who have supplied important information which might otherwise have been passed over. In particular I thank B. Amberg, R. Baer, P. Hall, K. A. Hirsch, K. W. Gruenberg, O. J. Kegel, J. E. Roseblade, S. E. Stonehewer and B. A. F. Wehrfritz. It was at Professor Baer's suggestion that this work was undertaken and I greatly appreciate the immense care and interest which he has taken in reading the manuscript: this has led to numerous improvements. In addition Dr. Roseblade has made many perceptive remarks on the manuscript. My debt to Professor Hall is that of student to teacher; it was through his lectures at Cambridge that I first became interested in the theory of groups and from them that I gained much of my knowledge.

My thanks are due to Springer-Verlag whose patience and cooperation have been all that one would expect of them. In addition I thank the National Science Foundation of the United States and the Department of Mathematics of the University of Illinois in Urbana for support and assistance during the period of writing.

This book is dedicated to my wife Gabriele in recognition of her advice, assistance and often-tried patience, all of which have greatly lightened the labour inherent in such a project as this.

Urbana, Illinois, January 1972 Derek Robinson

Contents

Contents of Part 1

Index of Notation

Page numbers in brackets refer to the original definition in the text. The symbols I and II refer to Part I and Part II.

(i) Classes and functions

$\mathfrak{X}, \mathfrak{Y}, \mathfrak{Z}, \ldots$	classes of groups (I: 1)
$\mathfrak{X} \leq \mathfrak{Y}$	$\mathfrak{X}$ is a subclass of $\mathfrak{Y}$
$\mathfrak{X}_1 \cdots \mathfrak{X}_n$	extension class (I: 2)
$\mathfrak{X}\mathfrak{Y}_1^{(i_1)} \cdots \mathfrak{Y}_r^{(i_r)}$	(II: 168)
χ	a subgroup theoretical class or property (I: 9)

Special classes of groups

$\mathfrak{F}, \mathfrak{F}_\pi, \mathfrak{P}, \mathfrak{G}, \mathfrak{C}, \mathfrak{A},$ $\mathfrak{N}, \mathfrak{S}, \mathfrak{I}, \mathfrak{D}$	(I: 1)
$\mathfrak{B}(W)$	(I: 7)
$\mathfrak{B}, \mathfrak{B}_i$	(I: 173)
$\mathfrak{I}$	(I: 174)
$\mathfrak{E}, \mathfrak{E}_n$	(II: 42)
$\mathfrak{U}_s, \mathfrak{U}_{s,n}$	(II: 69)
$\mathfrak{S}_0, \mathfrak{S}_1$	(II: 128, 137)
$\hat{\mathfrak{M}}, \check{\mathfrak{M}}$	(II: 165—166)
Max-f, Min-f	maximal and minimal conditions on f-subgroups (I: 37)
$\boldsymbol{A}, \boldsymbol{B}, \boldsymbol{C}, \ldots$	operations on classes of groups (I: 3)
$\boldsymbol{A} \leq \boldsymbol{B}$	$\boldsymbol{A}\mathfrak{X} \leq \boldsymbol{B}\mathfrak{X}$ for all $\mathfrak{X}$
$\boldsymbol{AB}$	product of operations $\boldsymbol{A}$ and $\boldsymbol{B}$ (I: 3)
$\boldsymbol{A}^\alpha$	ordinal power of $\boldsymbol{A}$ (I: 13)
$\langle \boldsymbol{A}_\lambda : \lambda \in \Lambda \rangle$	closure operation generated by operations $\boldsymbol{A}_\lambda$ (I: 5)

Special operations

$\boldsymbol{I}, \boldsymbol{U}, \boldsymbol{S}, \boldsymbol{S}_n, \boldsymbol{H}, \boldsymbol{P}, \boldsymbol{D},$ $\boldsymbol{D}_0, \boldsymbol{N}, \boldsymbol{N}_0, \boldsymbol{R}, \boldsymbol{R}_0$	(I: 4)
$\boldsymbol{L}, \boldsymbol{L}_c$	(I: 5)
$\grave{\boldsymbol{P}}, \acute{\boldsymbol{P}}, \hat{\boldsymbol{P}}, \grave{\boldsymbol{P}}_{sn}, \acute{\boldsymbol{P}}_{sn}, \hat{\boldsymbol{P}}_{sn},$ $\grave{\boldsymbol{P}}_n, \acute{\boldsymbol{P}}_n, \hat{\boldsymbol{P}}_n$	(I: 12—13)
$\acute{\boldsymbol{N}}$	(I: 19)
$\boldsymbol{M}, \overline{\boldsymbol{M}}$	(I: 36, 62)

$\mathfrak{X}^{A}$, $\mathfrak{X}^{-A}$	largest A-closed classes contained in and disjoint from $\mathfrak{X}$ (I: 6—7)
Rad $\mathfrak{X}$, Res $\mathfrak{X}$	the radical and residual classes generated by $\mathfrak{X}$ (I: 20, 23)
Cat $\mathfrak{X}$, $\mathfrak{X}^{(\alpha)}$	classes of groups with an $\mathfrak{X}$-category (I: 36—37)

(ii) Elements and groups

$x, y, z, \ldots$	group elements
1_G, 1	identity element of a multiplicative group
x^y	$y^{-1}xy$
x^{-y}	$(x^y)^{-1}$
x^{y+z}	$x^y x^z$
$[x_1, \ldots, x_n]$	(left normed) commutator (I: 42)
$[x, {}_n y]$	$[x, \underbrace{y, \ldots, y}_{\leftarrow n \rightarrow}]$
$G, H, \ldots$	groups, rings, sets, etc.
$H \leq G$, $H < G$	H is a subgroup and H is a proper subgroup of the group G
$H \lhd G$	H is a normal subgroup of the group G
$H \lhd^{\Sigma} G$	(I: 12)
H *sn* G, H *asc* G, H *desc* G and H *ser* G	H is a subnormal, ascendant, descendant or serial subgroup of G (I 12)
$\langle X_\lambda : \lambda \in \Lambda \rangle$	subgroup generated by the subsets X_λ
$\langle X : R \rangle$	group with generators X and relators R
$X_1 \cdots X_n$, $\prod_{\lambda \in \Lambda} X_\lambda$	products of subsets of a (multiplicative) group
$X_1 \cdots X_n$	additive subgroup generated by all $x_1 \cdots x_n$, $x_i \in X_i$, where X_i is a subset of a ring
X^{-1}	set of all x^{-1}, $x \in X$, where X is a subset of a group
G^n	subgroup generated by all nth powers g^n where $g \in G$ and G is a group
R^n	additive subgroup generated by all $r_1 \cdots r_n$, $r_i \in R$, where R is a ring
$\|G\|$	cardinality of G
$\|G : H\|$	index of H in G
$C_G(H)$, $N_G(H)$	centralizer and normalizer of H in G, — where H is a G-operator group
Core$_G$ H	core of H in G (I 60), — where H is a G-operator group
Aut$_G$ H	group of automorphisms of H induced by elements of G — where H is a G-operator group
Aut G, Inn G	automorphism and inner automorphism group
End G	set of all endomorphisms of G
Hom(G, H)	set of all homomorphisms of G into H
R^*	multiplicative group of units of a ring R
RG	group ring of a group G over a ring R
$H \times K$, $H \oplus K$, $\mathrm{Dr}_{\lambda \in \Lambda} H_\lambda$	direct products or sums
$H \otimes K$	tensor product
$H * K$, $\mathrm{Fr}_{\lambda \in \Lambda} H_\lambda$	free products
$H \wr K$, $\mathrm{Wr}_{\lambda \in \Lambda} H_\lambda$	wreath products (II: 18ff.)
$[X_1, \ldots, X_n]$,	commutator subgroup (I: 43)
$\gamma X Y^n$	$[X, \underbrace{Y, \ldots, Y}_{\leftarrow n \rightarrow}]$

X^Y	normal closure of X in Y (I: 42)
X^y	$y^{-1}Xy$
$X^{G,\alpha}$	αth term of the series of successive normal closures of X in G (I: 173)
$\{\Lambda_\sigma, V_\sigma: \sigma \in \Sigma\}$	series of order type Σ (I: 9–10)

(iii) Special subgroups

$G' = [G, G]$	derived subgroup of G
$G^{(\alpha)}$	αth term of the derived series of G (I: 45)
$\gamma_\alpha(G)$, $\gamma_\alpha^\chi(G)$	αth terms of the lower central and lower χ-central series (I: 29)
$\zeta(G)$	centre of G
$\zeta_\alpha(G)$, $\zeta_\alpha^\chi(G)$	αth terms of the upper central and upper χ-central series (I: 28)
$\bar{\gamma}(G)$, $\bar{\gamma}^\chi(G)$	hypocentre and χ-hypocentre (I: 29)
$\bar{\zeta}(G)$, $\bar{\zeta}^\chi(G)$	hypercentre and χ-hypercentre (I: 28)
$\varrho_{\mathfrak{X}}(G)$, $\varrho^*_{\mathfrak{X}}(G)$	$\mathfrak{X}$-radical and $\mathfrak{X}$-residual (I: 18)
$W(G)$, $W^*(G)$	verbal and marginal subgroup determined by a word or set of words W (I: 8, 9)
$L(G)$, $\bar{L}(G)$	sets of left and bounded left Engel elements of G (II: 40)
$R(G)$, $\bar{R}(G)$	sets of right and bounded right Engel elements of G (II:40)
$\varrho(G)$, $\bar{\varrho}(G)$	(II: 57)
$\omega(G)$	Wielandt subgroup of G (I: 177)
$\sigma(G)$	upper-finite radical of G (II: 180)

(iv) Miscellaneous

$GF(q)$	Galois field with q elements
$M(n, R)$	ring of all $n \times n$ matrices over the ring R
$GL(n, R)$	general linear group of all invertible $n \times n$ matrices over the ring R
$U(n, R)$	group of all upper unitriangular $n \times n$ matrices over the ring R
$M(\Lambda, F)$	McLain's group (II: 14)
$B(m, n)$	the Burnside group with m generators and exponent n (I: 35)
$\mathfrak{A}(J, \pi)$	a special class of J-modules (II: 146)
$\binom{n}{r}$	binomial coefficient
$[x]$	integral part of x
π'	complementary set of primes to π

Note to the Reader

A decimal classification is used throughout this work. Thus, for example, Theorem 6.29 and Theorem 6.29.1 are respectively the ninth and tenth results in the second section of Chapter 6. References to the bibliography are indicated by numbers in square brackets. As a general rule where the work cited contains a dozen or more pages, a page, section or theorem number is given with the reference. When a result is not ascribed to any author, either it is common knowledge or I am unaware of a source in the literature. Group theoretical classes are denoted by Gothic capitals unless customary usage dictates otherwise. Operations on classes appear as boldface capitals.

Chapter 6

Generalized Nilpotent Groups

Let $\mathfrak{X}$ be a class of groups: by *a class of generalized $\mathfrak{X}$-groups* we mean a class of groups $\mathfrak{Y}$ such that every $\mathfrak{X}$-group is a $\mathfrak{Y}$-group and such that every finite $\mathfrak{Y}$-group is an $\mathfrak{X}$-group, i.e.

$$\mathfrak{F} \cap \mathfrak{Y} \leqq \mathfrak{X} \leqq \mathfrak{Y}.$$

Thus a class of generalized finite groups is a class of groups satisfying a finiteness condition. In this chapter we will study *classes of generalized nilpotent groups* or, equivalently, group theoretical properties that are possessed by all nilpotent groups and which for finite groups imply nilpotence.

6.1 A Survey of Classes of Generalized Nilpotent Groups

We begin by describing the main classes of generalized nilpotent groups and the relationships which subsist between these classes. Most of the classes either contain or are contained in the class of locally nilpotent groups and we divide them accordingly into two categories, beginning with those that are subclasses of the class of locally nilpotent groups.

I. Classes of Locally Nilpotent Groups

(a) Fitting Groups, Groups with an Abelian Category, Baer Groups and Gruenberg Groups

We recall that

$$\boldsymbol{N}\mathfrak{A} \quad \text{and} \quad \acute{\boldsymbol{N}}\mathfrak{A}$$

are respectively the classes of *Baer groups* and *Gruenberg groups.* By Lemma 2.34 a group is a Baer (Gruenberg) group if and only if each finitely generated subgroup is subnormal (ascendant). Every Baer group is clearly a Gruenberg group and every Gruenberg group is locally nilpotent by the $\acute{\boldsymbol{N}}$-closure of the class $\boldsymbol{L}\mathfrak{N}$ (Theorem 2.31). One proves without difficulty that a *countable locally nilpotent group is a Gruenberg group*; however, there exist uncountable locally nilpotent groups which have trivial Gruenberg radical (Theorem 6.27).

The basic properties of Baer groups were obtained by Baer in [20] and those of Gruenberg groups by Gruenberg in [3]: see also Kemhadze [2] and [3], Plotkin and Kemhadze [1], Rips [1].

At the end of Chapter 2 it was shown that *the class of Baer groups is identical with the class Cat* $\mathfrak{A}$ *of groups with an abelian category.*

A group G which coincides with its Fitting subgroup $\varrho_{\mathfrak{N}}(G)$ is called a *Fitting group*. By the $\mathbf{N}_0$-closure of the class $\mathfrak{N}$, the Fitting groups are precisely the locally normal and nilpotent groups, and they are also the groups which can be covered by normal nilpotent subgroups. The class of Fitting groups is intermediate between the classes of nilpotent groups and Baer groups.

It was shown in Section 2.3 (p. 64) that

$$\mathfrak{N}_c \leqq \mathfrak{A}^{(c)}, \tag{1}$$

a result due to Kontorovič [7]. Equation (1) shows at once that *every Fitting group is an* $\mathfrak{A}^{(\omega)}$*-group.*

(b) Hypercentral Groups, Groups in which All Subgroups are Ascendant or Subnormal

A group is called an *N-group* or group with *the normalizer condition* if every proper subgroup is distinct from its normalizer. Let G be an N-group and let $H \leqq G$; we define an ascending series $\{H_\alpha\}$ from H to G by the rules

$$H_0 = H, \; H_{\alpha+1} = N_G(H_\alpha) \quad \text{and} \quad H_\lambda = \bigcup_{\lambda<\alpha} H_\alpha$$

where α is an ordinal and λ is a limit ordinal. Clearly this series reaches G after a finite or infinite number of steps, so H is ascendant in G. Conversely, a proper ascendant subgroup is evidently always distinct from its normalizer.

Consequently, *a group G is an N-group if and only if every subgroup is ascendant in G* (Baer [7], Theorem 4.13). Consequently every N-group is a Gruenberg group, and is, in particular, locally nilpotent (Plotkin [2], [6]; see also Hirsch [7]). By a remark in Section 2.1 (pp. 49—50), *every hypercentral group is an N-group*, so the class of N-groups lies between the class of hypercentral groups and the class of Gruenberg groups.

A group is said to be an N_1*-group* if every subgroup is subnormal. It is obvious that the class of N_1-groups is contained in the class of N-groups and in the class of Baer groups, and it is well-known that every nilpotent group is an N_1-group (Part 1, p. 49).

We mention briefly some related classes of groups. In [1] Mann studies groups which have dense ascendant or subnormal subgroups:

a group G has *dense ascendant (subnormal) subgroups* if whenever $H < K \leqq G$ and H is not maximal in K, there is an ascendant (subnormal) subgroup of G lying strictly between H and K. Obviously every N-group has dense ascendant subgroups and every N_1-group has dense subnormal subgroups: Mann proves that *every infinite, but not every finite, group with dense ascendant subgroups is a Gruenberg group* (Mann [1], Theorems 1 and 2).

In [33] Černikov studies a weaker form of the normalizer condition: every infinite proper subgroup is distinct from its normalizer; he shows that an infinite, non-periodic group with this property is an N-group. A similar property is studied by Garaščuk in [2].

Diagram of classes of groups

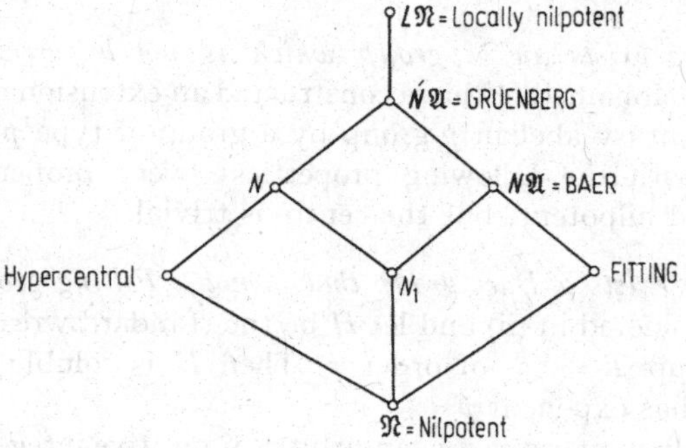

The following question remains unsettled: *is every N_1-group a Fitting group?*

Theorem 6.11. Apart from the doubtful case mentioned above, there exist no inclusions among the eight classes of groups except those indicated in the diagram.

Proof

(i) There exists a hypercentral group that is not a Baer group. For example, the locally dihedral 2-group is hypercentral with hypercentral length $\omega + 1$ (the least possible), but it has a subgroup of order 2 which is not subnormal. In fact for each prime p there exists a Černikov (and therefore hypercentral) p-group which is not a Baer group.

(ii) There exists a Fitting group that is not an N-group. Let G be the standard wreath product of a group of prime order p with a countably infinite, elementary abelian p-group X. The base group B of G is an

elementary abelian p-group. Hence, if $x \in X$,

$$[B, \underbrace{x, \ldots, x}_{p}] = B^{(x-1)^p} = B^{x^p-1} = 1,$$

so that $\langle x, B\rangle$ is nilpotent. Also $G' \leq B$, so $\langle x, B\rangle \lhd G$. Since G is the product of all the subgroups $\langle x, B\rangle$ where $x \in X$, the group G is a Fitting group. Now $C_B(X) = 1$ since X is infinite and an element of $C_B(X)$ must have all its X-components equal. Therefore $N_G(X) = X$ and G is not an N-group.

Notice that G is a metabelian group of exponent p^2 with trivial centre: other examples of metabelian p-groups with trivial centre are in papers of Baer ([7], p. 412) and O. J. Schmidt ([4], p. 369). An example of a torsion-free metabelian Fitting group with trivial centre has been given by Sesekin ([3], p. 438).

(iii) There exists an N_1-group which is not hypercentral. In fact Heineken and Mohamed [1] have constructed an extension of a countably infinite, elementary abelian p-group by a group of type p^∞ (where p is any prime) with the following properties: every proper subgroup is subnormal and nilpotent, but the centre is trivial.

(iv) There exists a Baer group that is not a Fitting group. Let G be the group considered in (ii) and let H be the standard wreath product of G with a group $T = \langle t\rangle$ of order p. Then H is soluble with derived length 3 and has exponent p^3.

G is not nilpotent, so if n is an arbitrary positive integer, we can find a sequence of elements $a_1, a_2, \ldots, a_n$ in G such that $[a_1, \ldots, a_n] \neq 1$. Let $b(i)$ be the element of the base group B (of the wreath product $H = G \wr T$) whose 1-component equals a_i and whose other components are trivial. Let $c(i) = [b(i), t]^{-1}$; then the 1-component of $c(i)$ is a_i, so the 1-component of $[c(1), \ldots, c(n)]$ is $[a_1, \ldots, a_n]$. Since $c(i) \in t^H$, it follows that t^H is not nilpotent; hence H is not a Fitting group.

On the other hand, it well-known that a soluble p-group of finite exponent is a Baer group (Theorem 7.17), so H is a Baer group.

(v) There exists a locally nilpotent group which is not a Gruenberg group. This follows from Theorem 6.27 in the next section. □

We mention that Dark [2] has constructed an example of a countably infinite, Baer p-group which has no non-trivial normal abelian subgroups; such a group is not hyperabelian and, *a fortiori*, is not a Fitting group. In the other direction, it is not difficult to show that *a Baer group which is nilpotent-by-abelian is a Fitting group*: for example, this may be deduced from Lemma 3.15 of Robinson [11].

Rips [1] has constructed for each positive integer n an example of a Baer p-group which cannot be generated by its subnormal nilpotent subgroups with subnormal index $\leq n$.

$\boldsymbol{N_0}$, $\boldsymbol{N}$ and $\boldsymbol{\acute{N}}$-closure

The class of locally nilpotent groups and the class of Gruenberg groups are $\acute{\boldsymbol{N}}$-closed; the class of Baer groups is $\boldsymbol{N}$-closed but not $\acute{\boldsymbol{N}}$-closed (otherwise every Gruenberg group would be a Baer group). The class of Fitting groups is not even $\boldsymbol{N}_0$-closed—an example is given in Robinson [11], p. 114—although it is obviously closed under the formation of direct products. We recall from Section 2.1 that the class of nilpotent groups and the class of hypercentral groups are $\boldsymbol{N}_0$-closed.

Finally, we observe that neither the class of N-groups nor the class of N_1-groups is closed with respect to forming finite direct products. For let H be the example of Heineken-Mohamed [1] already mentioned and let $G = H \times H$ be its direct square. Then H is an N_1-group, but G is not even an N-group. For suppose that G is an N-group and let D be the diagonal subgroup consisting of all (h, h) where $h \in H$. There exists an element (h_1, h_2) in $N_G(D)$ but not in D. For any $h \in H$

$$(h, h)^{(h_1,h_2)} = (h^{h_1}, h^{h_2}) \in D.$$

Thus $h^{h_1} = h^{h_2}$ and $h_1 h_2^{-1} \in \zeta(H) = 1$. Hence $h_1 = h_2$ and we obtain the contradiction $(h_1, h_2) \in D$.

II. Classes of Non-Locally Nilpotent Groups

(a) Z-groups, $\bar{Z}$-groups, Hypocentral Groups and Residually Nilpotent Groups

We recall that a *Z-group* is a group which has a central series. Let us write

$$\bar{Z} = Z^{\boldsymbol{H}},$$

so that the $\bar{Z}$-groups form the largest $\boldsymbol{H}$-closed subclass of the class of Z-groups. Now it is easy to show that minimal normal subgroup of a Z-group lies in the centre of the group; this, together with the fact that every group has a chief-series, shows that *a group is a $\bar{Z}$-group if and only if every chief-factor is central.* The class of $\bar{Z}$-groups is not $\boldsymbol{S}$-closed, for examples of Merzljakov [1] and P. Hall [14] show that a $\bar{Z}$-group may have non-cyclic free subgroups: see also Kargapolov [3]. Consequently, if we define

$$\bar{\bar{Z}} = Z^{\langle \boldsymbol{H},\boldsymbol{S} \rangle} = Z^{\boldsymbol{HS}},$$

so that the $\bar{\bar{Z}}$-groups form the largest subclass of Z which is closed with respect to forming sections, $\bar{Z}$ is a weaker property than $\bar{\bar{Z}}$. The classes Z, $\bar{Z}$ and $\bar{\bar{Z}}$ are all $\boldsymbol{L}$-closed (Mal'cev [2]), as will be demonstrated in Section 8.2: for $\bar{\bar{Z}}$-groups this has already been established (Theorem 5.27). Thus *every locally nilpotent group is a $\bar{\bar{Z}}$-group.*

The *hypocentral* (or ZD-) groups form another special class of Z-groups; these are the groups which have a descending central series or, equivalently, the groups whose lower central series reaches the identity subgroup after a finite or infinite number of steps.

Every residually nilpotent group is hypocentral since a group G is residually nilpotent if and only if the ωth term of its lower central series is trivial. It is known that there exist hypocentral groups with arbitrary hypocentral (i.e. lower central) length (Mal'cev [5], pp. 363—365: see also Theorem 6.22 below). Hence the residually nilpotent groups constitute a proper subclass of the class of hypocentral groups.

By a well-known theorem of Iwasawa [1], *every free group is residually a finite p-group*; thus in particular *all free groups are residually nilpotent* (Magnus [1], p. 269). Hence Z is not $\boldsymbol{H}$-closed and therefore contains Z properly. A theorem of Hall and Hartley ([1], Lemma 13) asserts that any group can be embedded in a suitable normal product of two free groups (this is Theorem 8.19.2 below). Hence the product of two normal residually nilpotent groups need not even be a Z-group. Kargapolov [3] has shown that Z is inherited by direct products and Merzljakov [1] has given an example to show that $\bar{Z}$ is *not* inherited by cartesian products and so is not $\boldsymbol{R}$-closed.

(b) Residually Central Groups

A group G is said to be *residually central* if for each non-trivial element x of G there is a normal subgroup N such that $x \notin N$ and xN lies in the centre of G/N. Residually central groups were introduced by Durbin [3] and Ayoub [2]. It is immediate that *a group G is residually central if and only if*

$$x \notin [G, x] \tag{2}$$

for each non-trivial element x of G.

Let G be a Z-group and suppose that $\{\Lambda_\sigma, V_\sigma : \sigma \in \Sigma\}$ is a central series in G. Let $1 \neq x \in G$; then $x \in \Lambda_\sigma \setminus V_\sigma$ for some $\sigma \in \Sigma$ and xV_σ is a non-trivial element of the centre of G/V_σ. Hence *every Z-group is residually central.* To complete the proof that the residually central groups form a class of generalized nilpotent groups we need to show that a finite residually central group is nilpotent: this is a consequence of the next theorem.

From (2) we see that the class of residually central groups is ***S***, ***L*** and ***R***-closed, but it is clearly not ***H***-closed. There is however the following weak form of ***H***-closure.

Theorem 6.12 (cf. Ayoub [2], Newell [2]). Let G be a residually central group and let N be a normal subgroup of G contained in the hypercentre. Then G/N is residually central.

We prove first a very simple result.

Lemma 6.13. Let Z be a subgroup of the centre of a group G and let x be an element in G such that $x \in Z[G, x]\backslash Z$. Then there is a non-trivial element c of $[G, x]$ such that $c \in [G, c]$.

Proof. By hypothesis we can write $x = zc$ where $z \in Z$ and $c \in [G, x]$: since $x \notin Z$, we see that $c \neq 1$. Let g be any element of G; then

$$[g, x] = [g, zc] = [g, c].$$

Hence $c \in [G, x] = [G, c]$. ∎

Proof of Theorem 6.12. Let $Z_\alpha = \zeta_\alpha(G)$. Since N is contained in the hypercentre of G, it is sufficient to prove that $G/N \cap Z_\alpha$ is residually central for every ordinal α. Suppose that α is the first ordinal for which this is false. Then $\alpha > 0$ and there is an element x such that $x \in (N \cap Z_\alpha)\,[G, x]$ but $x \notin N \cap Z_\alpha$. Assume that α is not a limit ordinal. Then $(N \cap Z_{\alpha-1})\,x$ does not belong to $N \cap Z_\alpha/N \cap Z_{\alpha-1}$, which is a subgroup of the centre of $G/N \cap Z_{\alpha-1}$, but it does belong to

$$(N \cap Z_\alpha/N \cap Z_{\alpha-1})\,[G/N \cap Z_{\alpha-1}, (N \cap Z_{\alpha-1})\,x].$$

Lemma 6.13 may therefore be applied to the group $G/N \cap Z_{\alpha-1}$ and we conclude that this group is not residually central. By this contradiction α is a limit ordinal and $x \in (N \cap Z_\beta)\,[G, x]$ for some $\beta < \alpha$. But $G/N \cap Z_\beta$ is residually central, so $x \in N \cap Z_\beta \leqq N \cap Z_\alpha$, our final contradiction. ∎

Corollary (Durban [3], Ayoub [2]). The residually central groups which satisfy Min-n are precisely the hypercentral Černikov groups.

Proof. Let G be a residually central group which satisfies Min-n and let H be the hypercentre of G. If $H = G$, then G is a Černikov group, by Corollary 2 to Theorem 5.21, and Theorem 6.12 shows that we may assume that $H = 1$. Let M be a minimal normal subgroup of G and let $1 \neq a \in M$. Since G has trivial centre, there is an element g in G such that $c = [g, a] \neq 1$. Now $c \in M$, so $M = c^G \leqq [G, a]$ and hence $a \in [G, a]$. This contradicts the residual centrality of G, so $G = 1$. ∎

Elements with Relatively Prime Orders in Residually Central Groups

Theorem 6.14

(i) Elements of a residually central group which have relatively prime, finite orders commute.

(ii) If G is a locally residually nilpotent group, the subgroup generated by the elements of finite order is the direct product of the subgroups G_p generated by the elements whose order is a power of the prime p. Moreover, each element of G_p has order ∞ or a power of p (Schenkman [3]).

Proof

(i) Let a and b be elements of a residually central group G with relatively prime, finite orders m and n, and assume that $c = [a, b] \neq 1$. There exists a normal subgroup N of G such that $c \notin N$ and $cN \in \zeta(G/N)$. Then

$$c^m \equiv [a^m, b] \equiv 1 \equiv [a, b^n] \equiv c^n \bmod N.$$

Hence $c \in N$, which is a contradiction.

(ii) We can assume G residually nilpotent. G_p is the subgroup generated by all elements of order a power of the prime p. Let T be the subgroup generated by all the elements with finite order; then certainly T is the product of the normal subgroups G_p. Suppose that x is an element which belongs to both G_p and the product of the G_q for $q \neq p$. If $x \neq 1$, there exists an $N \lhd G$ such that $x \notin N$ and G/N is nilpotent. Hence xN lies in the p-component *and* the p'-component of G/N, and this can only mean that $x \in N$. Thus T is the direct product of the G_p. Finally, G_p cannot contain an element of prime order q different from p, since such an element would have to lie in $G_p \cap G_q$. □

Corollary. *A periodic locally residually nilpotent group is the direct product of its Sylow subgroups* (cf. Baer [9], p. 147).

Whether the corollary is true for periodic hypocentral groups or for periodic Z-groups is doubtful. Since a locally finite Z-group is clearly locally nilpotent, a counterexample would have to be periodic but not locally finite.

It is usual to use the symbol

$$S$$

to denote the property "*every Sylow subgroup is normal*". Hence *a group G is an S-group if and only if the elements of finite order form a subgroup which is a direct product of p-groups.* A finite S-group is nilpotent and every locally nilpotent group is an S-group; hence the S-

groups form a class of generalized nilpotent groups. S-groups form a very large class, containing all torsion-free groups and all p-groups for example.

By the Corollary to Theorem 6.14 *every periodic, locally residually nilpotent group is an S-group.* However the elements of finite order in a residually nilpotent group do not in general form a subgroup, as the infinite dihedral group shows. Hence not every residually nilpotent group is an S-group. The example of Golod [1] shows that a residually finite p-group, and so, *a fortiori*, a periodic residually nilpotent group, may not be locally finite.

(c) Baer-Nilpotent Groups, Engel Groups

A group is said to be *Baer-nilpotent* if every finite section of the group is nilpotent (Baer [17], p. 87). This is a large class of generalized nilpotent groups, which is obviously $\boldsymbol{S}$, $\boldsymbol{H}$ and $\boldsymbol{L}$-closed; however $\boldsymbol{N}_0$-closure is doubtful.

Let x and y be elements of a group G and define

$$[x, {}_0y] = x \quad \text{and} \quad [x, {}_{i+1}y] = [[x, {}_iy], y]$$

for $i = 0, 1, 2, \ldots$ A group G is called an *Engel group* if for each pair of elements x and y of G there is an integer $n = n(x, y) \geqq 0$ such that

$$[x, {}_ny] = 1.$$

The class of Engel groups is denoted by

$$\mathfrak{E}.$$

It is a well-known theorem (see p. 52) that *a finite Engel group is nilpotent.* Since the class $\mathfrak{E}$ is $\boldsymbol{S}$ and $\boldsymbol{H}$-closed, it follows that *every Engel group is Baer-nilpotent.* On the other hand, it is evident that every locally nilpotent group, and even every weakly nilpotent group, is an Engel group: here a group is termed *weakly nilpotent* if each 2-generator subgroup is nilpotent (Vilyacer [1]). The example of Golod [1] shows that Engel groups need not be locally nilpotent. Generalizations of the Engel property are studied in Baer [40]. A detailed discussion of Engel groups is reserved for Chapter Seven.

If $n > 1$ and p is a large prime, the Burnside group $B(n, p)$ is Baer-nilpotent but not Engel since its abelian subgroups are finite: see Corollary, Theorem 7.21 and Part 1, p. 35. $\bar{Z}$-groups need not be Baer-nilpotent: indeed the groups constructed by Merzljakov [1] and P. Hall [14] are $\bar{Z}$-groups but are not Baer-nilpotent since they have non-cyclic free subgroups. Baer-nilpotent groups are not necessarily residually central because of the existence of finitely generated, infinite simple p-groups (see Section 5.1); for clearly a simple, residually central group is cyclic of prime order.

The difficulty involved in the study of Baer-nilpotent groups may be gauged by the fact that an infinite group with every proper non-trivial subgroup of prime order p—if such a group exists—is Baer-nilpotent. In this connection see the discussion of Schmidt's Problem in Section 3.4.

(d) Groups in which every Subgroup is Serial

Let us denote by

$$\tilde{N}$$

the property "every subgroup is serial"; this means that every subgroup occurs as a term in some series in the group.

Theorem 6.15. (Kuroš and Černikov [1], § 8). A group G is an $\tilde{N}$-group if and only if H is normal in K whenever H is a maximal subgroup of a subgroup K of G.

Proof. Let G be an $\tilde{N}$-group; if H is a maximal subgroup of a subgroup K of G, then H *ser* G and hence H *ser* K. By maximality of K we obtain $H \lhd K$. Conversely, let G have this property and let L be a subgroup of G. We consider the set of all chains which are refinements of the chain $1 \leqq L \leqq G$—these need not be series at present—and we order them by inclusion. By Zorn's Lemma there is a maximal refinement. If H and K are two consecutive terms of this chain, then H is maximal in K, so $H \lhd K$ by hypothesis. Hence the chain has each term normal in its successor; it is clearly a complete series in G containing L. ▯

In this second form the property $\tilde{N}$ has been studied by Baer [7]. It is clear that $\tilde{N}$ is $\boldsymbol{S}$ and $\boldsymbol{H}$-closed. A finite $\tilde{N}$-group is nilpotent because each maximal subgroup is normal (Wielandt [1]). Therefore *every $\tilde{N}$-group is Baer-nilpotent*. The proof of Theorem 5.38 showed that $\tilde{N}$ is $\boldsymbol{L}$-closed (Baer [7], Theorem 4.1); hence *every locally nilpotent group is an $\tilde{N}$-group*. Clearly the $\tilde{N}$-groups form a class of generalized nilpotent groups.

An $\tilde{N}$-group cannot be a finitely generated, infinite simple group, for such a group has non-normal maximal subgroups. Hence the existence of finitely generated, infinite simple p-groups shows that the $\tilde{N}$-groups form a proper subclass of the class of Baer-nilpotent groups.

On the other hand, it is unknown whether $\tilde{N}$-groups are locally nilpotent or even residually central. It is also open whether periodic $\tilde{N}$-groups are S-groups.

A natural subclass of the class of $\tilde{N}$-groups is formed by the *N_2-groups* or groups in which every subgroup is descendant. It is easy to prove that *a group G is an N_2-group if and only if $H < K \leq G$ always implies that $H^K < K$.* Locally nilpotent groups need not be N_2-groups, as the locally dihedral 2-group shows.

(e) The V-Groups and U-Groups of Baer

One of the best known properties of nilpotent groups is the commutativity of elements with relatively prime, finite orders, and in Theorem 6.14 we saw that even residually central groups have this property. Let

$$V$$

denote the following property: in every section elements with relatively prime, finite orders commute. A finite V-group is just the direct product of its Sylow subgroups and hence is nilpotent. Therefore *every V-group is Baer-nilpotent.* Clearly every p-group and every weakly nilpotent group is a V-group.

Another easily proved result is the following: *a periodic group is a V-group if and only if its elements with relatively prime orders commute.* Hence a periodic residually central group is a V-group. On the other hand, it is an open question whether a periodic V-group is a S-group.

V-groups were introduced in 1953 by Baer [17]. In the same paper Baer studies the property

$$U$$

defined as follows: a group G is a *U-group* if each finitely generated chief factor of a subgroup H of G is central in H. It is clear from the definitions that *every $\bar{\bar{Z}}$-group is a U-group* and that a finite U-group is nilpotent.

The position of the class of U-groups with respect to the classes of V-groups and $\tilde{N}$-groups is established by

Theorem 6.16. (i) Every $\tilde{N}$-group is a U-group (Baer [17], Satz 6). (ii) Every U-group is a V-group (Baer [16], § 3, Lemma 1).

In order to prove this let us first notice the following useful result (of which the case $\mathfrak{X} = \mathfrak{F}$ has already been mentioned on p. 171 of Part 1).

Lemma 6.17. Let $\mathfrak{X}$ be a class of finitely presented groups and let G be a finitely generated group which is not in $\mathfrak{X}$. Then G has a homomorphic image which is not in $\mathfrak{X}$ but all of whose proper homomorphic images are in $\mathfrak{X}$.

Proof. Let $\{N_\alpha: \alpha \in A\}$ be a chain of normal subgroups of G such that $G/N_\alpha \notin \mathfrak{X}$ for each $\alpha \in A$. Let N be the union of the N_α. If $G/N \in \mathfrak{X}$, then G/N is finitely presented and, since G is finitely generated, Lemma 1.43 shows that $N = a_1^G \cdots a_n^G$ for some finite subset $\{a_1, \ldots, a_n\}$.

But this implies that $N = N_\alpha$ for some $\alpha \in A$. Therefore $G/N \notin \mathfrak{X}$ and by Zorn's Lemma there is a normal subgroup K of G which is maximal subject to $G/K \notin \mathfrak{X}$. Then $H = G/K$ has the required property. □

Proof of Theorem 6.16. First of all let G be a U-group: to prove that G has V it is enough by the $\boldsymbol{S}$ and $\boldsymbol{H}$-closure of the property U to prove that if x and y are elements with relatively prime, finite orders, then $[x, y] = 1$. Suppose that this is false; then $S = \langle x, y\rangle$ is not abelian. Taking $\mathfrak{X}$ in Lemma 6.17 to be the class of finitely generated nilpotent groups—these are finitely presented by the Corollary to Lemma 1.43—we see that S can be assumed to have all its proper factor groups nilpotent. Each proper factor group of S is finite and cyclic, being a nilpotent group generated by two elements with relatively prime, finite orders. Let I denote the intersection of all the non-trivial normal subgroups of S; then $1 \neq [x, y] \in I$, so S/I is finite cyclic and Theorem 1.41 shows that I is finitely generated. But I is evidently the unique minimal normal subgroup of S, so the property U implies that I is central in S. It follows that S is abelian, which is not the case. Hence G is a V-group.

Now let G be an $\tilde{N}$-group and let H be a finitely generated minimal normal subgroup of G. To show that G has the property U, it suffices to prove that H lies in the centre of G since a subgroup or factor group of an $\tilde{N}$-group is an $\tilde{N}$-group. H is finitely generated, so it has a maximal subgroup M and $M \lhd H$ in view of the property $\tilde{N}$; hence $|H : M|$ is finite. Now H has just a finite number of normal subgroups with the same index as M: for there are only finitely many distinct homomorphisms of H onto a group of order $|H : M|$. It follows that H has a proper characteristic subgroup L of finite index; but $L \lhd G$, so $L = 1$ and H is finite, by the minimality of H. As a finite $\tilde{N}$-group H is nilpotent and, being characteristically simple, it is an elementary abelian p-group for some prime p.

Let $x \in G$; then for some positive integer m the element x^m centralizes H, so that $\langle x^m\rangle \lhd \langle x, H\rangle$ and consequently

$$\langle x, H\rangle/\langle x^m\rangle$$

is a finite $\tilde{N}$-group and is therefore nilpotent. Hence the endomorphism $a \to [a, x]$ of H is nilpotent. It follows that if i is a sufficiently large positive integer,

$$a^{(x^{p^i}-1)} = a^{(x-1)^{p^i}} = [a, \underbrace{x, \ldots, x}_{p^i}] = 1$$

for all $a \in H$, since H is an elementary abelian p-group. Therefore $\mathrm{Aut}_G H$ is finite p-group.

Let X be the holomorph of H by $\mathrm{Aut}_G H$ and observe that X is a finite p-group. Hence X is nilpotent and H, being a minimal normal subgroup of X, lies in the centre of X. This implies that H lies in the centre of G. □

We remark that the existence of finitely generated, infinite simple p-groups indicates that V is a weaker property than U.

It is obvious that V is $\boldsymbol{L}$-closed; to conclude let us prove that U is also $\boldsymbol{L}$-closed.

Theorem 6.18. The class of U-groups is $\boldsymbol{L}$-closed.

Proof. Suppose that G has the property U locally; since U is inherited by homomorphic images and subgroups, it suffices to show that a finitely generated minimal normal subgroup N of G lies in the centre of G. Now N is a U-group and, being finitely generated, it has a maximal normal subgroup M; by the property U the group N/M must be abelian. Hence $N' \leqq M$ and $N' = 1$ by minimality of N. Therefore N is a finitely generated, characteristically simple, abelian group, which means that N is a finite elementary abelian p-group, of order p^r, say, for some prime p. Now let X be an arbitrary finitely generated subgroup of G containing N. Then X is a U-group and, since N has order p^r and $N \lhd X$, it follows that $N \leqq \zeta_r(X)$. Hence $N \leqq \zeta_r(G)$, which implies that $N \leqq \zeta(G)$. □

Diagram of Classes of Groups

The known inclusions between the main classes of groups which we have been discussing in II are exhibited in the following diagram.

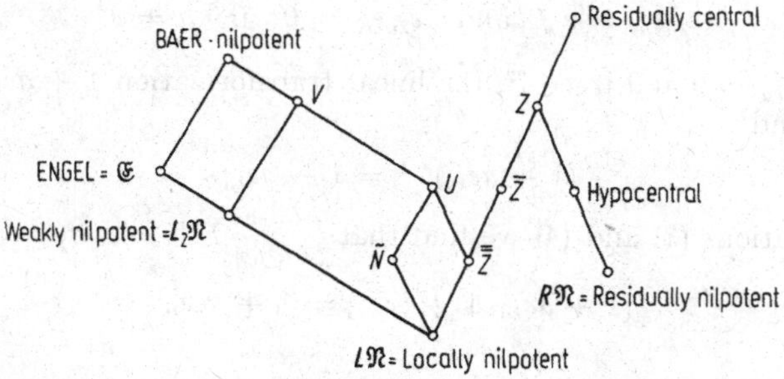

There are numerous unsettled questions regarding these classes: we mention some of the more interesting ones.

(i) Is every Baer-nilpotent group a V-group?

(ii) Is every residually central group a Z-group?

(iii) Does there exist an $\tilde{N}$-group which is not locally nilpotent?

(iv) Does the property U imply either $\tilde{N}$ or residual centrality?

(v) Does every weakly nilpotent group have one of the properties U and Z?

6.2 Some Constructions for Locally Nilpotent Groups

In this section we will construct examples of locally nilpotent groups which are, in one sense or another, far from being nilpotent; for example perfect locally nilpotent groups with trivial centre will be found; we will also obtain locally nilpotent groups which have trivial Baer or Gruenberg radicals.

McLain's Locally Nilpotent Groups

In [1] and [2] McLain has introduced a useful method of constructing locally nilpotent groups which are essentially generalizations of groups of unitriangular matrices.

The ingredients for this construction are a linearly ordered set Λ and a field F. Let V be a vector space over F having as basis a set of elements

$$\{v_\lambda : \lambda \in \Lambda\}.$$

For each pair of elements λ and μ from Λ such that $\lambda < \mu$ we define a linear transformation $e_{\lambda\mu}$ of V by the rules

$$(v_\lambda)\, e_{\lambda\mu} = v_\mu \quad \text{and} \quad (v_\nu)\, e_{\lambda\mu} = 0 \quad \text{if} \quad \nu \neq \lambda.$$

Evidently these linear transformations obey the laws

$$e_{\lambda\mu} e_{\mu\nu} = e_{\lambda\nu} \quad \text{and} \quad e_{\lambda\mu} e_{\nu\zeta} = 0 \quad \text{if} \quad \mu \neq \nu. \tag{3}$$

Thus $e_{\lambda\mu}^2 = 0$ and if $a \in F$, the linear transformation $1 + ae_{\lambda\mu}$ is non-singular and

$$(1 + ae_{\lambda\mu})^{-1} = 1 - ae_{\lambda\mu}. \tag{4}$$

From equations (3) and (4) we find that

$$\left.\begin{aligned} &[1 + ae_{\lambda\mu}, 1 + be_{\mu\nu}] = 1 + abe_{\lambda\nu} \\ \text{and} \\ &[1 + ae_{\lambda\mu}, 1 + be_{\nu\zeta}] = 1 \quad \text{if} \quad \mu \neq \nu \quad \text{and} \quad \lambda \neq \zeta, \end{aligned}\right\} \tag{5}$$

where a and b belong to F.

The McLain group determined by Λ and F is the group of linear transformations generated by all the $1 + ae_{\lambda\mu}$, that is

$$M(\Lambda, F) = \langle 1 + ae_{\lambda\mu} : \lambda, \mu \in \Lambda, \lambda < \mu, a \in F \rangle.$$

It is obvious from the definition that the elements $e_{\lambda\mu}$ are linearly independent over F, so each element of $M(\Lambda, F)$ has a unique representation as a sum of the form

$$1 + \sum_{\lambda<\mu} a_{\lambda\mu} e_{\lambda\mu} \tag{6}$$

where $a_{\lambda\mu} \in F$, and, of course, only finitely many of the $a_{\lambda\mu}$ are different from 0. Conversely, by repeated use of the identity

$$1 + u + v = (1 + u)(1 + v) - uv$$

we can show that every element of the form (6) belongs to $M(\Lambda, F)$.

If Λ is a finite set with n elements, then $M(\Lambda, F)$ is isomorphic with the group of all $n \times n$ (upper) unitriangular matrices over F and hence is nilpotent of class $n - 1$. Thus, to obtain interesting examples we will require Λ to be infinite.

Theorem 6.21 (McLain [1], [2])

(i) $M(\Lambda, F)$ is the product of its normal abelian subgroups and hence is a Fitting group.

(ii) Let p be the characteristic of F. Then $M(\Lambda, F)$ is either torsion-free or a p-group according as p is 0 or a prime.

(iii) Let N be a non-trivial normal subgroup of $M(\Lambda, F)$; then there exist λ and μ in Λ such that $1 + ae_{\lambda\mu} \in N$ for some non-zero a in F; if Λ does not have both a first and a last element, then N contains $1 + ae_{\lambda\mu}$ for all $a \in F$ and a suitable λ and μ.

(iv) Suppose that Λ contains more than one element and has the following property: given $\lambda_1, \lambda_2, \mu_1, \mu_2$ in Λ such that $\lambda_1 < \mu_1$ and $\lambda_2 < \mu_2$, there exists an order-automorphism* α of Λ for which $\lambda_1\alpha = \lambda_2$ and $\mu_1\alpha = \mu_2$. Then $M(\Lambda, F)$ is characteristically simple.

Proof. Let $M = M(\Lambda, F)$. The commutator relations (5) show that $(1 + ae_{\lambda\mu})^M$ lies in the subgroup generated by all elements of the form $1 + be_{\nu\zeta}$ where $b \in F$ and $\nu \leqq \lambda < \mu \leqq \zeta$ and that all such elements commute with each other. Hence $(1 + ae_{\lambda\mu})^M$ is abelian and M is the product of these normal subgroups.

Let $x \in M$; then there exist elements $\lambda_1, \dots, \lambda_n$ in Λ such that $\lambda_1 < \lambda_2 < \cdots < \lambda_n$ and x belongs to the subgroup

$$H = \langle 1 + ae_{\lambda_i\lambda_{i+1}} : i = 1, \dots, n - 1, a \in F \rangle. \tag{7}$$

But H is isomorphic with the group of all $n \times n$ unitriangular matrices over F and hence is torsion-free or a p-group according as p is 0 or a prime. So far (i) and (ii) have been established.

---

* An *order-automorphism* α of a linearly ordered set Λ is a one-one mapping of Λ onto itself which preserves the ordering of Λ, i.e. $\lambda < \mu$ implies that $\lambda\alpha < u\alpha$.

Now let $1 \neq N \lhd M$ and let $1 \neq x \in N$. Then x belongs to a subgroup H of the form (7); hence $1 \neq N \cap H \lhd H$, which implies that N intersects the centre of H non-trivially since H is nilpotent. But the centre of H consists of all $1 + ae_{\lambda_1\lambda_n}$ where $a \in F$, so N contains an element $1 + ae_{\lambda_1\lambda_n}$ where $a \neq 0$.

If Λ does not have a first element, there exists a $\mu \in \Lambda$ such that $\mu < \lambda_1$. Then N contains the element

$$[1 + ba^{-1}e_{\mu\lambda_1}, 1 + ae_{\lambda_1\lambda_n}] = 1 + be_{\mu\lambda_n}$$

for all $b \in F$. If Λ has no last element, the argument is similar. Thus (iii) is established.

Finally, assume that Λ has the property specified in (iv) and let N be a non-trivial characteristic subgroup of M. Clearly we may assume that Λ has more than two elements, and in this case Λ cannot have either a first or a last element: by (iii) there exist λ and μ in Λ such that N contains all $1 + ae_{\lambda\mu}$ where $a \in F$. Let λ' and μ' be elements of Λ such that $\lambda' < \mu'$. By hypothesis there is an order-automorphism α of Λ such that $\lambda\alpha = \lambda'$ and $\mu\alpha = \mu'$. The mapping $v_\zeta \to v_{\zeta\alpha}$ extends to a non-singular linear transformation α^* of V. Moreover

$$(\alpha^*)^{-1} (1 + ae_{\zeta\nu})\, \alpha^* = 1 + ae_{\zeta\alpha\, \nu\alpha},$$

so that transformation by α^* induces an automorphism in M. Hence N contains $1 + ae_{\lambda'\mu'}$ for all $a \in F$. Therefore $N = G$ and G is characteristically simple. $\square$

A natural choice for Λ in Theorem 6.21 (iv) is the set Q of all rational numbers in their natural order. Thus

$$M(Q, F)$$

is characteristically simple; since it is obviously not abelian, it is perfect and has trivial centre (McLain [2]). If F is the field $GF(p)$, then $M(Q, F)$ is a locally finite p-group.

The first example of an infinite locally finite p-group with trivial centre was given by Kuroš [5]; subsequently Schmidt ([5], p. 156) and Ado [1] found infinite perfect locally finite p-groups.

It is easy to prove that $M(\Lambda, F)$ *is perfect if and only if* Λ *is dense* and that $M(\Lambda, F)$ *has trivial centre if and only if* Λ *does not have both a first and a last elementor* $|\Lambda| = 1$.

Hypocentral Groups with Arbitrary Hypocentral Length

Theorem 6.22. Let F be an arbitrary field and let Λ be a well-ordered set with ordinal type $\alpha > 0$. Then $M(\Lambda, F)$ is a hypocentral group with hypocentral length exactly $-1 + \alpha$. Consequently there exist hypocentral groups with arbitrary hypocentral length.

Proof. Let $M = M(\Lambda, F)$. We shall take Λ to be the set of ordinal numbers preceding α, in their natural order. If β is an ordinal such that $\beta \leqq \alpha$, we will prove that $\gamma_\beta(M)$ consists of all

$$1 + \sum_{\lambda<\mu} a_{\lambda\mu} e_{\lambda\mu}$$

where

$$a_{\lambda\mu} = 0 \quad \text{if} \quad \mu < \lambda + \beta.$$

From this it follows at once that $\gamma_\alpha(M) = 1$ and that if $\beta < \alpha$, then $1 + e_{0\beta} \in \gamma_\beta(M)$, so $\gamma_\beta(M) \neq 1$.

Let β be the first ordinal for which the assertion of the last paragraph is false. Then $\beta > 1$ and β cannot be a limit ordinal because $\gamma_\beta(M)$ would be the intersection of all preceding terms of the lower central series and the $e_{\lambda\mu}$ are linearly independent. If

$$x = 1 + \sum_{\mu \geqq \lambda+\beta-1} a_{\lambda\mu} e_{\lambda\mu}$$

and $y \in M$, then a short computation shows that

$$[x, y] = 1 + \sum_{\mu \geqq \lambda+\beta} b_{\lambda\mu} e_{\lambda\mu}$$

for certain $b_{\lambda\mu}$ in F. Thus it remains only to prove that if

$$z = 1 + \sum_{\mu \geqq \lambda+\beta} c_{\lambda\mu} e_{\lambda\mu},$$

then $z \in \gamma_\beta(M)$. First of all observe that by repeated use of the identity

$$1 + u + v = (1 + u)(1 + v) - uv,$$

we can write z as a product

$$\prod_{\mu \geqq \lambda+\beta} (1 + d_{\lambda\mu} e_{\lambda\mu})$$

for suitable $d_{\lambda\mu} \in F$, only finitely many being non-zero. Now if $\mu \geqq \lambda + \beta$,

$$1 + d_{\lambda\mu} e_{\lambda\mu} = [1 + d_{\lambda\mu} e_{\lambda\,\lambda+\beta-1}, 1 + e_{\lambda+\beta-1\,\mu}] \in [\gamma_{\beta-1}(M), M] = \gamma_\beta(M),$$

since $\beta > 1$. Hence $z \in \gamma_\beta(M)$. □

The existence of hypocentral groups of arbitrary hypocentral length was first established by Mal'cev ([5], p. 363—365). In the same paper Mal'cev shows that there exist hypoabelian groups whose derived series has arbitrary ordinal type, a result that can also be obtained by inspecting the group $M = M(\Lambda, F)$ of Theorem 6.22: for this group has the property

$$M^{(\beta)} = \gamma_{2^\beta}(M).$$

In [1] (Theorem B1) Hartley constructs an abelian-by-cyclic group which is hypocentral with arbitrary infinite hypocentral length and is also hypercentral with hypercentral length $\omega + 1$. The McLain construction can also be used to obtain hypercentral groups with arbitrary hypercentral length, but here it is necessary to choose for Λ a suitable partially ordered set; for details the reader is referred to McLain [4]. See also the paper of Gluškov [5] on this question.

For further information on McLain's groups see Roseblade [2], Hall and Hartley [1] (2.7) and Robinson [9] (4.2).

Constructions by Means of Wreath Products*

The *standard wreath product* or wreath product of a pair of groups in their regular representations has been used several times to construct groups with special properties. For our present purpose it is necessary to deal with wreath products of linearly ordered sets of permutation groups. Accordingly we will give an account of the simpler properties of these general wreath products.

The first systematic account of wreath products of finite sets of permutation groups is contained in four papers of Kalužnin and Krasner [1]–[4]: for some later developments see Brumberg [1], P. Hall [4], Houghton [1] and P. M. Neumann [1]. The treatment which follows is drawn from a paper of P. Hall [12]. Wreath products of partially ordered sets of permutation groups have been studied by Holland [1]: see also McCleary [1].

Wreath Products in General

Let

$$\{H_\lambda : \lambda \in \Lambda\}$$

be a set of transitive permutation groups and let X_λ be the set upon which the group H_λ acts. In each X_λ we single out an element and denote it by 1_λ. Let X be the *restricted* set product of all the X_λ, that is to say X consists of all ordered sets

$$x = (x_\lambda)_{\lambda \in \Lambda}$$

where $x_\lambda \in X_\lambda$ and $x_\lambda = 1_\lambda$ for all but a finite number of λ. In particular X contains the element

$$1 = (1_\lambda)_{\lambda \in \Lambda}.$$

From now on we assume that Λ is a linearly ordered set. If x and y are elements of X and $\lambda \in \Lambda$, we will use

$$x \equiv y \bmod \lambda$$

* All wreath products are restricted unless the contrary is stated.

to mean that

$$x_\mu = y_\mu \text{ for all } \mu > \lambda.$$

We now represent H_λ faithfully as a group of permutations of X in the following manner. Let $\xi \in H_\lambda$ and $x \in X$; then a permutation ξ_X of X is defined thus. If $x \equiv 1 \bmod \lambda$,

$$(x\xi_X)_\lambda = x_\lambda \xi \quad \text{and} \quad (x\xi_X)_\mu = x_\mu, \quad \text{if} \quad u \neq \lambda; \tag{8}$$

on the other hand if $x \not\equiv 1 \bmod \lambda$,

$$x\xi_X = x. \tag{9}$$

So only elements congruent to 1 modulo λ are affected by ξ_X and these only in their λ-coordinate. Clearly the mapping $\xi \to \xi_X$ is a faithful representation of H_λ as a group J_λ of permutations of X.

The *wreath product* of the groups H_λ, $\lambda \in \Lambda$, is by definition the group of permutations of X generated by the J_λ:

$$W = \underset{\lambda \in \Lambda}{\mathrm{Wr}}\, H_\lambda = \langle J_\lambda : \lambda \in \Lambda \rangle.$$

If the H_λ are all isomorphic *as permutation groups* with a given transitive group H, we write

$$W = \mathrm{Wr}\, H^\Lambda$$

and call W the Λth *wreath power* of H.

Naturally the wreath product depends on the groups H_λ and on the way in which H_λ acts on X_λ, but W *does not depend on the choice of the distinguished elements* 1_λ. For if $\bar{1}_\lambda$ is any other element of X_λ, there is a permutation τ_λ in H_λ which maps 1_λ to $\bar{1}_\lambda$—here we use the transitivity of H_λ. Let $\bar{X}$ be the set of all restricted sequence $(\bar{x}_\lambda)_{\lambda \in \Lambda}$ and let $\bar{W}$ be the wreath product of the H_λ which arise from the choice of the $\bar{1}_\lambda$ as distinguished elements. The rule $(x\tau)_\lambda = x_\lambda \tau_\lambda$ defines a one-one mapping of X onto $\bar{X}$. If $\xi \in H_\lambda$, let $\bar{\xi} = \tau_\lambda^{-1} \xi \tau_\lambda$; then

$$\tau^{-1} \xi_X \tau = (\bar{\xi})_{\bar{X}}.$$

Hence W and $\bar{W}$ are isomorphic as permutation groups and consequently there is no dependence on the choice of 1_λ.

The next point to observe is that W *is a transitive permutation group on* X. To prove this it is enough to show that given any $x \in X$ there is an element ξ in W such that $1\xi = x$. Let $\lambda_1, \ldots, \lambda_r$ be the elements of Λ such that $x_{\lambda_i} \neq 1_{\lambda_i}$ and suppose that $\lambda_1 < \cdots < \lambda_r$. Since H_{λ_i} is transitive on X_{λ_i}, there is a ξ_i in H_{λ_i} such that $1_{\lambda_i} \xi_i = x_{\lambda_i}$. Now let

$$\xi = (\xi_1)_X \cdots (\xi_r)_X.$$

Then $\xi \in W$ and $1\xi = x$ by (8) and (9).

Segments, Segmentations and Bisections

We adhere to the previous notation. If Γ is a subset of Λ, define

$$J_\Gamma = \langle J_\gamma : \gamma \in \Gamma \rangle \tag{10}$$

and recall that J_γ is the group of permutations of X that represents H_γ. Also let

$$W_\Gamma = \underset{\gamma \in \Gamma}{Wr}\, H_\gamma$$

where the ordering of Γ is that inherited from Λ. The notation

$$\Gamma \ll \Delta$$

means that Γ and Δ are subsets of Λ and that $\gamma \in \Gamma$ and $\delta \in \Delta$ imply that $\gamma < \delta$.

A subset Γ of Λ is called a *segment* of Λ if for each $\lambda \in \Lambda$ exactly one of relations $\lambda \in \Gamma$, $\{\lambda\} \ll \Gamma$ and $\Gamma \ll \{\lambda\}$ holds. Γ is a *lower segment (upper segment)* of Λ if $\Gamma \ll \{\lambda\}$ (respectively $\{\lambda\} \ll \Gamma$) for all $\lambda \in \Lambda \backslash \Gamma$. Notice that a segment cannot be empty.

A *segmentation* of Λ is a decomposition of Λ,

$$\Lambda = \bigcup_{i \in I} \Gamma_i, \tag{11}$$

into mutually disjoint segments Γ_i of Λ. Clearly, if i and i' are two distinct elements of I, then either $\Gamma_i \ll \Gamma_{i'}$ or $\Gamma_{i'} \ll \Gamma_i$. In other words the segments Γ_i, and hence the set I, can be linearly ordered according to the relation $\ll$: thus $i < i'$ if and only if $\Gamma_i \ll \Gamma_{i'}$.

We will show next how each segmentation (11) of Λ leads to another decomposition of W as a wreath product. Let us write

$$J^{(i)} = J_{\Gamma_i} \quad \text{and} \quad W^{(i)} = W_{\Gamma_i} \tag{12}$$

and let $X^{(i)}$ be the set of all restricted sequences of the form

$$x^{(i)} = (x_\lambda)_{\lambda \in \Gamma_i}$$

where we use the original distinguished elements 1_λ. If $x \in X$, it is natural to identify x with the ordered set $(x^{(i)})_{i \in I}$. When this has been done, X is simply the restricted set product of the $X^{(i)}$, $i \in I$; here the distinguished element of $X^{(i)}$ is $1^{(i)} = (1_\lambda)_{\lambda \in \Gamma_i}$.

If $i \in I$, we define $x \equiv y \bmod i$ to mean that $x^{(j)} = y^{(j)}$ for all $j > i$. We can now form the wreath product

$$\overline{W} = \underset{i \in I}{Wr}\, W^{(i)}$$

using the above ordering of the set I.

Let $\mu \in \Gamma_i$, let $\xi \in H_\mu$ and let $\eta \in W^{(i)}$; denote by $\eta \to \eta^*$ the representation of $W^{(i)}$ as a permutation group on X obtained during the formation of $\overline{W}$. The definitions of ξ_X, $\xi_{X(i)}$ and η^* show that

$$(\xi_{X(i)})^* = \xi_X.$$

Hence

$$J^{(i)} = (W^{(i)})^* \simeq W^{(i)}, \quad \text{for all } i \in I,$$

and $\overline{W} = W$. Thus

$$W = \underset{i \in I}{Wr}\, W^{(i)}. \tag{13}$$

Equation (13) expresses the generalized associative law for wreath products.

The simplest type of segmentation is a *bisection* of Λ consisting of one lower and one upper segment,

$$\Lambda = \Gamma_1 \cup \Gamma_2$$

where

$$\Gamma_1 \ll \Gamma_2.$$

In this case I consists of the integers 1 and 2 in their natural order and (13) becomes

$$W = \underset{i=1,2}{Wr}\, W^{(i)} = W^{(1)} \wr W^{(2)} = \langle J^{(1)}, J^{(2)} \rangle.$$

The Base Group

We conclude this general discussion by identifying the base group of the wreath product

$$W = H_1 \wr H_2$$

and exhibiting W as a semi-direct product of the base group by J_2.

As usual H_1 and H_2 act on sets X_1 and X_2 respectively and X denotes the set product of X_1 and X_2. Let $x \in X_1$ and $y \in X_2$ and let $\xi \in H_1$ and $\eta \in H_2$. Then by definition

$$(x, 1_2)\, \xi_X = (x\xi, 1_2) \quad \text{and} \quad (x, y)\, \xi_X = (x, y) \quad \text{if} \quad y \neq 1_2$$

while

$$(x, y)\eta_X = (x, y\eta).$$

If $1_2\, \eta = y$, then $\eta_X^{-1} \xi_X \eta_X$ maps (x, y) to $(x\xi, y)$ and fixes (x, z) if $z \neq y$. Consequently $\eta_X^{-1} \xi_X \eta_X$ depends only on ξ and y and we may write

$$\xi(y) = \eta_X^{-1} \xi_X \eta_X \tag{14}$$

and

$$J_1(y) = \eta_X^{-1} J_1 \eta_X$$

where $y = 1_2 \eta$. It follows that if $y \neq y'$, then $\xi(y)$ and $\xi(y')$ do not affect the same element of X; consequently

$$J_1^W = J_1^{J_2} = \underset{y \in Y}{Dr}\, J_1(y).$$

This is called *the base group* B of W (cf. Ore [2], p. 17). Since $(J_1(y))^\eta = J_1(y\eta)$, the elements of J_2 permute the conjugates of J_1 in the same manner as the elements of Y are permuted by the elements of H_2. The group J_2 acts on B as a group of automorphisms and W is the semi-direct product of B by J_2.

Applications

Theorem 6.23 (P. Hall [12], Theorem B). Let Λ be a linearly ordered set without a first element and let $\{H_\lambda : \lambda \in \Lambda\}$ be a set of non-trivial transitive permutation groups. Let

$$W = \underset{\lambda \in \Lambda}{Wr}\, H_\lambda.$$

Then W contains no non-trivial finitely generated subnormal subgroups.

For example, let $C(p)$ be a cyclic group of prime order p in its regular representation and let Z^- be the set of negative integers in their natural order. Then the wreath power

$$W = Wr\, C(p)^{Z^-}$$

is a countably infinite, locally finite p-group with trivial Baer radical. Since it is countable, W is a Gruenberg group. Another such example is in a paper of Kuroš ([11], p. 924): an example of a torsion-free countably infinite, locally nilpotent group with trivial Baer radical has been given by Levič [3].

We will need the following lemma in the proof of Theorem 6.23.

Lemma 6.24 (P. Hall [12], Lemma 6). If H is a transitive permutation group containing an element of order m and K is a transitive permutation group containing an element of order n, then $W = H \wr K$ has an element of order mn.

Proof. Let H and K act on the sets X and Y respectively. Let $\xi \in H$ have order m and let $\eta \in K$ have order n. Then there must exist a finite set of cycles of η

$$(y_1, y_2, \ldots, y_{n_1}), (z_1, z_2, \ldots, z_{n_2}), \ldots$$

such that the least common multiple of $n_1, n_2, \ldots$ is n. Let

$$\zeta = \eta \xi(y_1)\, \xi(z_1) \cdots$$

where $\xi(y)$ is defined in equation (14). We will show that ζ has order mn. Now the $\xi(y_i)$, $\xi(z_i)$, etc. lie in different direct factors of the base group, so they commute among themselves; also $\xi(y_i)^\eta = \xi(y_{i+1})$, $\xi(z_i)^\eta = \xi(z_{i+1})$ etc.; by induction on the positive integer k we obtain

$$\zeta^k = \eta^k \left(\prod_{i=1}^{k} \xi(y_i)\right)\left(\prod_{i=1}^{k} \xi(z_i)\right)\cdots.$$

Here it is to be understood that, for example, the suffix i in $\xi(y_i)$ may have to be reduced modulo n_1. Thus $\zeta^k = 1$ if and only if $\eta^k = 1$ and

$$1 = \prod_{i=1}^{k} \xi(y_i) = \prod_{i=1}^{k} \xi(z_i) = \cdots. \tag{15}$$

This will be the case if $k = mn$ since, for example,

$$\xi(y_1) \cdots \xi(y_{mn}) = (\xi(y_1)^m \cdots \xi(y_{n_1})^m)^{\frac{n}{n_1}} = 1.$$

Now suppose that $\zeta^k = 1$. Then $k = rn$ for some positive integer r. Also, n_i divides n, so we may write $n = n_i s_i$ and $k = rn_i s_i$ for $i = 1, 2, \ldots$ Thus (15) becomes

$$1 = \left(\prod_{i=1}^{n_1} \xi(y_i)\right)^{rs_1} = \left(\prod_{i=1}^{n_2} \xi(z_i)\right)^{rs_2} = \cdots.$$

Since $y_1, \ldots, y_{n_1}$ are all different, it follows that $\xi(y_1)^{rs_1} = 1$, so that m divides rs_1; similarly m divides rs_2, rs_3, etc. But $s_1, s_2, \ldots$ are relatively prime because n is the least common multiple of $n_1, n_2, \ldots$ Hence m divides r and therefore mn divides $rn = k$. It follows that ζ has order mn. □

Proof of Theorem 6.23. Let S be a non-trivial finitely generated, subnormal subgroup of W. Then $S \leq \langle J_{\lambda_1}, \ldots, J_{\lambda_n} \rangle$ for some finite subset $\{\lambda_1, \ldots, \lambda_n\}$ of Λ: here as usual J_λ is the canonical image of H_λ. Since Λ has no first element, there is a bisection

$$\Lambda = \Gamma_1 \cup \Gamma_2$$

such that Γ_1 is infinite and each λ_i belongs to Γ_2. Setting $H = W^{(1)}$ and $K = W^{(2)}$, we have $W = H \wr K$; for convenience of notation we will identify K with $J^{(2)}$, so that $S \leq K$.

Let $1 \neq \eta \in S$ and let $(\ldots y_{-1}, y_0, y_1, \ldots)$ be a cycle of η with length greater than 1. If ξ is any element of H and s is any positive integer, then an induction on s shows that

$$[\xi(y_0), \underbrace{\eta, \ldots, \eta}_{s}] = \prod_{i=0}^{s} \xi(y_i)^{(-1)^{s-i}\binom{s}{i}}. \tag{16}$$

(Here we may have to reduce the suffixes i modulo the length of the cycle if the latter is finite). If we take s to be the subnormal index of S in W, then

$$[\xi(y_0), \underbrace{\eta, \ldots, \eta}_{s}] \in B \cap S \leqq B \cap K = 1, \tag{17}$$

where B is the base group of $W = H \wr K$. Therefore by (16) the elements $y_0, y_1, \ldots, y_s$ cannot all be different; consequently the cycle $(\ldots y_{-1}, y_0, y_1, \ldots)$ has finite length $n \leqq s$. Let us change the notation slightly and write $(y_0, y_1, \ldots, y_{n-1})$ for this cycle.

Let p be a prime greater than s. Then, by (16) and (17) with p in place of s,

$$\prod_{i=0}^{p} \xi(y_i)^{(-1)^{p-i}\binom{p}{i}} = 1.$$

Taking the y_0-component of both sides, we obtain $\xi(y_0)^d = 1$ and hence

$$\xi^d = 1,$$

where

$$d = (-1)^p + (-1)^{p-n}\binom{p}{n} + (-1)^{p-2n}\binom{p}{2n} + \cdots.$$

Since $p > s \geqq n > 1$, the binomial coefficient $\binom{p}{in}$ is divisible by p if $i > 0$. Hence $d \equiv \pm 1 \bmod p$, so $d \neq 0$. Since $\xi^d = 1$ for all $\xi \in H$, the subgroup H has finite exponent dividing d. But H is the wreath product of an infinite collection of non-trivial groups and Lemma 6.24 implies that H has infinite exponent. Thus we have a contradiction. □

Characteristically Simple, Locally Nilpotent Groups with Trivial Baer Radical

A characteristically simple group must either be a Baer group or have trivial Baer radical. McLain's construction leads to a characteristically simple group which is even a Fitting group: our next objective is the opposite extreme, a characteristically simple, locally nilpotent group with trivial Baer radical.

Let Λ be a linearly ordered set and let H be a transitive permutation group on a set X with distinguished element 1. Form the wreath power

$$W = Wr\, H^{\Lambda}$$

acting on the set V of all *restricted* sequences $v = (v_\lambda)_{\lambda \in \Lambda}$ where $v_\lambda \in X$: by this we mean that $v_\lambda = 1$ for all but a finite number of the λ. If each $v_\lambda = 1$, let us write (1) for v.

Let A be a group of order-automorphisms of Λ. Then A can be made into a group of permutations of V by defining

$$(va)_\lambda = v_{\lambda a^{-1}}, \quad (\lambda \in \Lambda, v \in V, a \in A).$$

Let $\xi \in H$ and $\lambda \in \Lambda$; denote by $\xi^{(\lambda)}$ the image of ξ in the canonical isomorphism of H with J_λ. Let $v \in V$ and $a \in A$, and assume that $v \equiv (1) \bmod \lambda a$, so that $va^{-1} \equiv (1) \bmod \lambda$. Then

$$(va^{-1}\xi^{(\lambda)}a)_{\lambda a} = v_{\lambda a}\xi$$

and

$$(va^{-1}\xi^{(\lambda)}a)_{\mu a} = v_{\mu a}$$

if $\mu \neq \lambda$. On the other hand, if $v \not\equiv (1) \bmod \lambda a$, then

$$va^{-1}\xi^{(\lambda)}a = v.$$

In other words $a^{-1}\xi^{(\lambda)}a$ affects only the v in V such that $v \equiv (1) \bmod \lambda a$ and those only in their λa-component, upon which it acts like ξ. Therefore $a^{-1}\xi^{(\lambda)}a = \xi^{(\lambda a)} \in J_{\lambda a}$ and

$$(J_\lambda)^a = J_{\lambda a}. \tag{18}$$

The groups J_λ generate W, so $W^a = W$ for each a in A. Hence A normalizes W and equation (18) shows that $C_A(W) = 1$, so A is faithfully represented as a group of automorphisms of W. Now form the semidirect product of W by A

$$G = WA,$$

so that

$$W \lhd G \quad \text{and} \quad W \cap A = 1.$$

The group of order-automorphisms A is said to be *irreducible* (see P. M. Cohn [4]) if, given any pair of elements (λ, μ) from Λ, there is an element a of A such that

$$\mu < \lambda a.$$

Concerning the group G there is the following theorem.

Theorem 6.25 (P. Hall [12], Theorem D). Let A be an irreducible group of order-automorphisms of a linearly ordered set Λ, let H be a non-trivial transitive permutation group and let $W = Wr\, H^\Lambda$. If G denotes the semidirect product of W by A, then G is monolithic with monolith W'. Thus, in particular, W' is characteristically simple.

For example, let Z be the set of all integers in their natural order and let $A = \langle a \rangle$ where a is the order-automorphism in which $n \to n+1$; obviously A is irreducible. Let $C(p)$ be a cyclic group of prime order p. If

$$W = Wr\, C(p)^Z,$$

then W' is a countably infinite, locally finite p-group which is characteristically simple and has trivial Baer radical, by Theorems 6.23 and 6.25.

W is generated by the groups J_n, so by (18)

$$G = \langle J_0, a\rangle = \langle b_0, a\rangle$$

where $J_0 = \langle b_0\rangle$. Evidently W is a Gruenberg group; hence G is an SN^*-group, by Lemma 2.35. G/W' is a finitely generated metabelian group, so it satisfies Max-n by Theorem 5.34; also W' is minimal normal in G, therefore G satisfies Max-n.

Corollary (P. Hall [12], p. 174). There exists a 2-generator monolithic SN^*-group satisfying Max-n whose monolith has trivial Baer radical. This group is therefore neither subsoluble nor an SI-group.

The same group G shows that *a cyclic extension of a locally nilpotent group need not even be locally soluble.*

To prove Theorem 6.25 we need a further simple property of finite wreath products: in what follows we shall identify K with its canonical image in $W = H \wr K$.

Lemma 6.26 (P. Hall [12], Lemma 4). Let H and K be transitive permutation groups and let $W = H \wr K$. If $N \lhd W$ and $N \cap K \neq 1$, then N contains the derived subgroup of the base group of W.

Proof. Let B denote the base group of W and suppose that K acts on a set Y. Then B is the direct product of groups $H(y)$, $y \in Y$, where $H \simeq H(y)$. Let $1 \neq \eta \in N \cap K$. Then $y \neq y\eta = y'$ for some $y \in Y$. Let ξ_1 and ξ_2 be two elements of $H(y)$; then $\xi_1^\eta \in H(y')$, so ξ_1^η commutes with ξ_1^{-1} and ξ_2 and

$$[\xi_1^{-1}, \xi_2] = [\xi_1^{-1}\xi_1^\eta, \xi_2] = [\xi_1, \eta, \xi_2],$$

which belongs to N. Therefore $H(y)' \leqq N$ and conjugation by elements of K (which is transitive) leads to $B' \leqq N$. ▯

Proof of Theorem 6.25. Let N be a non-trivial normal subgroup of G contained in W'. Let $\eta \in W'$, so that $\eta \in S'$ where

$$S = \langle J_{\mu_1}, \dots, J_{\mu_m}\rangle,$$

$\mu_1 < \mu_2 < \dots < \mu_m$ and $\mu_i \in \Lambda$. If $1 \neq \xi \in N$, then

$$\xi \in T = \langle J_{\lambda_1}, \dots, J_{\lambda_n}\rangle$$

where $\lambda_1 < \lambda_2 < \dots < \lambda_n$. Since Λ is irreducible, there is an element a of Λ such that $\mu_m < \lambda_1 a$. Consider the bisection

$$\Lambda = \Gamma_1 \cup \Gamma_2$$

where

$$\Gamma_1 = \{\mu : \mu \in \Lambda, \mu < \lambda_1 a\}.$$

Since $T^a = \langle J_{\lambda_1 a}, \ldots, J_{\lambda_n a}\rangle$ and $\lambda_1 a < \lambda_2 a < \cdots < \lambda_n a$, it follows that T^a is contained in $J^{(2)}$. Hence $1 \neq \xi^a \in N \cap J^{(2)}$. Since $N \lhd W$, Lemma 6.26 shows that $N \geqq B'$ where B is the base group of the wreath product $W = W^{(1)} \wr W^{(2)}$. But $\eta \in S'$ and $S \leqq J^{(1)} \leqq B$, so $\eta \in B' \leqq N$. Since η was an arbitrary element of W', it follows that $N \geqq W'$ and W' is a minimal normal subgroup of G.

To prove that W' is the monolith of G it is enough to show that an arbitrary non-trivial normal subgroup M of G intersects W' non-trivially. Suppose that this is false and $M \cap W' = 1$. Then

$$[M \cap W, W] \leqq M \cap W' = 1,$$

so that $M \cap W$ lies in the centre of W. But the latter is trivial by Theorem 6.23, for the irreducibility of Λ obviously implies that Λ has no first (or last) element. Consequently, $M \cap W = 1$ and $[M, W] = 1$.

Let $1 \neq x \in M$ and write $x = wa$ where $w \in W$ and $a \in A$. There is a bisection

$$\Lambda = \Gamma_1 \cup \Gamma_2$$

such that $w \in J^{(1)}$ and Γ_2 has no last element. Thus

$$W = W^{(1)} \wr W^{(2)} = \langle J^{(1)}, J^{(2)}\rangle.$$

Suppose that a fixes each element of Γ_2; then a centralizes $J^{(2)}$. Since $x = wa$ centralizes W, it follows that $[J^{(2)}, w] = 1$. However $w \in J^{(1)}$ and $J^{(2)}$ is transitive, so $w = 1$. Hence a centralizes W and this implies that $a = 1$ and $x = 1$.

Consequently there exists a λ in Γ_2 such that $\lambda a \neq \lambda$. Let ξ be an element of H such that $1\xi \neq 1$: such a ξ exists because H is non-trivial and transitive. The only "non-trivial" component of $(1)\xi^{(\lambda)}$ is the λ-component, which equals 1ξ. Since $w \in J^{(1)}$, the λa-component of $(1)\xi^{(\lambda)}x$ equals 1ξ. On the other hand, $\lambda \neq \lambda a$, so the λa-component of $(1)x\xi^{(\lambda)}$ is $((1)x)_{\lambda a} = ((1)\,w)_\lambda = 1$, since $w \in J^{(1)}$ and $\lambda \in \Gamma_2$. Hence $\xi^{(\lambda)}x \neq x\xi^{(\lambda)}$, which contradicts $[M, W] = 1$. ∎

Locally Nilpotent Groups with Trivial Gruenberg Radical

The first example of a locally nilpotent group which is not a Gruenberg group was found independently by Kargapolov [4] and by L. G. Kovács and B. H. Neumann (unpublished). Such a group must of course be uncountable.

The construction here also involves wreath products. Let ϱ denote the first uncountable ordinal and for each ordinal $\alpha < \varrho$ let H_α be any non-trivial group. An ascending chain of groups $\{W_\alpha : \alpha \leqq \varrho\}$ is defined by the rules

$$W_0 = 1,\ W_{\alpha+1} = H_{\alpha+1} \wr W_\alpha \text{ and } W_\lambda = \bigcup_{\beta<\lambda} W_\beta \text{ for}$$

for each ordinal $\alpha < \varrho$ and each limit ordinal $\lambda \leqq \varrho$: here the wreath product is the standard one and W_α is embedded in $W_{\alpha+1}$ in the natural way. Now define

$$W = W_\varrho.$$

Theorem 6.27. Let S be an SN^*-subgroup of W; then $S \leqq W_\alpha$ for some $\alpha < \varrho$. Also the Gruenberg radical of W is trivial.

The proof depends on the following lemma.

Lemma 6.28. Let W be the standard wreath product of two groups H and K and let S be an infinite subgroup of K. Then $N_W(S) = N_K(S)$.

Proof. Let $z \in N_W(S)$ and let B be the base group of W. We write $z = bk$ where $b \in B$ and $k \in K$. If $s \in S$, then $s^{bk} \in S$, so $s^b \in K$ and consequently $[s, b] \in B \cap K = 1$ for all $s \in S$. Hence $b \in C_B(S)$ and it will be enough to prove that $C_B(S) = 1$.

Let $b \in C_B(S)$ and denote by b_x the x-component of b where $x \in K$. If $s \in S$, then $b = b^s$ and, equating x-components, we obtain $b_x = b_{xs^{-1}}$ for all $x \in K$. Hence b_x remains constant as x varies over a left coset of S in K. But S is infinite and only finitely many components of b are nontrivial. Therefore $b_x = 1$ for all $x \in K$, and $b = 1$. □

Proof of Theorem 6.27. Let S be an SN^*-subgroup of W: we must show that $S \leqq W_\alpha$ for some $\alpha < \varrho$ and obviously we may assume that S is infinite. The first step is to prove that S has a countably infinite, ascendant subgroup K. Now there is a strictly ascending abelian series in S, say

$$1 = S_0 < S_1 < \cdots S_\sigma = S;$$

let S_τ be the first infinite term. Then certainly $\tau \leqq \omega$. If $\tau = \omega$, we can take K to be S_ω, for this is countable, infinite and ascendant in S. If τ is finite, then $\tau \geqq 1$ and $S_{\tau-1}$ is finite while $S_\tau/S_{\tau-1}$ is an infinite abelian group; let $K/S_{\tau-1}$ be a countably infinite subgroup of $S_\tau/S_{\tau-1}$; then $K \lhd S_\tau$, so K is ascendant in S and it is clearly countably infinite.

Since ϱ is the first uncountable ordinal, $K \leqq W_\alpha$ for some $\alpha < \varrho$. Let

$$K = K_0 \lhd K_1 \lhd \cdots K_\beta = S$$

be an ascending series from K to S. We will show that $K_\gamma \leqq W_\alpha$ for all $\gamma \leqq \beta$; from this it will follow that $S \leqq W_\alpha$. If this is not the case, let γ be the first ordinal such that $K_\gamma \not\leqq W_\alpha$; then γ is not a limit ordinal and $K_{\gamma-1} \leqq W_\alpha$. Let $x \in K_\gamma$ and put $L = \langle x, K_{\gamma-1}\rangle$; then $L \leqq W_{\alpha_0}$ for some least $\alpha_0 < \varrho$. Suppose that $\alpha < \alpha_0$. Now α_0 cannot be a limit ordinal; for if it were, then $x \in W_{\alpha_1}$ for some $\alpha_1 < \alpha_0$ and $L \leqq W_{\alpha_2}$ where $\alpha_2 = \max\{\alpha, \alpha_1\} < \alpha_0$, which contradicts our choice of α_0. Hence

$$W_{\alpha_0} = H_{\alpha_0} \wr W_{\alpha_0 - 1}$$

and $K_{\gamma-1} \leqq W_\alpha \leqq W_{\alpha_0-1}$. But x belongs to W_{α_0} and normalizes the infinite group $K_{\gamma-1}$ since $K_{\gamma-1} \lhd K_\gamma$. Lemma 6.28 now assures us that $x \in W_{\alpha_0-1}$, which implies that $L \leqq W_{\alpha_0-1}$. This contradiction to the minimality of α_0 can only mean that $\alpha_0 \leqq \alpha$. Hence $x \in W_\alpha$ and consequently $K_\gamma \leqq W_\alpha$, the required contradiction.

Now let R be the Gruenberg radical of W. Then R is an SN^*-group by Lemma 2.35, and by the first part of the theorem $R \leqq W_\alpha$ for some $\alpha < \varrho$. Let B denote the base group of the wreath product

$$W_{\alpha+1} = H_{\alpha+1} \wr W_\alpha.$$

Since $R \lhd W_{\alpha+1}$, we have

$$[B, R] \leqq B \cap R \leqq B \cap W_\alpha = 1,$$

so that $R \leqq C_{W_\alpha}(B)$. But $C_{W_\alpha}(B) = 1$ since $H_{\alpha+1}$ is non-trivial; therefore $R = 1$. ▯

For example, if each H_α is cyclic with prime order p, the group W is a locally finite p-group with trivial Gruenberg radical. To obtain a characteristically simple example we form the wreath power

$$W^* = W \mathrm{r}\, W^Z$$

where Z is the set of integers in the natural order. $(W^*)'$ is clearly a locally finite p-group and by Theorem 6.25 it is characteristically simple. Now $(W^*)'$ cannot be a Gruenberg group since it has subgroups isomorphic with W'. Hence the Gruenberg radical of $(W^*)'$ is trivial. We state this as a

Corollary. For each prime p there exists a characteristically simple, locally finite p-group with trivial Gruenberg radical.

Levič and Tokarenko [1] have shown that there exist torsion-free locally nilpotent groups which are not Gruenberg groups.

6.3 The Effect of Finiteness Conditions on Generalized Nilpotent Groups

We consider the effect on the hierarchy of classes of generalized nilpotent groups when finiteness conditions are imposed. In the first place it is clear that *a locally finite Baer-nilpotent group is locally nilpotent*; also *a locally finite, residually central group is locally nilpotent*, by the Corollary to Theorem 6.12. Hence for locally finite groups the properties Baer-nilpotent, residually central and locally nilpotent coincide, as must all intervening properties. On the other hand, p-groups need not be locally nilpotent by Golod's example.

Next we consider the effect of imposing minimal conditions. Recall that the residually central groups that satisfy Min-n are precisely the hypercentral Černikov groups (Corollary to Theorem 6.12); also, under Min-n the classes of Baer groups, Fitting groups and N_1-groups coincide with the class of nilpotent Černikov groups (see Part 1, p. 155).

However the structure of Baer-nilpotent groups and p-groups satisfying Min is unknown. For example, it is an open question whether a Baer-nilpotent group satisfying Min is an S-group, i.e. a direct product of p-groups. In the positive direction it is known that the Engel groups which satisfy Min are just the hypercentral Černikov groups (Vilyacer [1]; see also Section 7.2). In addition there is the following result.

Theorem 6.31. The U-groups which satisfy Min are precisely the hypercentral Černikov groups.

Proof. Suppose that there is a U-group which satisfies Min but which is not a Černikov group. Since the class of U-groups is $\boldsymbol{S}$-closed, there is a U-group G which is a minimal non-Černikov group. If G is not finitely generated, then each finitely generated subgroup is proper and hence is a Černikov group; thus G is locally finite and therefore locally nilpotent. However, this is a contradiction because a locally nilpotent group which satisfies Min is a Černikov group (Corollary 2 to Theorem 5.27). Hence G must be finitely generated. It now follows via Zorn's Lemma that G has a maximal normal subgroup N and N is a Černikov group. G/N is a finitely generated chief-factor of G, so, by definition of a U-group, G/N is abelian and hence is finite. Thus G is a Černikov group. ▯

We turn now to maximal conditions. By Theorem 5.37 a locally nilpotent group that satisfies Max-n is finitely generated and nilpotent. Hence all classes between $\boldsymbol{L}\mathfrak{N}$ and $\mathfrak{N}$ coincide under Max-n. For the classes of generalized nilpotent groups that contain the class $\boldsymbol{L}\mathfrak{N}$ the effect of even Max is far from clear. An Engel group satisfying Max is nilpotent by a result of Baer ([28], Satz N: see Section 7.2). On the other hand it is unknown if a Baer-nilpotent group or even an $\tilde{N}$-group that satisfies Max is nilpotent. Nor is it known whether a residually nilpotent group satisfying Max is polycyclic. It is easy to see that *a U-group satisfying Max and a Z-group satisfying Max-n are hypocentral*. It is open whether a residually central group satisfying Max-n is hypocentral.

Linearity. Imposition of the finiteness condition linearity has a more decisive effect. We consider first locally nilpotent linear groups.

Theorem 6.32. Let G be a locally nilpotent linear group over a field F.

(i) If S is the torsion-subgroup of G, then G/S is nilpotent.

(ii) If p is a prime not equal to the characteristic of F, the p-component of G is a Černikov group.

(iii) If the characteristic of F is a prime p, the p-component of G lies in $\zeta_m(G)$ for some finite m and has finite exponent (but need not be a Černikov group).

(iv) G is hypercentral. (Garaščuk [1] Kegel [2] p. 542).

Corollary 1 (Wehrfritz [5]). A locally nilpotent linear group is nilpotent if it contains no quasicyclic subgroups.

Corollary 2 (Suprunenko and Medvedeva [1]). A locally nilpotent linear group over the field of rational numbers is nilpotent.

Corollary 3 (cf. Wehrfritz [5], Corollary 2). A locally nilpotent linear group satisfies the maximal condition on normal abelian subgroups if and only if it is nilpotent and satisfies Max.

Proofs. Corollary 1 follows at once from the theorem while Corollary 2 follows from Corollary 1 and Lemma 5.29.1.

Let G be a locally nilpotent linear group satisfying Max-*nab*, the maximal condition on normal abelian subgroups. By parts (ii) to (iv) of the theorem, the torsion-subgroup of G lies in some $\zeta_l(G)$ where l is finite, and by (i) G is nilpotent. To complete the proof let us show that *if H is any nilpotent group satisfying Max-nab, then H satisfies Max.* Let A be a maximal normal abelian subgroup of H. A subgroup lying between $A \cap \zeta_i(H)$ and $A \cap \zeta_{i+1}(H)$ is abelian and normal in H; hence

$$A \cap \zeta_{i+1}(H)/A \cap \zeta_i(H)$$

satisfies Max and, since H is nilpotent, A satisfies Max. We note that $A = C_H(A)$ and that Theorem 3.27 implies that H/A satisfies Max. Hence H satisfies Max. The converse is obvious. □

In connection with this last result it is worth noting that by Theorem 6.27 there exist non-trivial locally nilpotent groups with trivial Gruenberg radical; such groups satisfy even the maximal condition on ascendant abelian subgroups but do not satisfy Max.

To prove Theorem 6.32 we need two auxiliary results.

Lemma 6.33. A torsion-free locally nilpotent group which is a periodic extension of a nilpotent group of class c is itself nilpotent of class c.

Proof. Let G be a torsion-free locally nilpotent group and suppose that N is a normal subgroup of G such that N is nilpotent of class c

and G/N is periodic. We can assume that $c > 0$. Let $a \in \zeta(N)$ and $g \in G$; then $[a, g^m] = 1$ for some $m > 0$. Hence $g^m \in \zeta(H)$ where $H = \langle a, g \rangle$, a finitely generated, torsion-free nilpotent group. But $H/\zeta(H)$ is torsion-free by Theorem 2.25, so $g \in \zeta(H)$ and $[a, g] = 1$. It follows that $\zeta(N) \leqq \zeta(G)$; the same argument proves that $G/\zeta(G)$ is torsion-free. Hence $G/\zeta(G)$ is nilpotent of class $\leqq c - 1$ by induction on c, and G is nilpotent of class c. ▯

Lemma 6.34 (G. Baumslag [3]). A p-group G which is an extension of a nilpotent group of finite exponent by a finite group is nilpotent.

Proof. Let $N \lhd G$ where N is a nilpotent group of finite exponent and G/N is finite. Suppose that H and K are normal subgroups of G contained in N such that $H \leqq K$ and K/H is an elementary abelian group. Let x be an element of G with order p^m and let $k \in K$; then

$$[k, \underbrace{x, \ldots, x}_{p^m}] \equiv [k, x^{p^m}] \equiv 1 \bmod H.$$

Since N, and consequently G, has finite exponent, it follows that

$$[N, \underbrace{x, \ldots, x}_{i}] = 1$$

for some i, and this implies that $\langle x, N \rangle$ is nilpotent*. Let $\{x_1, \ldots, x_n\}$ be a transversal to N in G. If G/N is abelian, $\langle x_i, N \rangle \lhd G$ and G is the product of the normal nilpotent subgroups $\langle x_i, N \rangle$. By Fitting's theorem G is nilpotent in this case. Suppose that G/N has nilpotent class $c > 1$; then Nx_i^G/N has nilpotent class at most $c - 1$. By induction on c each x_i^G is nilpotent. Since $G = Nx_1^G \ldots x_n^G$, the group G is nilpotent. ▯

Proof of Theorem 6.32. Let G be a group of $n \times n$ matrices over a field F, which we may suppose to be algebraically closed. Assume that G is locally nilpotent. Then G is soluble by Zassenhaus' theorem (3.23), and by Mal'cev's theorem on soluble linear groups (Theorem 3.21) G contains a normal subgroup T with finite index consisting of upper triangular matrices. Let U be the subgroup of all unitriangular matrices in T; then $U \lhd G$.

We show first that $U \leqq \zeta_{n-1}(T)$. Let e_{ij} denote the $n \times n$ matrix with 1 in its (i, j)th position and 0 elsewhere: let U_k consist of all matrices in U which have the form

$$a = 1 + \sum_{j-i \geqq k} \alpha_{ij} e_{ij} \quad (\alpha_{ij} \in F).$$

Then $1 = U_n \leqq U_{n-1} \leqq \cdots \leqq U_1 = U$; it will be enough to prove that $[U_k, T] \leqq U_{k+1}$. If $a \in U_k$, we can write

$$a = 1 + \sum_{i=1}^{n-k} \alpha_{i\,i+k} e_{i\,i+k} + \cdots$$

---

* Use induction on the nilpotent class of N.

where we are omitting terms in e_{ij} for which $j - i > k$. Let $x \in T$ and let $d_1, d_2, \ldots, d_n$ be the coefficients on the principal diagonal of x. Then

$$[a, x] = 1 + \sum_{i=1}^{n-k} (d_i^{-1} d_{i+k} - 1)\, \alpha_{i\,i+k} e_{i\,i+k} + \cdots. \tag{19}$$

Now the subgroup $\langle a, x\rangle$ is nilpotent, with positive class c say. Hence

$$1 = [a, \underbrace{x, \ldots, x}_{c}] = 1 + \sum_{i=1}^{n-k} (d_i^{-1} d_{i+k} - 1)^c\, \alpha_{i\,i+k} e_{i\,i+k} + \cdots.$$

Consequently $(d_i^{-1} d_{i+k} - 1)^c\, \alpha_{i\,i+k} = 0$ and therefore

$$(d_i^{-1} d_{i+k} - 1)\, \alpha_{i\,i+k} = 0.$$

It follows from (19) that $[a, x]$ (which surely belongs to U) is in U_{k+1}, and this proves that $[U_k, T] \leqq U_{k+1}$.

Now T/U is isomorphic with a group of diagonal matrices, so it is abelian. Hence T is nilpotent and G is nilpotent-by-finite. Let S be the torsion-subgroup of G; then G/S is nilpotent-by-finite, as well as being torsion-free and locally nilpotent. By Lemma 6.33 the group G/S is nilpotent.

Let p be a prime and let P be the p-component of G. Suppose that p does not equal the characteristic of the field F. Then U is either torsion-free or a q-group where q is a prime other than p, so $P \cap U = 1$ and $P \simeq PU/U$. Now G/T is finite and T/U is isomorphic with a subgroup of the direct product of n copies of F^*, the multiplicative group of F; also, it is well-known that the torsion-subgroup of F^* is a direct product of r^∞-groups, one for each prime r not equal to the characteristic of F (see Fuchs [3], Lemma 76.1). Therefore P is a Černikov group.

Now suppose that the characteristic of F equals p; then U is a p-group with exponent dividing p^{n-1} and U is contained in P, the p-component of G. Let

$$R_k = U \cap \zeta_{k+1}(T)/U \cap \zeta_k(T);$$

then $T \leqq C_G(R_k)$ and G induces in R_k a finite group of automorphisms G_k. Now R_k is an abelian p-group with finite exponent and G is locally nilpotent: it is easy to deduce from these facts that G_k is a finite p-group. The semi-direct product of R_k by G_k is therefore nilpotent, by Lemma 6.34. But $U \leqq \zeta_{n-1}(T)$, so $U \leqq \zeta_{m_0}(G)$ for some finite m_0. Since F^* has no elements of order p, the group P/U is finite. We deduce by way of Theorem 4.38 that $P/U \leqq \zeta_\omega(G/U)$ and hence that $P/U \leqq \zeta_{m_1}(G/U)$ for some finite m_1; therefore $P \leqq \zeta_m(G)$ where $m = m_0 + m_1$.

Finally, by what we have already shown, the torsion-subgroup S of G lies in the FC-hypercentre of G and G/S is nilpotent. Theorem 4.38 shows that G is hypercentral. ▯

On the other hand, a *locally nilpotent linear group need not be nilpotent*: for example the group generated by the matrices

$$a_i = \begin{pmatrix} \xi_i & \cdot \\ \cdot & 1 \end{pmatrix}, \quad b_i = \begin{pmatrix} 1 & \cdot \\ \cdot & \xi_i \end{pmatrix}, \quad t = \begin{pmatrix} \cdot & 1 \\ 1 & \cdot \end{pmatrix},$$

where $i = 1, 2, \ldots$, and ξ_i is a primitive complex 2^ith root of unity, is hypercentral with length $\omega + 1$ (Gruenberg [6], p. 292).

Wehrfritz [5] has shown that the length of the upper central series of a linear group of degree n is at most $\omega + \frac{1}{2}(n!)$ and he also obtains upper bounds—which are best possible—for the hypercentral length of a locally nilpotent linear group over an algebraically closed field: see also Gruenberg [6], [7] and Wehrfritz [9] (§ 8).

Theorem 6.35 (cf. Suprunenko and Garaščuk [2]). A linear Baer group is nilpotent.

Proof. Let G be a linear Baer group over a field F. Let p be a prime not equal to the characteristic of F and denote the p-component of G by P. By Theorem 6.32 the group P is Černikov, so P has a characteristic subgroup R with finite index which is also a radicable abelian p-group. Evidently it is sufficient to prove that R is contained in $\zeta_r(G)$ for some finite r, by Theorem 6.32. Let X be a finitely generated subgroup of G. Then X is nilpotent and subnormal in G. If we define $R_0 = R$ and $R_{i+1} = [R_i, X]$, then for some integer m

$$1 = R_m < R_{m-1} < \cdots < R_1 < R_0 = R$$

since $R \lhd G$. Suppose that R_i is radicable: if $x \in X$, then $[R_i, x]$, being an endomorphic image of R_i, is radicable, so $R_{i+1} = \prod_{x \in X} [R_i, x]$ is radicable. Hence by induction on i each R_i is radicable. Thus R_{i+1} is a direct factor of R_i. If R has rank r, the impossibility of expressing R as a direct product of more than r groups of type p^∞ shows that $m \leqq r$ and $R_r = 1$. This is true for all X, so

$$[R, \underbrace{G, \ldots, G}_{r}] = 1$$

and $R \leqq \zeta_r(G)$. □

In [5] (Theorem 0) Gruenberg has shown how to generalize Theorem 6.32 (iv) and Theorem 6.35 to linear groups over a commutative Noetherian ring R, that is to say to groups that can be faithfully represented as groups of R-automorphisms of finitely generated R-modules. This is not possible if R is merely a commutative ring by examples of Tokarenko [1].

Baer-nilpotent Linear Groups

As Baer has pointed out in [36] (p. 57), *a Baer-nilpotent linear group is hypercentral.* We only sketch the proof. Let G be a finitely generated Baer-nilpotent linear group of degree n. By a result of Mal'cev ([1], Theorems VII and VIII) G is residually a finite linear group of degree n. Zassenhaus' theorem (3.23) now implies that G is soluble. But a soluble Baer-nilpotent group is locally nilpotent (Baer [18], § 2, Satz 2: this will be proved in Section 10.5). Hence G is nilpotent and the assertion now follows from Theorem 6.32.

On the other hand every free group is linear (Part 1, p. 40) and Magnus's theorem asserts that all free groups are residually nilpotent. Consequently residually nilpotent, linear groups need not be locally nilpotent.

Locally Nilpotent Groups with Finite Abelian Subgroup Rank

Let A be an abelian group. The *0-rank* or *torsion-free rank* of A is the cardinal of a maximal independent set consisting of elements of A with infinite order: if p is a prime, the *p-rank* of A is the cardinal of a maximal independent set of non-trivial elements of A each of which has order a power of p. It is well-known that p-rank and 0-rank are invariants of A (Fuchs [3], § 8). We recall that *the rank* of a group, on the other hand, is the minimal number of elements required to generate a finitely generated subgroup if this is finite and otherwise is ∞ (Section 1.4, pp. 33–34).

It is easy to see that A has finite 0-rank if and only if the factor group of A by its torsion-subgroup is a subgroup of the direct product of finitely many copies of the additive group of rational numbers, and that A has finite p-rank if and only if the p-component of A is the direct product of finitely many cyclic or quasicyclic groups.

Following Baer ([44], p. 95) we say that a group G has *finite abelian subgroup rank* if each abelian subgroup has finite 0-rank and finite p-rank for all primes p. Obviously G has finite abelian subgroup rank if it satisfies either Max-*ab* or Min-*ab*.

Our main objective is the following theorem:

Theorem 6.36 (Mal'cev [7], Theorem 5). Let G be a locally nilpotent group. Then the abelian subgroups of G have finite 0-rank if and only if the factor group of G by its torsion-subgroup is a torsion-free nilpotent group of finite rank.

It is convenient to precede this with a lemma.

Lemma 6.37 (cf. Čarin [9], Theorem 9). Let G be a locally nilpotent group and let N be a normal subgroup of G. If N is torsion-free and abelian of finite rank r, then $N \leqq \zeta_r(G)$ and $G/C_G(N)$ is a torsion-free nilpotent group of finite rank.

Proof. Let X be a free abelian subgroup of N with maximal rank, so that X has rank r and N/X is periodic. Let Y be any finitely generated subgroup of G; then $\langle X, Y\rangle$ is nilpotent, of class c say. If $a \in N$, then $a^m \in X$ for some positive integer m. Since N is abelian,

$$1 = [a^m, \underbrace{Y, \ldots, Y}_{c}] = [a, \underbrace{Y, \ldots, Y}_{c}]^m;$$

but N is torsion-free, so

$$[N, \underbrace{Y, \ldots, Y}_{c}] = 1.$$

Through its action on N the group Y may therefore be represented by a group of $r \times r$ unitriangular matrices over Q, the field of rational numbers. Consequently

$$[N, \underbrace{Y, \ldots, Y}_{r}] = 1$$

and it follows that N lies in $\zeta_r(G)$. Also, $G/C_G(N)$ is isomorphic with a group of $r \times r$ unitriangular matrices over Q, so it is torsion-free and nilpotent. Finally, the group of $r \times r$ unitriangular matrices over Q has finite rank since each upper central factor is a direct sum of finitely many copies of Q. □

Proof of Theorem 6.36. The sufficiency of the condition is obvious. Let G be a locally nilpotent group in which each abelian subgroup has finite 0-rank.

We assume first of all that G is torsion-free. If G is not nilpotent, then clearly it contains a countable subgroup which is not nilpotent. Thus we can assume that G is countable and that there is a sequence of nilpotent subgroups $N_1, N_2, \ldots$ such that

$$1 < N_1 < N_2 < \cdots$$

and G is the union of the N_i. Let Z_i be the centre of N_i and let Z be the subgroup generated by the Z_i. Then Z is abelian and torsion-free, so by hypothesis it has finite rank. Consequently, if F is a free abelian subgroup of Z with maximal rank, F is finitely generated and Z/F is periodic. Since F satisfies Max, there is an integer i such that

$$F \cap Z_j \leqq \prod_{k=1}^{i} (F \cap Z_k)$$

for all j. Thus if $j \geqq i$, we have $F \cap Z_{j+1} \leqq N_i \leqq N_j \leqq N_{j+1}$, which implies that $F \cap Z_{j+1} \leqq F \cap Z_j$. Now N_{j+1}/Z_{j+1} is torsion-free, by Theorem 2.25, and

$$F \cap Z_j / F \cap Z_{j+1} \simeq (F \cap Z_j)\, Z_{j+1}/Z_{j+1} \leqq N_{j+1}/Z_{j+1}.$$

Hence $F \cap Z_{j+1}$ either has smaller rank than $F \cap Z_j$ or coincides with it. Therefore there is an integer $j \geqq i$ such that

$$F \cap Z_j = F \cap Z_{j+1} = \text{etc.}$$

Consequently $F \cap Z_j \leqq \zeta(G)$. Also $F \cap Z_j = 1$ would imply that $Z_j \simeq Z_j F/F$, which is periodic; since G is torsion-free, this yields the contradiction $Z_j = 1$. It follows that

$$\zeta(G) \neq 1.$$

Now consider the group $G/\zeta(G)$. Suppose $x\zeta(G)$ is an element with finite order m; then for all sufficiently large i we have $x \in N_i$ and therefore $x^m \in Z_i$. But N_i/Z_i is torsion-free, so $x \in Z_i$ and hence $x \in \zeta(G)$. Consequently $G/\zeta(G)$ is torsion-free. Next let $A/\zeta(G)$ be an abelian subgroup of $G/\zeta(G)$, so A is nilpotent of class at most 2. Let M be a maximal normal abelian subgroup of A. Then $M = C_A(M)$ by Lemma 2.19.1 and A/M has finite rank by Lemma 6.37. Hence A has finite rank.

We have proved that $G/\zeta(G)$ inherits the properties of G. Consequently the first part of the proof shows that

$$\zeta_i(G) < \zeta_{i+1}(G)$$

for each integer i. Now define $L = \zeta_\omega(G)$, a hypercentral group, and let N be a maximal normal abelian subgroup of L. Then $N = C_L(N)$ and N has finite rank. Hence, by Lemma 6.37, the group L is nilpotent and has finite rank—here we recall that to have finite rank is a $\boldsymbol{P}$-closed property by Lemma 1.44. It follows that we can choose a finitely generated subgroup X of L such that $\zeta_{i+1}(G)X/\zeta_i(G)X$ is periodic for each positive integer i. Thus for any $x \in L$ there is a positive integer m such that $x^m \in X$. But X is finitely generated, so $X \leqq \zeta_i(G)$, for some integer i. Since $L/\zeta_i(G)$ is torsion-free by Theorem 2.25, it follows that $L = \zeta_i(G)$, which is a contradiction.

We return now to the general case and denote by T the torsion-subgroup of G. To complete the proof we have only to show that an abelian subgroup of G/T has finite 0-rank. Suppose that $x_1T, x_2T, \ldots$ generate a free abelian subgroup of G/T with countably infinite rank. Let i and j be two positive integers and let $J = \langle x_i, x_j \rangle$. Then J is a finitely generated nilpotent group, so its torsion-subgroup S is finite, being a finitely generated, periodic nilpotent group. Since $J' \leqq J \cap T = S$, the group J/S is abelian. For some positive integer r we have $[S, x_j^r] = 1$. Since $[x_i, x_j^r] \in S$, it follows that

$$[x_i, x_j^r, x_j^r] = 1. \qquad (20)$$

Let s be the order of $[x_i, x_j^r]$. Then (20) implies that $[x_i, x_j^{rs}] = 1$. We have therefore found for each pair of positive integers i and j a positive integer m_{ij} such that

$$[x_i, x_j^{m_{ij}}] = 1.$$

It follows that the subgroup

$$H = \langle x_1, x_2^{d_2}, x_3^{d_3}, \ldots\rangle$$

is abelian where $d_i = m_{1i} m_{2i} \ldots m_{i-1i}$ if $i \geqq 2$. Hence H has finite 0-rank and there is a non-trivial relation between the x_i and therefore between the $x_i T$. This contradicts our choice of the $x_i T$. ▯

Corollary 1. A locally nilpotent group G has finite abelian subgroup rank if and only if the torsion-subgroup is a direct product of Černikov p-groups and its factor group in G is torsion-free and nilpotent of finite rank. Such a group is hypercentral.

Proof. Let G have finite abelian subgroup rank. Then each primary component of G satisfies Min-*ab* and is therefore a Černikov group by Theorem 3.32. By Corollary 1, Theorem 4.38, the torsion-subgroup of G lies in the hypercentre; hence G is hypercentral. The rest is clear. ▯

Corollary 2 (Mal'cev [7], Theorem 5; Myagkova [1], § 4, Theorem 1). If G is a locally nilpotent group, the following properties of G are equivalent.

(i) G has finite rank.

(ii) Each abelian subgroup of G has finite rank.

(iii) If T is the torsion-subgroup of G, then T is a direct product of Černikov p-groups of bounded rank for different primes p and G/T is a torsion-free nilpotent group of finite rank.

We remark that *if G is a Baer group with finite rank r, then all quasicyclic subgroups of G lie in $\zeta_r(G)$*: the proof is similar to that of Theorem 6.35 From this it follows easily that *a Baer group of finite rank with elements of only finitely many distinct prime orders is nilpotent* (Gruenberg [4], Theorem 1.3, Čan Van Hao [3]: for Fitting groups this was found by Mal'cev ([7], Theorem 6).

However in general a Baer group with finite rank need not even be soluble (see Section 10.3, p. 179).

Nilpotent and locally nilpotent torsion-free groups in which a maximal abelian subgroup has finite rank are studied in two paper of Sesekin [2], [3]. We mention also that Vilyacer [1] has shown that *a weakly nilpotent group with finite rank is locally nilpotent and hence hypercentral.*

Further properties of torsion-free locally nilpotent groups relating to their subgroups of finite rank may be found in a paper of Gluškov [4]: see also Plotkin [12], § 10.

Theorem 6.38 (Čarin [9], Theorem 1). Let N be a normal subgroup of the locally nilpotent group G. Then N lies in the hypercentre of G if and only if N has an ascending G-admissible series whose factors are abelian with finite rank.

Proof. The condition is trivially necessary if N is to lie in the hypercentre of G; for we can intersect N with the upper central series of G term by term and refine the resulting series to an ascending G-admissible series in N with cyclic factors.

Conversely, let N satisfy the condition and assume that N is not contained in $\bar{\zeta}(G)$, the hypercentre of G. Then it is easy to show that $G/\bar{\zeta}(G)$ has a non-trivial normal abelian subgroup which is either finite or torsion-free of finite rank. In each case we obtain the contradiction that $G/\bar{\zeta}(G)$ has non-trivial centre by using either Theorem 4.38 or Lemma 6.37. □

Corollary (Čarin [9] Theorem 2). A locally nilpotent group is hypercentral if and only if it has an ascending normal series whose factors are abelian with finite rank.

For locally finite p-groups the result of this corollary appears in a much earlier paper of Baer ([7], p. 410).

Chapter 7

Engel Groups

7.1 Left and Right Engel Elements

Let x and y be elements of a group G and let n be a non-negative integer. The commutator

$$[x, {}_n y]$$

is defined by the rules

$$[x, {}_0 y] = x \quad \text{and} \quad [x, {}_{n+1} y] = [[x, {}_n y], y].$$

x is called a *right Engel element* of G if for each g in G there is an integer $n = n(x, g) \geqq 0$ such that

$$[x, {}_n g] = 1.$$

Observe that the variable element g is on the *right* here. If n can be chosen independently of g, then x is called a *right n-Engel element* or simply a *bounded right Engel element of G*. The sets of right Engel elements and bounded right Engel elements of G are denoted by

$$R(G) \quad \text{and} \quad R(G)$$

respectively.

An element x is called a *left Engel element* of G if for each g in G there exists an integer $n = n(x, g) \geqq 0$ such that

$$[g, {}_n x] = 1.$$

Here the variable g appears on the *left*. If n can be chosen independently of g, then x is called a *left n-Engel element* or simply a *bounded left Engel element of G*. The sets of left Engel elements and bounded left Engel elements of G are denoted by

$$L(G) \quad \text{and} \quad \bar{L}(G).$$

It is an unsolved problem whether these four subsets are always subgroups; it is evident they are normal subsets of G. The following relation holds between left and right Engel elements.

Theorem 7.11 (Heineken [2]). In any group G the inverse of a right Engel element is a left Engel element and the inverse of a right n-Engel element is a left $(n+1)$-Engel element. Thus

$$R(G)^{-1} \subseteq L(G) \quad \text{and} \quad \bar{R}(G)^{-1} \subseteq \bar{L}(G).$$

Proof. Let x be a right n-Engel element and let g be any element of G. Then x^g is also a right n-Engel element of G, so

$$1 = [x^g, {}_n x^{-1}]^{x^{-1}} = [[x\,[x, g], x^{-1}], {}_{n-1}x^{-1}]^{x^{-1}}$$
$$= [[x, g], {}_n x^{-1}]^{x^{-1}} = [g, {}_{n+1}x^{-1}]$$

since $[x, g]^{x^{-1}} = [g, x^{-1}]$. This argument also shows that

$$R(G)^{-1} \subseteq L(G). \quad \square$$

However it is not known whether a right Engel element is necessarily a left Engel element.

Let G be an arbitrary group, let x belong to the Hirsch-Plotkin radical $\varrho_{L\mathfrak{N}}(G)$ and let g be any element of G. Then $[g, x] \in \varrho_{L\mathfrak{N}}(G)$ and consequently the subgroup $\langle x, [g, x]\rangle$ is nilpotent. Hence $[g, {}_n x] = 1$ for some $n \geqq 0$. It follows that $L(G)$ contains the Hirsch-Plotkin radical of G. Now let x belong to the Baer radical $\varrho_{N\mathfrak{A}}(G)$ and let $g \in G$. Then $\langle x\rangle \lhd^n G$ for some integer n and this implies that $[g, {}_n x] \in \langle x\rangle$; hence $[g, {}_{n+1}x] = 1$. Thus x is a left $(n+1)$-Engel element of G and $\bar{L}(G)$ contains the Baer radical of G.

We state these results as the first part of

Lemma 7.12. Let G be any group.

(i) $L(G)$ contains the Hirsch-Plotkin radical of G and $\bar{L}(G)$ contains the Baer radical of G.

(ii) $R(G)$ contains the hypercentre $\bar{\zeta}(G)$ of G and $\bar{R}(G)$ contains $\zeta_\omega(G)$, the ω-hypercentre of G.

The proof of the second part of the lemma is also straightforward and we omit it.

The major goals of Engel theory can be stated as follows: to find conditions on a group G which will ensure that $L(G)$, $\bar{L}(G)$, $R(G)$ and $\bar{R}(G)$ are subgroups and, if possible, coincide with the Hirsch-Plotkin radical, the Baer radical, the hypercentre and the ω-hypercentre respectively.

We must observe that the celebrated example of Golod [1] is a finitely generated, infinite p-group in which each element is left Engel: this shows that in general $L(G)$ is larger than the Hirsch-Plotkin radical of G. It is also known that $R(G)$ and $\bar{R}(G)$ can differ from $\bar{\zeta}(G)$ and $\zeta_\omega(G)$ (see p. 57): whether the corresponding phenomenon can occur for $\bar{L}(G)$ is unknown.

Engel and n-Engel Groups

If G is a group, it is obvious that $G = L(G)$ and $G = R(G)$ are equivalent statements; a group with this property is called an *Engel group* (Gruenberg [1]) and the class of Engel groups is denoted by

$$\mathfrak{E}.$$

By an *n-Engel group* we mean a group G such that $[x, {}_n y] = 1$ for all x and y in G, so that every element of G is both a left and a right n-Engel element. The class of n-Engel groups is written

$$\mathfrak{E}_n.$$

Clearly

$$L\mathfrak{N} \leqq \mathfrak{E} \quad \text{and} \quad \mathfrak{N}_n \leqq \mathfrak{E}_n,$$

and Golod's example shows that there exist non-locally nilpotent Engel groups. However it is not known if every $\mathfrak{E}_n$-group is locally nilpotent.

Left and Right Normed Commutators

The element $[x, {}_n y]$ is a *left-normed* commutator. Some authors employ *right-normed* commutators $[{}_n y, x]$ defined by

$$[{}_0 y, x] = x \quad \text{and} \quad [{}_{n+1} y, x] = [y, [{}_n y, x]],$$

and define terms such as left and right Engel in terms of these. For example, in the right normed theory an element x of a group G is right Engel if, given g in G, there is an integer $n \geqq 0$ such that $[{}_n g, x] = 1$. Thus it is important to establish the relation between these types of commutators.

Let y be an element of a group G and let the mappings $\lambda(y)$ and $\varrho(y)$ be defined by the rules

$$\lambda(y): x \to [y, x] \quad \text{and} \quad \varrho(y): x \to [x, y].$$

If $\tau(y)$ denotes the inner automorphism of G induced by transformation by y, one verifies at once that

$$\varrho(y) = \lambda(y^{-1})\, \tau(y).$$

Since $\lambda(y^{-1})$ and $\tau(y)$ commute,

$$(\varrho(y))^n = (\lambda(y^{-1}))^n\, (\tau(y))^n = (\lambda(y^{-1}))^n\, \tau(y^n)$$

for each non-negative integer n. Consequently

$$[x, {}_ny] = [{}_ny^{-1}, x]^{y^n}.$$

Hence $[x, {}_ny] = 1$ and $[{}_ny^{-1}, x] = 1$ are equivalent. This means that the term right Engel (right n-Engel) is the same in both theories, but that a left Engel (left n-Engel) element in the left-normed theory is the inverse of a left Engel (left n-Engel) element in the right-normed theory. Allowance for this fact must be made in interpreting theorems about left Engel elements. On the other hand, the classes $\mathfrak{E}$ and $\mathfrak{E}_n$ are the same in both theories.

We remark that the semigroups generated by the mappings $\lambda(g)$ and $\varrho(g)$ for g an element of a group G are the subject of four papers by N. D. Gupta [1]—[4] and three papers by Heineken [8], [9], [10]. In [2] Gupta has proved that *if G is a group without elements of order* 3 *and if $\varrho(g)^n = \varrho(g)^{n+1}$ for all g in G, then G is an n-Engel group.*

Right 2-Engel Elements and 2-Engel Groups

Obviously a 0-Engel group has order 1 and the 1-Engel groups are the abelian groups. For less trivial reasons 2-Engel groups also have a relatively simple structure: 2-Engel groups have been studied by Burnside [4], Hopkins [1], Levi [2] and Kappe [1], [2].

Theorem 7.13 (Levi [2], Kappe [1]). Let a be a right 2-Engel element and let x, y and z be arbitrary elements of a group G.

(i) a is a left 2-Engel element.

(ii) a^G is abelian and its elements are right (and therefore left) 2-Engel elements.

(iii) $[a, x, y] = [a, y, x]^{-1}$.

(iv) $[a, [x, y]] = [a, x, y]^2$.

(v) $[a, x, y, z]^2 = 1$.

(vi) $[a, [x, y], z] = 1$.

Proof. Since a is a right 2-Engel element,

$$1 = [a, ax, ax] = [a, x, ax] = [a, x, a]^x,$$

so $[x, a, a] = 1$ and a is a left 2-Engel element of G.

In view of (i), we have

$$[a^x, a^y] = [a^{xy^{-1}}, a]^y = [a\,[a, xy^{-1}], a]^y = 1,$$

so a^G is abelian. Denote by

$$\varrho(x)$$

the mapping $b \to [b, x]$, $(b \in a^G)$. Then, since a^G is abelian, $\varrho(x)$ is an endomorphism of a^G. Also $\varrho(x)^2$ maps a^y onto $[a^y, x, x] = [a, x^{y^{-1}} x^{y^{-1}}]^y = 1$. Hence

$$\varrho(x)^2 = 0 \tag{1}$$

and a^G consists of right 2-Engel elements.

The formula $[b, xy] = [b, y]\,[b, x]\,[b, x, y]$ and the commutativity of a^G show that

$$\varrho(xy) = \varrho(x) + \varrho(y) + \varrho(x)\,\varrho(y). \tag{2}$$

Also, from $[b, x^{-1}] = [b, x]^{-x^{-1}}$ we obtain

$$\varrho(x^{-1}) = -\varrho(x), \tag{3}$$

using (1).

Since $(xy)^{-1}$ commutes with $[b, xy]$ for each b in a^G, we have

$$0 = \varrho(xy)\,\varrho(y^{-1}x^{-1}) = (\varrho(x) + \varrho(y) + \varrho(x)\,\varrho(y))\,(-\varrho(y) - \varrho(x) + \varrho(y)\varrho(x))$$
$$= -\varrho(x)\,\varrho(y) - \varrho(y)\,\varrho(x),$$

using (1). Therefore

$$\varrho(x)\,\varrho(y) = -\varrho(y)\,\varrho(x) \tag{4}$$

and consequently $[a, x, y] = [a, y, x]^{-1}$. Next (2) shows that

$$\varrho([x, y]) = \varrho((yx)^{-1}) + \varrho(xy) + \varrho((yx)^{-1})\,\varrho(xy).$$

Expanding the right side by means of (2) and (3), and using (4), we obtain

$$\varrho([x, y]) = 2\varrho(x)\,\varrho(y), \tag{5}$$

so that $[a, [x, y]] = [a, x, y]^2$.

By (4)

$$\varrho(x)\,\varrho(yz) + \varrho(yz)\,\varrho(x) = 0.$$

Using (2) to simplify this, we obtain

$$0 = \varrho(x)\,\varrho(y)\,\varrho(z) + \varrho(y)\,\varrho(z)\,\varrho(x) = 2\varrho(x)\,\varrho(y)\,\varrho(z). \tag{6}$$

Thus $[a, x, y, z]^2 = 1$. Next (5) and (6) show that

$$\varrho([x, y])\,\varrho(z) = 2\varrho(x)\,\varrho(y)\,\varrho(z) = 0,$$

so $[a, [x, y], z] = 1$, as required. □

Corollary 1 (Kappe [1]). The right 2-Engel elements of a group form a characteristic subgroup.

Proof. Let a and b be right 2-Engel elements of G and let x be any element of G. Then

$$\begin{aligned}[ab^{-1}, x, x] &= [[a, x]^{b^{-1}} [b, x]^{-b^{-1}}, x] \\ &= [[a, x] [b, x]^{-1}, x[x, b]]^{b^{-1}} \\ &= 1,\end{aligned}$$

by parts (iii) and (vi) of the theorem. Hence ab^{-1} is a right 2-Engel element of G. ▯

We remark that the left 2-Engel elements of a group do not form a subgroup in general. For example, the standard wreath product of a group of order 2 with an elementary abelian group of order 4 is generated by left 2-Engel elements, but does not *consist* of such elements.

Corollary 2 (Macdonald and Neumann [1]). In an arbitrary group a right 2-Engel element of odd order lies in the third term of the upper central series.

This follows from (ii) and (v) of Theorem 7.13. Notice that Corollary 2 is not true for right 2-Engel elements in general: for in the standard wreath product of a group of order 2 with a countably infinite, elementary abelian 2-group the base group is the set of right 2-Engel elements but the hypercentre is trivial.

It is easy to show that the *norm* or intersection of all normalizers of subgroups of a group G consists of right 2-Engel elements. For comparison with Corollary 2 we mention that *the norm of a group always lies in the second term of the upper central series*: this is due to Schenkman [6] and improves an earlier result of Wos [1]. Recently Cooper ([1], Theorem 2.2.1) has generalized Schenkman's theorem as follows: *an automorphism which leaves every subgroup of a group G invariant induces the trivial automorphism in the central factor group $G/\zeta(G)$*.

Corollary 3 (Levi [2]; see also Hopkins [1]). A 2-Engel group is nilpotent of class at most 3.

This follows from Theorem 7.13 (vi). The number 3 is the least that will do here: for example, the Burnside group $B(3,3)$ has order 3^7 and is nilpotent of class exactly 3 (Hopkins [1], Levi and van der Waerden [1]; see also M. Hall [2], p. 323); by Theorem 7.14 the group $B(3,3)$ is 2-Engel.

Several extensions of Corollary 3 are known. For example, Heineken [5] has proved that *a group in which the identity*

$$[x, g_1, g_1, g_2, g_2, \ldots, g_n, g_n] = 1$$

is valid is nilpotent of class at most 3n provided G has no elements of order 2. Metabelian groups subject to more general types of identical relations are studied by Gupta and Newman in [1].

Corollary 4 (Levi [2]). The commutator operation on a group G is associative if and only if G is nilpotent of class $\leqq 2$.

Proof. If G is nilpotent of class $\leqq 2$, then all commutators of weight three are trivial and the associative law is obviously valid. Conversely, let G satisfy this associative law. Then

$$[x, y, y] = [x, [y, y]] = 1,$$

so $G \in \mathfrak{E}_2$. By Theorem 7.13 (iv)

$$[x, y, z] = [x, [y, z]] = [x, y, z]^2,$$

so $[x, y, z] = 1$ and G is nilpotent of class $\leqq 2$. □

There is a well-known relation between 2-Engel groups and groups of exponent 3.

Theorem 7.14 (Levi [2]). Every group of exponent 3 is a 2-Engel group. If G is a 2-Engel group, then $[G', G]^3 = 1$.

Proof. Suppose that G has exponent 3 and let x and y belong to G. Then $(xy^{-1})^2 = (xy^{-1})^{-1} = yx^{-1}$, so that, multiplying by y^2, we have

$$\begin{aligned} xy^{-1}xy = yx^{-1}y^2 = y^{-2}x^{-1}y^{-1} &= y^{-1}(y^{-1}x^{-1}y^{-1}) \\ = y^{-1}x(x^{-1}y^{-1})^2 = y^{-1}x(x^{-1}y^{-1})^{-1} &= y^{-1}xyx. \end{aligned}$$

Hence x commutes with x^y and consequently

$$1 = [x^y, x] = [x, y, x].$$

Hence $G \in \mathfrak{E}_2$.

Conversely let $G \in \mathfrak{E}_2$ and let x, y and z belong to G. Consider the commutator identity

$$[x, y^{-1}, z]^y \, [y, z^{-1}, x]^z \, [z, x^{-1}, y]^x = 1$$

(see Part 1, p. 42). By (iii) of Theorem 7.13 this becomes

$$[x, y, z]^3 = 1.$$

Finally $[G', G] \leqq \zeta(G)$ by (vi) of Theorem 7.13, so $[G', G]^3 = 1$. □

Macdonald has generalized both the second part of Theorem 7.14 and Corollary 3 to Theorem 7.13 as follows. *If a group G satisfies the identity $[x, y_1, \ldots, y_n, x] = 1$, then G is nilpotent of class at most $n + 2$ and $(\gamma_{n+2}(G))^3 = 1$* (Macdonald [3], Theorem 1).

Characterizations of 2-Engel Groups

Theorem 7.15 (Kappe [1]). The following properties of a group G are equivalent.

(i) G is a 2-Engel group.
(ii) If $x \in G$, then $C_G(x) \lhd G$.
(iii) Each maximal abelian subgroup is normal in G.
(iv) Each 2-generator subgroup of G is nilpotent of class $\leqq 2$.
(v) The identity $[x, [y, z]] = [x, y, z]^2$ holds in G.
(vi) The identity $[x, y, z] = [x, z, y]^{-1}$ holds in G.

Proof

(i) *implies* (ii). Let $G \in \mathfrak{E}_2$, let $g \in G$ and let $c \in C_G(g)$. Then for any x in G we have

$$[g, c^x] = [g, [c, x]] = [g, c, x]^2 = 1$$

by Theorem 7.13 (iv). Hence $c^x \in C_G(g)$ and $C_G(g) \lhd G$.

(ii) *implies* (iii). Let A be a maximal abelian subgroup of G. Then $A = C_G(A)$, so A is the intersection of all the $C_G(a)$ with $a \in A$. Hence $A \lhd G$.

(iii) *implies* (iv). Let a and b belong to G and let A and B be maximal abelian subgroups of G containing a and b respectively. Then $A \lhd G$ and $B \lhd G$, so AB is nilpotent of class $\leqq 2$ and the same is true of $\langle a, b\rangle$.

(iv) *implies* (i). This is obvious.

It remains to show that (i), (v) and (vi) are equivalent: we see by Theorem 7.13 that (v) and (vi) follow from (i). If G satisfies (v) and x and y belong to G, then

$$1 = [y, y, x]^2 = [y, [y, x]],$$

so $[x, y, y] = 1$ and G is a 2-Engel group. Similarly (i) follows from (vi). ∎

If each 2-generator subgroup of a group G is nilpotent of class at most 2, then G is nilpotent of class at most 3. This is clear from Corollary 3 to Theorem 7.13. We mention in passing that Heineken [4] has proved that *if $n > 2$, a group in which each n-generator subgroup is nilpotent of class at most n is itself nilpotent of class at most n*: of course this is false for $n = 2$. More recently N. D. Gupta [7] has shown that *a group whose n-generator subgroups are nilpotent of class at most $c \geqq 3$ is locally nilpotent if there exist positive integers c_1 and c_2 such that*

$$n \geqq c_1 + c_2 + 1 \quad \text{and} \quad c \leqq c_1c_2 + c_1 + c_2.$$

On this subject the reader may also consult Macdonald and Neumann [1] and H. Neumann's book [1] (Chapter 3, Section 4).

n-Engel Groups when $n \geqq 3$

The study of these groups is much harder and little is known of them except when $n = 3$. Heineken ([3], Hauptsätze 1 and 2) has shown that *a 3-Engel group is locally nilpotent, and is even nilpotent of class at most 4 if there are no elements of order 2 or 5 present*; whether every n-Engel group is locally nilpotent when $n \geqq 4$ is unknown. Notice, however, that 3-Engel groups need not be nilpotent *: the standard wreath product of a group of order 2 and countably infinite elementary abelian 2-group is a non-nilpotent 3-Engel group. 3-Engel groups are also studied in papers of Širšov [1] and Ivanjuta [1].

Engel Conditions and p-Groups of Finite Exponent

It has been known for some time that there are connections between p-groups of finite exponent and Engel conditions: see for example Levi's Theorem 7.14.

A group G is said to satisfy *the nth Engel congruence* if

$$[x, {}_n y] \in \gamma_{n+2}(G)$$

for all x and y in G. For example, the commutator collection process of P. Hall ([1], § 33) may be applied to show that a group of prime exponent p satisfies the $(p-1)$th Engel congruence (see M. Hall [2], § 18.4).

More generally Gupta and Newman [2], improving an earlier result of Sanov ([1]; see also Glauberman, Krause and Struik [1], and Struik [1], [2]), have proved that *if p is a prime, a group of exponent p^e satisfies the nth Engel congruence where*

$$n = ep^e - 1 - \sum_{i=0}^{e-1} p^i + e.$$

This goes some way towards proving a conjecture of Bruck [1] that n may be taken to be

$$ep^e - 1 - (e-1)\, p^{e-1},$$

and indeed Gupta and Newman verify Bruck's conjecture when $e = 2$. Also Krause [1] has shown that *a group of exponent 8 satisfies the 14th Engel congruence*. For further results see Kostrikin [5].

On the other hand a group of high prime exponent p need not be an Engel group. If $n \geqq 2$ and $p \geqq 4381$, the Burnside group $B(n, p)$ has finite abelian subgroups (Novikov and Adjan [2]) but is not an Engel group; for if it were, it would be finite by the Corollary to Theorem 7.21 in the next section. In the other direction there is the following simple consequence of Kostrikin's positive solution of the restricted Burnside problem for prime exponent (Kostrikin [4]).

* Recently Bachmuth and Mochizuki have Constructed a 3-Engel group of exponent 5 which is insoluble.

Theorem 7.16 (Černikov [28]). An SN-group of prime exponent is locally finite and hence is an Engel group.

Proof. Let G be a finitely generated SN-group with prime exponent p and let R be the finite residual of G. By Kostrikin's theorem there is an upper bound for the order of a finite factor group of G. Since G has only finitely many normal subgroups of given index, it follows that G/R is finite. Thus R is a finitely generated SN-group with no proper subgroups of finite index. Suppose that $R \neq 1$ and let $\{\Lambda_\sigma, V_\sigma : \sigma \in \Sigma\}$ be a series in R with abelian factors. Let $\{x_1, \dots, x_m\}$ be a set of non-trivial generators of R; then $x_i \in \Lambda_{\sigma_i} \backslash V_{\sigma_i}$ where $\sigma_i \in \Sigma$. Let σ be the last of $\sigma_1, \dots, \sigma_n$ in the ordering of Σ. Then $V_\sigma < \Lambda_\sigma = R$. Since $V_\sigma \lhd \Lambda_\sigma$, the group R/V_σ is a non-trivial finite elementary abelian p-group. By this contradiction $R = 1$ and G is finite. □

Soluble p-Groups with Finite Exponent

Theorem 7.17. Let G be a non-trivial soluble p-group of finite exponent and let l be the minimum length of a normal series of G with elementary abelian factors. Then

$$[G, \underbrace{\langle g\rangle, \dots, \langle g\rangle}_{r}] = 1$$

for all g in G where

$$r = 1 + p + p^2 + \cdots + p^{l-1}.$$

In particular G is a Baer group and an r-Engel group.

Proof. Let $1 = G_0 < G_1 < \cdots < G_l = G$ be a normal series of G with elementary abelian factors and let $g \in G$. If $l = 1$, then G is abelian and the result is obvious, so let $l > 1$. Then by induction on l

$$[G, \underbrace{\langle g\rangle, \dots, \langle g\rangle}_{r'}] \leq G_1 \tag{7}$$

where $r' = 1 + p + p^2 + \cdots + p^{l-2}$. Now $g^{p^{l-1}} \in G_1$, so if $a \in G_1$,

$$1 = [a, g^{p^{l-1}}] = a^{g^{p^{l-1}}-1} = a^{(g-1)^{p^{l-1}}} = [a, {}_{p^{l-1}}g]$$

since G_1 is an elementary abelian p-group. Hence

$$[G_1, \underbrace{\langle g\rangle, \dots, \langle g\rangle}_{p^{l-1}}] = 1. \tag{8}$$

Since $r' + p^{l-1} = r$, the result follows from (7) and (8). □

A metabelian group of exponent p^2 (where p is a prime) need not be nilpotent, as we see from the standard wreath product of a group of order p and a countably infinite elementary abelian p-group: notice that this is a $(p+1)$-Engel group. However, for soluble groups of exponent p the situation is quite different.

Theorem 7.18 (N. D. Gupta [6]). A soluble group of prime exponent p and derived length d is nilpotent with class not exceeding

$$1 + (p-1) + (p-1)^2 + \cdots + (p-1)^{d-1}.$$

Taking $d = 2$ we obtain a well-known result of Meier-Wunderli [2].

Corollary. A metabelian group of prime exponent p is nilpotent of class at most p.

Theorem 7.18 with the less exact bound p^{d-1} was given in an earlier paper of Tobin [1].

The proof of the theorem leans heavily on

Lemma 7.19. Let X be an abelian group of automorphisms of an elementary abelian p-group A and assume that for each x in X the endomorphism $\sum_{i=0}^{p-1} x^i$ is zero. Then

$$[A, \underbrace{X, \ldots, X}_{p-1}] = [A, \underbrace{X, \ldots, X}_{p}].$$

Proof. If $x \in X$, denote by $\xi(x)$ the endomorphism $x - 1$ of A in which $a \to [a, x] = a^{-1}a^x$; then $x = 1 + \xi(x)$. By hypothesis

$$0 = \sum_{i=0}^{p-1} x^i = \sum_{i=0}^{p-1} (1 + \xi(x))^i$$

$$= p + \binom{p}{2}\xi(x) + \cdots + \binom{p}{p-1}\xi(x)^{p-2} + \xi(x)^{p-1}.$$

Since $\binom{p}{i} \equiv 0 \bmod p$ if $0 < i < p$, this leads to

$$\xi(x)^{p-1} = 0. \tag{9}$$

Let $x_1, x_2, \ldots$ be elements of X and let $\xi_i = \xi(x_i)$. It will be sufficient to prove that

$$[A, \underbrace{X, \ldots, X}_{p-1}] = 1$$

under the assumption

$$[A, \underbrace{X, \ldots, X}_{p}] = 1. \tag{10}$$

Since X is abelian, this will follow if we can show that

$$\xi_1^{m_1} \cdots \xi_r^{m_r} = 0$$

whenever the positive integers m_i satisfy $m_1 + \cdots + m_r = p - 1$. If $r = 1$, then $m_1 = p - 1$ and the result is true by (9). Let $r > 1$ and assume that the result is true for $r - 1$. Write $\eta = \xi(x_1^i x_2)$ where i is a non-negative integer. Then

$$\eta = (1 + \xi_1)^i (1 + \xi_2) - 1 = i\xi_1 + \xi_2 + \cdots$$

where the terms of degree 2 or more in ξ_1 and ξ_2 have been omitted. Let $m = m_1 + m_2$. Then by induction on r

$$0 = \eta^m \xi_3^{m_3} \cdots \xi_r^{m_r} = (i\xi_1 + \xi_2)^m \xi_3^{m_3} \cdots \xi_r^{m_r}, \tag{11}$$

since all products of p or more ξ's are zero by the hypothesis (10). We conclude that there exist endomorphisms α_j, independent of i, such that

$$\alpha_0 + i\alpha_1 + \cdots + i^m \alpha_m = 0$$

for $i = 0, 1, \ldots, m$. The coefficient matrix of this set of $m + 1$ linear equations in the α_i is of the Vandermonde type, so its determinant does not vanish and we conclude that $\alpha_0 = \alpha_1 = \cdots = \alpha_m = 0$. But a direct inspection of equation (11) reveals that

$$\alpha_{m_1} = \binom{m}{m_1} \xi_1^{m_1} \xi_2^{m_2} \cdots \xi_r^{m_r}$$

since $m = m_1 + m_2$. Now $m < p$, so p does not divide $\binom{m}{m_1}$ and consequently $\xi_1^{m_1} \xi_2^{m_2} \cdots \xi_r^{m_r} = 0$. □

Proof of Theorem 7.18. G is a soluble group with derived length d and exponent p. Without loss of generality we can assume that G is finitely generated and hence finite. Let $d > 1$ and assume by induction on d that G' is nilpotent of class at most

$$c' = 1 + (p - 1) + \cdots + (p - 1)^{d-2}.$$

Let H/K be a lower central factor of G'; let $a \in H$ and $x \in G$. Then, since G has exponent p,

$$1 = (xa)^p \equiv x^p a^{1+x+\cdots+x^{p-1}} \equiv a^{1+x+\cdots+x^{p-1}} \bmod K.$$

Also $C = C_G(H/K) \geq G'$, so G/C is abelian. Apply Lemma 7.19 with $A = H/K$ and $X = G/C$; since G is nilpotent (being a finite p-group), we conclude that

$$[H, \underbrace{G, \ldots, G}_{p-1}] \leq K.$$

It follows that G is nilpotent of class at most

$$c = 1 + (p - 1)\, c' = 1 + (p - 1) + (p - 1)^2 + \cdots + (p - 1)^{d-1}. \quad \square$$

Further bounds for the nilpotent class of a soluble group of prime exponent are derived in the paper [6] of N. D. Gupta. Bounds for the nilpotent class of finite metabelian p-groups have been obtained by several authors: the relevant papers are Bachmuth, Heilbronn and Mochizuki [1], Bachmuth and Mochizuki [1], N. D. Gupta [8], Gupta and Gupta [1], Gupta, Newman and Tobin [1] and Gupta and Tobin [1]: see also Bryce [1].

It is noteworthy that Bachmuth, Mochizuki and Walkup [1] have found a locally nilpotent group of exponent 5 which is insoluble.

7.2 Engel Structure in Groups Satisfying Finiteness Conditions

The following problem has been the object of much study. To find suitable finiteness conditions on a group G which will ensure that $L(G)$, $\overline{L}(G)$, $R(G)$ and $\overline{R}(G)$ are subgroups and coincide with the Hirsch-Plotkin radical, the Baer radical, the hypercentre and the ω-hypercentre respectively. The first finiteness condition we consider is Max-*ab*; observe that if G satisfies Max-*ab*, the following radicals coincide and satisfy Max; the Hirsch-Plotkin radical, the Baer radical, the Fitting radical. This is by Theorem 3.31. By the same result the hypercentre of G coincides with some finite term of the upper central series.

Theorem 7.21 (Peng [1]). Let G be a group which satisfies Max-*ab*. Then $L(G)$ and $\overline{L}(G)$ coincide with the Fitting subgroup of G, and $R(G)$ and $\overline{R}(G)$ coincide with the hypercentre of G.

The first part of the theorem is due to Plotkin [14]. Previously this theorem had been proved for groups satisfying Max by Baer ([28], p. 257) and for finite groups by Schenkman [1]. In particular, a finite Engel group is nilpotent (Zorn [1], Zassenhaus [2], p. 5).

The bulk of the work resides in the proof of the next lemma.

Lemma 7.22 (Peng [1]). Let G be a group which satisfies Max-*ab* and let a be an element of G such that for each g in G there is an integer $n = n(g)$ for which

$$[a^g, {}_n a] = 1.$$

Then a^G is nilpotent and satisfies Max.

Proof. We shall employ the notation

$$\{a^G\}$$

for the set of conjugates of a in G; this should be distinguished from a^G, the *subgroup* generated by all such conjugates. We will call a subgroup

X of G *a-generated* if it can be generated by those conjugates of a which it contains, i.e. if

$$X = \langle X \cap \{a^G\}\rangle.$$

A technical result will now be established.

If X and Y are nilpotent a-generated subgroups of G and $X < Y$, then $N_Y(X)$ contains a conjugate of a which does not lie in X.

Since Y is nilpotent, X *sn* Y and there is a series of finite length $X = X_0 \lhd X_1 \lhd \cdots \lhd X_s = Y$. Since $X \neq Y$ and Y is a-generated, there is a conjugate of a in Y but not in X. It follows that there is an integer i satisfying $0 \leqq i < s$ such that

$$X \cap \{a^G\} = X_i \cap \{a^G\} \subset X_{i+1} \cap \{a^G\}.$$

Let y belong to $X_{i+1} \cap \{a^G\}$ but not to X_i. Since $X_i \lhd X_{i+1}$, we conclude that y normalizes X_i and

$$(X \cap \{a^G\})^y = X \cap \{a^G\}. \tag{12}$$

But X is a-generated, so (12) implies that y normalizes X and our assertion is proved.

The group G contains at least one a-generated nilpotent subgroup, namely $\langle a\rangle$. By Theorem 3.31 the property Max-*ab* implies the maximal condition on nilpotent subgroups. Hence every a-generated nilpotent subgroup of G is contained in a maximal a-generated nilpotent subgroup. If there is a unique subgroup of this kind, say M, then $M \lhd G$; also $\langle a\rangle \leqq M$, so $a^G \leqq M$ and a^G is nilpotent and satisfies Max.

Let us therefore suppose that U and V are two distinct maximal a-generated nilpotent subgroups of G. We can even choose U and V so that

$$I = \langle U \cap V \cap \{a^G\}\rangle$$

is maximal, for G satisfies the maximal condition on nilpotent subgroups. Here it is understood that $I = 1$ if $U \cap V \cap \{a^G\}$ should turn out to be empty. By definition I is a-generated and is properly contained in both U and V. Let

$$W = \langle N_U(I) \cap \{a^G\}\rangle. \tag{13}$$

By the first part of the proof $I < W$. For a similar reason $N_V(I) \cap \{a^G\}$ contains an element v which is not in I; clearly $v \notin U$.

Suppose that $v \in N_G(W)$. Since W is nilpotent and satisfies Max-*ab*, it is finitely generated and hence may be generated by a finite set of conjugates of a, say $\{a^{g_1}, \ldots, a^{g_m}\}$. Since v is a conjugate of a, there exists a positive integer n such that $[a^{g_i}, {}_n v] = 1$ for $i = 1, \ldots, m$. Let $H = \langle v, W\rangle$; then $W \lhd H$. Clearly H/W' is nilpotent; since W is nilpotent, it follows by a theorem of P. Hall that H is nilpotent (see Theorem 2.27). Now v is a conjugate of a and W is a-generated, so H is a-generated. Therefore H lies in a maximal a-generated nilpotent subgroup T of G. Since $v \in T \backslash U$, we see that $U \neq T$. Also $W \leqq H \leqq T$; thus we have $N_U(I) \cap \{a^G\} \subseteq U \cap T \cap \{a^G\}$ and therefore

$$\langle U \cap T \cap \{a^G\} \rangle \geqq W > I,$$

which contradicts the maximality of I.

By this contradiction $v \notin N_G(W)$. Since $I < W$, there is an element u in $N_U(I) \cap \{a^G\}$ which does not lie in I. By hypothesis $[v, {}_n u] = 1$ for some integer n because u and v are conjugates of a. Hence there is a least integer k such that $[v, {}_k u]$ normalizes W. Now $k > 0$ since $v = [v, {}_0 u] \notin N_G(W)$. Let $z = [v, {}_{k-1} u]$. Then $[z, u] = (u^z)^{-1} u \in N_G(W)$, and $u \in W$, so $u^z \in N_G(W)$. Just as in the previous paragraph we see that $K = \langle u^z, W \rangle$ is a nilpotent a-generated subgroup. Let S be a maximal nilpotent a-generated subgroup of G containing K. If $S \neq U$, then

$$\langle U \cap S \cap \{a^G\} \rangle \geqq W > I,$$

which contradicts the maximality of I. Hence $S = U$ and $u^z \in U$. By construction u and v belong to $N_G(I)$, so that $z = [v, {}_{k-1} u] \in N_G(I)$; also u^z is a conjugate of a. Therefore

$$u^z \in N_U(I) \cap \{a^G\} \subseteq W$$

by (13). But u was chosen to lie in $N_U(I) \cap \{a^G\}$, so $u \in W$ and $u^z \in W^z$. Thus

$$u^z \in U \cap W^z \cap \{a^G\}.$$

Now $I \leqq W$ and $I = I^z \leqq W^z$ because z belongs to $N_G(I)$. Thus $I \leqq \langle U \cap W^z \cap \{a^G\} \rangle$. Also $u^z \notin I$ because $u \notin I$. Hence

$$\langle U \cap W^z \cap \{a^G\} \rangle > I. \tag{14}$$

Suppose now that $W^z \nleqq U$. Then W^z lies in a maximal nilpotent a-generated subgroup other than U and by (14) this contradicts the maximality of I. Hence $W^z \leqq U$ and

$$(N_U(I) \cap \{a^G\})^z \subseteq N_U(I) \cap \{a^G\},$$

which shows that $W^z \leqq W$. Since $W^z \neq W$, this implies that

$$W < W^{z^{-1}} < W^{z^{-2}} < \cdots,$$

which is impossible by the maximal condition on nilpotent subgroups. ∎

Proof of Theorem 7.21. Let G satisfy Max-*ab*. If $a \in L(G)$, then Lemma 7.22 shows that a^G is nilpotent and hence is contained in the Fitting subgroup of G. By Lemma 7.12 (i) the subsets $L(G)$ and $\overline{L}(G)$ coincide with the Fitting subgroup of G.

Now let $a \in R(G)$. Then $a^{-1} \in L(G)$ by Theorem 7.11. Hence $(a^{-1})^G = a^G = A$ is nilpotent and finitely generated. Let $a^{g_1}, \ldots, a^{g_m}$ generate a^G and let g be any element of G. Then $[a^{g_i}, {}_n g] = 1$ for some integer n since $a^{g_i} \in R(G)$. Hence $\langle g, A\rangle / A'$ is nilpotent. But A is nilpotent, so it follows by Theorem 2.27 that $\langle g, A\rangle$ is nilpotent. Since A satisfies Max, its upper central series can be refined to a series of G-admissible subgroups of finite length whose factors are either free abelian groups of finite rank or finite elementary abelian p-groups. Let $F = H/K$ be a factor of this series with rank r. Consider the action of G upon the finite factor F/F^p where p is a prime. Since $\langle g, A\rangle$ is nilpotent, the group of automorphisms induced in F/F^p by G is a finite p-group. Thus the holomorph of F/F^p by this group of automorphisms is a finite p-group and hence is nilpotent. It follows that $H/H^pK \leqq \zeta_r(G/H^pK)$. If F is an elementary abelian p-group, then, of course, $H^p \leqq K$. Otherwise F is free abelian, so the intersection of all the F^p is trivial and $H/K \leqq \zeta_r(G/K)$. Hence in either case $H/K \leqq \zeta_r(G/K)$. (The reader will observe that this result could have been achieved more quickly if we had used the well-known theorem of Wedderburn [1] that a finite dimensional algebra over a field is nilpotent if it possesses a basis of nilpotent elements). Finally we deduce that $A \leqq \zeta_s(G)$ for some integer s, so that $R(G) \leqq \bar{\zeta}(G)$. Hence $\zeta_i(G) = \bar{R}(G) = R(G)$ for some integer i. ▯

Corollary (Plotkin [6]). An Engel group with all its abelian subgroups finite is a finite nilpotent group.

Engel Sets

A subset S of a group G will be called *an Engel set* if given x and y in S there is an integer $n = n(x, y)$ such that $[x, {}_n y] = 1$.

Theorem 7.23. Let the group G satisfy Max-*ab*. Then a normal subset of G is an Engel set if and only if it is contained in the Fitting subgroup of G.

Proof. Suppose that the normal subset S of G is an Engel set and let $a \in S$. If $g \in G$, then $a^g \in S$ and $[a^g, {}_n a] = 1$ for some integer n. Lemma 7.22 implies that a^G is nilpotent. Therefore S is contained in the Fitting subgroup of G. The converse is clear. ▯

Theorem 7.24 (Held [2], Satz 1). Let G be a group satisfying Min and suppose that the elements whose orders are powers of p form an Engel set for each prime p. Then G is a hypercentral Černikov group.

Proof. It is sufficient to prove that G is locally finite. For then G will be locally nilpotent, by Theorem 7.23, and the result follows by Corollary 2, Theorem 5.27. Hence we may assume that G is a minimal non-locally finite group. By the proof of Schmidt's theorem (3.17) G

has a homomorphic image H which is finitely generated and simple and whose maximal subgroups are disjoint. Let M be a maximal subgroup of H and let x be an element of prime order p in M: since H is periodic, such an x exists. If all the elements of order p lay in M, they would generate a proper non-trivial normal subgroup of H; this contradicts the simplicity of H, so we must be able to find an element y of order p which does not belong to M. The elements of order p form an Engel set, so there is a first integer n such that $[y, {}_n x] \in M$, and $n > 0$ since $y \notin M$. Thus $z = [y, {}_{n-1} x] \notin M$. Now

$$x^z = x[z, x]^{-1} \in M,$$

so

$$M \cap M^z \neq 1.$$

Hence $M = M^z$ and $z \in N_G(M)$. Thus $M < N_G(M)$ and consequently $N_G(M) = G$, which yields the contradiction $M \lhd G$. □

Corollary (Vilyacer [1]). The Engel groups which satisfy Min are precisely the hypercentral Černikov groups.

However a good deal more than this is known. For example Held [5], relying on earlier work of Plotkin [14], has shown that *in a group G which satisfies Min-ab the set* $\bar{L}(G)$ *coincides with the Fitting subgroup and* $\bar{R}(G)$ *coincides with the* ω*-hypercentre.* Recently J. Martin has shown that in addition $L(G)$ *equals the Hirsch-Plotkin radical and* $R(G)$ *equals the hypercentre* (unpublished).

In [6] Plotkin has proved that *an Engel group whose abelian subgroups have finite rank and finite torsion-subgroup is nilpotent of finite rank.* A further recent result of J. Martin states that *a torsion-free n-Engel group in which at least one maximal abelian subgroup has finite rank is nilpotent of finite rank.* However the Engel structure of groups with finite abelian subgroup rank is still unknown: for soluble groups see Gruenberg [4] (Theorem 1.3) and Wehrfritz [10] (Theorem E 1).

Gruenberg [6] has investigated the Engel structure of groups of R-automorphisms of a finitely generated R-module where R is a commutative Noetherian ring with unity: here the four sets of Engel elements coincide with the four corresponding subgroups: special cases of this result had previously been obtained by Garaščuk [1] and Suprunenko and Garaščuk [1], [2]. The Engel structure of the automorphism group of a finitely generated soluble group of finite rank has been studied by Wehrfritz ([10], Theorem E 2).

Further results about the Engel structure of groups subject to finiteness conditions are to be found in the following papers: Baer [36], [40]; Held [1], [3], [6]; Plotkin [12] (§ 12), [5], [6], [10], [14].

7.3 Engel Structure in Generalized Soluble Groups

In discussing the Engel structure of soluble groups it is necessary to modify our aims as expressed in Sections 7.1 and 7.2. The reason is that in an arbitrary soluble group G the set $R(G)$ may very easily be distinct from the hypercentre. For example, if G is the standard wreath product of a group of prime order p with a countably infinite elementary abelian p-group, then $G = R(G) = \bar{R}(G)$ but $\bar{\zeta}(G) = 1$. To remedy this we replace the hypercentre and ω-hypercentre by two larger subgroups which we will then endeavour to prove are the sets of right and bounded right Engel elements in favourable situations.

The Subgroups $\varrho(G)$ and $\bar{\varrho}(G)$

Let G be an arbitrary group. Following Gruenberg ([3], p. 159) we define

$$\varrho(G)$$

to consist of all elements a of G such that

$$\langle x \rangle \ asc \ \langle x, a^G \rangle$$

for every x in G: similarly,

$$\bar{\varrho}(G)$$

is defined to be the set of all a in G for which there is a positive integer $n = n(a)$ such that

$$\langle x \rangle \lhd^n \langle x, a^G \rangle$$

for all x in G.

The first point to settle is that $\varrho(G)$ and $\bar{\varrho}(G)$ are subgroups.

Lemma 7.31 (Gruenberg [3], Theorem 3). Let G be an arbitrary group. Then $\varrho(G)$ and $\bar{\varrho}(G)$ are characteristic subgroups of G satisfying

$$\bar{\zeta}(G) \leqq \varrho(G) \leqq R(G) \quad \text{and} \quad \zeta_\omega(G) \leqq \bar{\varrho}(G) \leqq \bar{R}(G).$$

In addition, $\varrho(G)$ is contained in the Gruenberg radical of G and $\bar{\varrho}(G)$ is contained in the Baer radical of G.

Proof. Let a and b belong to $\varrho(G)$. For an arbitrary x in G we have $\langle x \rangle \ asc \ \langle x, a^G \rangle$ and $\langle x \rangle \ asc \ \langle x, b^G \rangle$: since ascendance is preserved under homomorphisms, the second relation implies that $\langle x, a^G \rangle \ asc \ \langle x, a^G, b^G \rangle$; hence $\langle x \rangle \ asc \ \langle x, a^G, b^G \rangle$ and in particular $\langle x \rangle \ asc \ \langle x, (ab^{-1})^G \rangle$. Therefore $ab^{-1} \in \varrho(G)$ and $\varrho(G)$ is a subgroup, which is obviously characteristic in G. For $\bar{\varrho}(G)$ the proof is similar.

Now let $x \in G$ and observe that $\langle x, \zeta_\alpha(G)\rangle \lhd \langle x, \zeta_{\alpha+1}(G)\rangle$ for every ordinal α. Thus $\langle x\rangle \lhd^{x} \langle x, \zeta_x(G)\rangle$ and this implies at once that $\bar{\zeta}(G) \leqq \varrho(G)$ and $\zeta_\omega(G) \leqq \bar{\varrho}(G)$. Also, if $a \in \varrho(G)$ and $x \in G$, then $\langle x\rangle$ *asc* $\langle x, a^G\rangle$. Therefore $[a, {}_n x] = 1$ for some integer n; for otherwise, by examining those terms of the ascending series between $\langle x\rangle$ and $\langle x, a^G\rangle$ that contain the $[a, {}_n x]$, we should be able to find a set of ordinals without a first element. Hence $\varrho(G) \leqq R(G)$. It is equally easy to see that $\bar{\varrho}(G) \leqq \bar{R}(G)$. Finally, if $a \in \varrho(G)$, then $\langle a\rangle$ *asc* a^G, so $\langle a\rangle$ *asc* G; therefore $\varrho(G)$ is contained in the Gruenberg radical of G and for a similar reason $\bar{\varrho}(G)$ is contained in the Baer radical of G. □

The following simple result is frequently useful.

Lemma 7.32 (Gruenberg [3], Lemma 2). Let x be an element and let Y be a non-empty subset of a group G and put $S = \langle Y\rangle$. Suppose that $S^x \leqq S$ and that for each y in Y there is an integer $n = n(y)$ such that $[y, {}_n x] = 1$. Then $S^x = S$.

Proof. Let $y \in Y$: since $S^x \leqq S$ and

$$[y, {}_i x] = [y, {}_{i-1} x]^{-1} [y, {}_{i-1} x]^x, \quad (i > 0),$$

it follows by induction on i that $[y, {}_i x] \in S$, the case $i = 0$ being trivial. Now

$$y^{x^{-1}} = y[y, x^{-1}]$$

and

$$[[y, {}_i x], x^{-1}] = [y, {}_{i+1} x]^{-x^{-1}} = [[y, {}_{i+1} x], x^{-1}]^{-1} [y, {}_{i+1} x]^{-1}.$$

Since $[y, {}_n x] = 1$ for some $n = n(y)$, it follows that $y^{x^{-1}}$ is a product of elements of the form $[y, {}_i x]^{\pm 1}$. Therefore $y^{x^{-1}} \in S$ and consequently $S^{x^{-1}} \leqq S$. Thus $S \leqq S^x$ and $S = S^x$ as required. □

We will establish one further preliminary result.

Lemma 7.33 (Gruenberg [3], Proposition 3). A weakly nilpotent SN^*-group is a Gruenberg group.

Proof. Let G be weakly nilpotent (i.e. every 2-generator subgroup is nilpotent), and assume that G is an SN^*-group and $\{G_\alpha\}$ is an ascending series in G with abelian factors. Let $g \in G$; we have to prove that $\langle g\rangle$ *asc* G. Let

$$\bar{G}_\lambda = \bigcap_{i=0,\pm 1,\pm 2,\ldots} G_\lambda^{g^i};$$

in order to prove that $\{\bar{G}_\lambda\}$ is an ascending series in G we need only show that

$$\bar{G}_\lambda \leqq \bigcup_{\alpha<\lambda} \bar{G}_\alpha \tag{15}$$

where λ is a limit ordinal. Let $a \in \bar{G}_\lambda$; then $H = \langle a, g\rangle$ is a finitely generated nilpotent group, so it satisfies Max and a^H is finitely generated. Now $a \in \bar{G}_\lambda = (\bar{G}_\lambda)^H$, so $a^H \leqq \bar{G}_\lambda \leqq G_\lambda$. Hence $a^H \leqq G_\alpha$ for some $\alpha < \lambda$ and consequently $a^H \leqq \bar{G}_\alpha$. Thus (15) is established. Since

$$(\bar{G}_{\alpha+1})' \leqq \bigcap_{i=0,\pm 1,\pm 2,\ldots} (G'_{\alpha+1})^{g^i} \leqq \bar{G}_\alpha,$$

$\{\bar{G}_\alpha\}$ is an ascending series in G with $\langle g\rangle$-admissible terms and abelian factors.

Now let $A = \bar{G}_1 \neq 1$. If $1 \neq a \in A$, then $\langle a, g\rangle$ is nilpotent, so $A(1) = C_A(g) \neq 1$, and of course $\langle g\rangle \lhd \langle g, A(1)\rangle$. Next we define recursively

$$A(\alpha+1)/A(\alpha) = C_{A/A(\alpha)}(gA(\alpha)) \quad \text{and} \quad A(\lambda) = \bigcup_{\alpha<\lambda} A(\alpha)$$

where α is an ordinal and λ a limit ordinal. Then it is evident that $A = A(\beta)$ for some ordinal β and that $\langle g, A(\alpha)\rangle \lhd \langle g, A(\alpha+1)\rangle$. Thus $\langle g\rangle$ *asc* $\langle g, A\rangle$. When applied to the group $\langle g, \bar{G}_{\alpha+1}\rangle/\bar{G}_\alpha$, this argument shows that $\langle g, \bar{G}_\alpha\rangle$ *asc* $\langle g, \bar{G}_{\alpha+1}\rangle$. Hence $\langle g\rangle$ *asc* G as required. ▯

In particular *a locally nilpotent SN^*-group is a Gruenberg group*, a result which may also be found in the later papers of Baer [36] (Satz 3.3) and Kemhadze [3].

The Engel Structure of Radical Groups

Theorem 7.34. Let G be a radical group. Then $L(G)$ coincides with the Hirsch-Plotkin radical of G and $R(G)$ is a locally nilpotent subgroup. Furthermore, $R(G) = \varrho(G)$ if and only if $R(G)$ is a Gruenberg group.

The first part of Theorem 7.34 is due to Plotkin ([7], Theorem 9): special cases of the theorem are in papers of Gruenberg ([3], Proposition 3) and Plotkin [5], [6]. We remark that Plotkin [14] has also proved a more general result: *if a group has an ascending series whose factors satisfy Max locally, the set of left Engel elements coincides with the Hirsch-Plotkin radical and the set of right Engel elements is a subgroup.* See also Baer [36] (Satz 5.2).

Concerning the second part of Theorem 7.34, we remark that if G is a radical group, it is quite possible for $R(G)$ and $\varrho(G)$ to differ: for example, let G be one of the locally finite p-groups of Kargapolov-Kovács-Neumann (see Theorem 6.27); here $G = L(G) = R(G)$ but $\varrho(G) = 1$ because G has trivial Gruenberg radical.

On the other hand, Lemma 7.33 and Theorem 7.34 imply the following

Corollary 1. If G is an SN^*-group, then $R(G) = \varrho(G)$.

Another fact worthy of mention is

Corollary 2 (cf. Gruenberg [3], Lemma 14). In an arbitrary group the right Engel elements that lie in the final term of the upper Hirsch-Plotkin (i.e. locally nilpotent) series form a subgroup.

Proof of Theorem 7.34. G is a radical group. First of all we will show that $L(G)$ equals the Hirsch-Plotkin radical H of G. Let us assume for the present that G is countable.

Let $x \in L(G)$. Choose a finite subset $\{a_1, \ldots, a_m\}$ of H and define

$$Y = \{[a_i, {}_jx]\colon\ i = 1, \ldots, m,\ j = 0, 1, 2, \ldots\}. \tag{16}$$

Since $x \in L(G)$, the subset Y is finite. Let $S = \langle Y \rangle$. Then by (16) we have $S^x \leqq S$ and Lemma 7.32 may be applied to give $S^x = S$, so that $S \lhd T = \langle x, S \rangle$. Since S is a finitely generated subgroup of H, it is nilpotent. Also T/S' is nilpotent because $x \in L(G)$ and S/S' is finitely generated. The theorem of P. Hall (Theorem 2.27) now shows that T is nilpotent. Hence $\langle x, H \rangle$ is locally nilpotent and therefore a Gruenberg group, for G is countable. It follows that $\langle x \rangle$ *asc* $\langle x, H \rangle$. Let H_α denote the αth term of the upper Hirsch-Plotkin series of G. Then evidently our argument proves that $\langle x, H_\alpha \rangle$ *asc* $\langle x, H_{\alpha+1} \rangle$ for each ordinal α. But G is a radical group, so $G = H_\beta$ for some β and consequently $\langle x \rangle$ *asc* G. Hence x belongs to the Gruenberg radical of G (which is also the Hirsch-Plotkin radical of G in view of our hypothesis of countability).

Now we abandon the assumption that G is countable and denote by x a left Engel element as before. Let X be a countable subgroup of G containing x. Obviously $x \in L(X)$, so x^X is locally nilpotent by the first part of the proof. Hence x^G is locally nilpotent and x belongs to the Hirsch-Plotkin radical of G.

We turn now to the right Engel elements. By Theorem 7.11 $R(G)^{-1} \subseteq L(G)$, and $R(G)$ is a subset of the Hirsch-Plotkin radical of G by the first part of the theorem. We must prove that $R(G)$ is a subgroup. Let a and b belong to $R(G)$ and let x be any element of G. Define

$$Y = \{[a, {}_ix], [b, {}_ix]\colon\ i = 0, 1, 2, \ldots\}; \tag{17}$$

if $S = \langle Y \rangle$, then $S^x \leqq S$, so $S^x = S$ by Lemma 7.32. Now Y is finite because a and b are right Engel elements, so S is nilpotent, being a subgroup of the Hirsch-Plotkin radical. Just as before this leads to the nilpotence of the subgroup $\langle x, S \rangle$: of course the argument now hinges on the fact that a and b are right Engel elements of G. By (17) we have $\langle x, S \rangle = \langle x, a, b \rangle$; hence there is an integer n such that $[ab^{-1}, {}_nx] = 1$; consequently $ab^{-1} \in R(G)$ and $R(G)$ is a subgroup.

Finally, assume that $R(G)$ is a Gruenberg group and let $a \in R(G)$ and $x \in G$. Then $a^G \leq R(G)$ and a^G actually *consists* of right Engel elements. Let y belong to $\langle x, a^G \rangle$ and write $y = x^i b$ where i is an integer and $b \in a^G$. Then, for a sufficiently large integer n,

$$[y, {}_n x] = [b, {}_n x] = 1,$$

which implies that $x \in L(\langle x, a^G \rangle)$. Therefore x, as well as a^G, is contained in the Hirsch-Plotkin radical of $\langle x, a^G \rangle$, by the first part of the theorem. Hence $\langle x, a^G \rangle$ is locally nilpotent. Now by hypothesis a^G is a Gruenberg group, so $\langle x, a^G \rangle$ is an SN^*-group. Lemma 7.33 shows that $\langle x, a^G \rangle$ is a Gruenberg group and $\langle x \rangle$ *asc* $\langle x, a^G \rangle$. Thus $a \in \varrho(G)$ and $\varrho(G) = R(G)$. Conversely, if $R(G) = \varrho(G)$, then $R(G)$ is a Gruenberg group by Lemma 7.31. □

The Bounded Engel Structure of Soluble Groups

Theorem 7.35 (Gruenberg [3], Theorem 4). Let G be a soluble group. Then $\overline{L}(G)$ coincides with the Baer radical of G and $\overline{R}(G)$ coincides with $\overline{\varrho}(G)$.

Proof. Let G be a soluble group with derived length d. First of all let $x \in \overline{L}(G)$ and suppose that x is a left n-Engel element. We will show that

$$\langle x \rangle \lhd^{nd} G,$$

so that x lies in the Baer radical of G. If $d = 0$, this is clear, so let $d > 0$. By induction on d we have $\langle x, A \rangle \lhd^{n(d-1)} G$ where $A = G^{(d-1)}$, the least non-trivial term of the derived series of G. Let ξ be the mapping $a \rightarrow [a, x]$ where $a \in A$; since A is abelian, ξ is an endomorphism of A and since x is a left n-Engel element, $\xi^n = 0$. Hence $\langle x, A \rangle$ is nilpotent of class n at most. From this it follows that $\langle x \rangle \lhd^n \langle x, A \rangle$ and consequently that $\langle x \rangle \lhd^{nd} G$.

The second part of the proof is harder. The first point to establish is that $\overline{R}(G)$ is a subgroup. Let a and b belong to $\overline{R}(G)$ and suppose that a is a right m-Engel element and b is a right n-Engel element. Let $x \in G$; we will prove that $[a^{-1}b, {}_l x] = 1$ where the integer l depends on m, n and d only; this is enough to show that $\overline{R}(G)$ is a subgroup.

Let

$$Y = \{[a, {}_i x], [b, {}_i x]: \ i = 0, 1, 2, \ldots\}$$

and $S = \langle Y \rangle$. Then Y has at most $m + n$ elements. By Lemma 7.32 $S^x = S$ and $S \lhd \langle x, S \rangle$. The problem now is to show that $\langle x, S \rangle$ is nilpotent with class not exceeding a number depending only on m, n and d. The identity

$$[c, {}_{i+1}x] = [c, {}_i x]^{-1} [c, {}_i x]^x$$

shows that S can also be generated by the $m + n$ elements a^{x^i}, b^{x^j} where $i = 0, 1, \ldots, m - 1$ and $j = 0, 1, \ldots, n - 1$. Hence S is generated by $m + n$ right r-Engel elements where

$$r = \max\{m, n\}.$$

Let F be the free group on a set $\{y_0, y_1, \ldots, y_{m+n-1}\}$ and denote by N the smallest normal subgroup of F containing $F^{(d)}$ and the set

$$\{[y_i, {}_r f]\colon\ i = 0, 1, \ldots, m + n - 1, f \in F\}.$$

The mappings

$$y_i \to a^{x^i},\ (0 \leqq i < m), \quad \text{and} \quad y_j \to b^{x^{j-m}}, \quad (m \leqq j < m + n),$$

extend to a homomorphism of F onto S whose kernel contains N. Hence there is induced a homomorphism of F/N onto S. Now F/N is a finitely generated soluble group generated by right r-Engel elements, so it is nilpotent, by Theorem 7.34 or by the first part of the present theorem and Theorem 7.11. Moreover, if the nilpotent class of F/N is $c = c(m, n, d)$, then S is nilpotent of class at most c.

Let s denote a^{x^i} or b^{x^j}; then in either case $[s, {}_r x] = 1$ and it follows that $\langle x, S\rangle/S'$ is nilpotent of class at most r. Theorem 2.27 implies that $\langle x, S\rangle$ is nilpotent with class not exceeding.

$$l = \binom{c+1}{2} r - \binom{c}{2}.$$

Since $\langle x, S\rangle = \langle x, a, b\rangle$, it follows that $[a^{-1}b, {}_l x] = 1$. Finally l depends only on m, n and d.

It remains to prove that $\overline{R}(G) \leqq \overline{\varrho}(G)$. Let x be any element of G and let a be a right m-Engel element of G; then $a^G \leqq \overline{R}(G)$ since $\overline{R}(G)$ is a subgroup. Let e denote the derived length of a^G. It will be shown that there is an integer $t = t(m, e)$ depending only on m and e such that

$$[u, {}_t x] = 1$$

for all u in a^G. If $e = 0$, then $t(m, 0) = 0$ will suffice, so let $e > 0$. Define $A = (a^G)^{(e-1)}$; then by induction on e we may assume that an integer $t_1 = t(m, e - 1)$ has been found such that $[u, {}_{t_1} x] \in A$ for all u in a^G. Since $A \leqq \gamma_{2^{e-1}}(a^G)$, each v in A can be expressed in terms of commutators of the form $c = [z_1, z_2, \ldots, z_{2^{e-1}}]$ where z_i is a conjugate of a. Now c can be written in the form $(z')^{-1}z''$, where z' and z'' are conjugates of a and hence are right m-Engel elements. Therefore c is a right f-Engel element where $f = l(m, m, d)$, by what has already been proved. Since A is abelian, it actually consists of right f-Engel elements. Hence $[v, {}_f x] = 1$ for all v in A and consequently $[u, {}_t x] = 1$ for all u in a^G where $t = f + t_1$. Now let $t(m, e) = t$; then, as in the first paragraph of the proof, $\langle x\rangle \lhd^{te} \langle x, a^G\rangle$ since a^G is soluble with derived length e. □

Whether $\overline{L}(G)$ equals the Baer radical when G is a radical group is open. Nor is it known, apparently, whether $\overline{R}(G) = \overline{\varrho}(G)$ when G is an SN^*-group.

Inclusion diagram

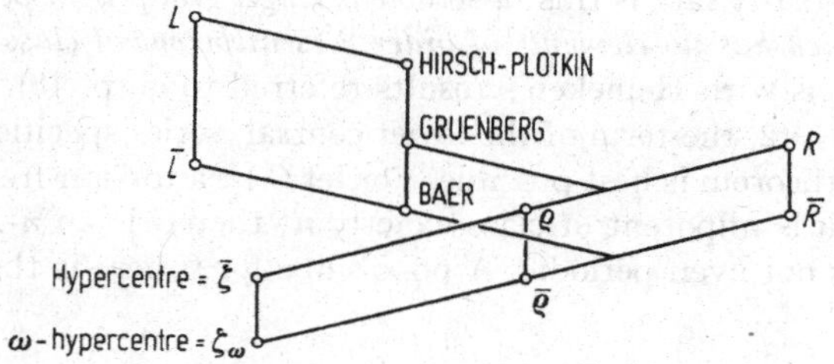

The accompanying diagram exhibits the inclusions existing in an arbitrary group between the sets of Engel elements and the seven subgroups under discussion. If G is a soluble group, then, by Theorems 7.34 and 7.35 and Lemma 7.33, the following coincide: $L(G)$, the Hirsch-Plotkin radical and the Gruenberg radical; $\overline{L}(G)$ and the Baer radical; $R(G)$ and $\varrho(G)$; $\overline{R}(G)$ and $\overline{\varrho}(G)$. The remaining six subgroups are distinct in general, as Gruenberg ([3], Theorem 5) shows by means of a countable metabelian group.

Soluble n-Engel Groups

For the remainder of this section we shall be concerned with proving the following theorem.

Theorem 7.36. Let G be a soluble n-Engel group with positive derived length d.

(i) The $(n^{d-1} + 1)$th term of the lower central series of G has finite exponent dividing a power of $(n + 1)((n - 1)!)$.

(ii) The $((n + 1)^{d-1} + 1)$th term of the lower central series of G has finite exponent dividing a power of $(n - 1)!$.

In its original form this theorem is due to Gruenberg ([4], Theorem 1.10): the present version incorporates certain refinements devised by Gupta and Newman in the course of their investigations of metabelian groups subject to more general identities than Engel conditions (Gupta and Newman [1]).

To begin with let us note some consequences of the theorem.

Corollary. Let G be a soluble n-Engel group with positive derived length d. If G has no elements with prime order less than n, then G is nilpotent of class at most $(n+1)^{d-1}$. If in addition G has no elements of order $n+1$, then the nilpotent class of G is at most n^{d-1}.

In particular *a torsion-free soluble n-Engel group is nilpotent*. Another noteworthy fact is this: *a soluble 3-Engel group with positive derived length d which has no elements of order 2 is nilpotent of class at most 3^{d-1}* (compare this with Heineken's results referred to on p. 48).

When $d = 2$, the term of the lower central series specified in the first part of the theorem is best possible. For let G be a torsion-free metabelian group which is nilpotent of class exactly n; then G is an n-Engel group, but $\gamma_n(G)$ is not even periodic. A possibility for G here is the group with generators

$$x, a_1, \ldots, a_n$$

and relations

$$[a_i, a_j] = 1, \; a_i^x = a_i a_{i+1} \; (i < n) \; \text{ and } \; a_n^x = a_n.$$

It is not possible to replace $(n+1)((n-1)!)$ by $n((n-1)!)$ in the statement of the theorem; for we saw in Section 7.1 that there exist 2-Engel 3-groups of nilpotent class 3 and by Theorem 7.13 (iii) and (vi) all 2-Engel groups are metabelian. Nor is it possible to replace $(n-1)!$ by $(n-2)!$ in part (iii); for example, let G be the standard wreath product of a group of prime order p and a countably infinite, elementary abelian p-group; G is a $(p+1)$-Engel group and $\gamma_{p+3}(G)$ does not have exponent dividing $(p-1)!$

The proof of Theorem 7.36 depends on three lemmas on metabelian groups. We precede these with two well-known remarks.

Let G be a metabelian group, let $a \in G'$ and let $x, y, z \in G$. Use of the standard commutator identity (3) on p. 42 of Part 1 shows that

$$[a, x, y] = [a, y, x] \tag{18}$$

and

$$[x, y, z]\,[y, z, x]\,[z, x, y] = 1. \tag{19}$$

Lemma 7.37 (Gupta and Newman [1]). Let G be a metabelian n-Engel group. Then $\gamma_{n+1}(G)/\gamma_{n+2}(G)$ has finite exponent dividing $(n+1)((n-1)!)$.

Proof. We will assume that $\gamma_{n+2}(G) = 1$. Let $x, y_1, \ldots, y_n$ be elements of G. Since $\gamma_{n+1}(G)$ lies in the centre of G, the commutator $[x, y_1, \ldots, y_n]$ is homomorphic in each argument; therefore

$$1 = [x, {}_n y_1 \ldots y_n] = \prod [x, y_{i_1}, \ldots, y_{i_n}], \tag{20}$$

the product being formed over all possible ordered sets $(i_1, \ldots, i_n)$ where $1 \leqq i_j \leqq n$. Set $y_1 = 1$ in (20); then the product of all the commutators which do not involve y_1 is 1 and, since G' is abelian, they can be dropped from equation (20). Similarly commutators not involving $y_2, y_3, \ldots$ or y_n can be dropped. Therefore (20) becomes

$$\prod_{\pi \in S_n} [x, y_{\pi(1)}, y_{\pi(2)}, \ldots, y_{\pi(n)}] = 1$$

where S_n is the symmetric group on $\{1, 2, \ldots, n\}$. Now use (18) to permute the symbols in positions 3 to n of the commutators in the last product. We get

$$([x, y_1, y_2, \ldots, y_n] [x, y_2, y_1, y_3, \ldots, y_n] \cdots [x, y_n, y_1, y_2, \ldots, y_{n-1}])^{(n-1)!} = 1. \tag{21}$$

Interchanging x and y_1, we obtain

$$([y_1, x, y_2, \ldots, y_n] [y_1, y_2, x, y_3, \ldots, y_n] \cdots [y_1, y_n, x, y_2, \ldots, y_{n-1}])^{(n-1)!} = 1.$$

Now invert this last equation and multiply the result together with (21), expanding by means of the equation

$$[y_1, y_i, x] = [x, y_1, y_i]^{-1} [x, y_i, y_1]$$

(see (19)). By further use of (18) we obtain

$$[x, y_1, \ldots, y_n]^{(n+1)((n-1)!)} = 1,$$

which proves the lemma. ☐

To simplify the notation in the sequel we will write

$$[x, {}_i y, {}_j z]$$

for

$$[[x, {}_i y], {}_j z].$$

Lemma 7.38. Let G be a metabelian group, let $a \in G'$ and let $x, y \in G$. Then for each positive integer r

$$[a, {}_r xy] = [a, {}_r x] [a, {}_{r-1} x, y]^r [a, {}_r x, y]^r c_r,$$

c_r being a possibly empty product of commutators of the form

$$[a, {}_i x, {}_j y]^{\pm 1} \tag{22}$$

where $i + j \geqq r$ and $j \geqq 2$.

Proof. Since

$$[a, xy] = [a, x]\,[a, y]\,[a, x, y], \tag{23}$$

G' being abelian, the result is true for $r = 1$ with $c_1 = 1$. Let $r > 1$ and assume that a suitable c_{r-1} has been found so that

$$[a, {}_{r-1}xy] = [a, {}_{r-1}x]\,[a, {}_{r-2}x, y]^{r-1}\,[a, {}_{r-1}x, y]^{r-1}\,c_{r-1}.$$

Since commutation by xy induces an endomorphism in G',

$$[a, {}_rxy] = [a, {}_{r-1}x, xy]\,[a, {}_{r-2}x, y, xy]^{r-1}\,[a, {}_{r-1}x, y, xy]^{r-1}\,[c_{r-1}, xy].$$

Now apply the identity (23) to the right side of this equation. In the resulting expression certain commutators are seen to involve two or more y's and by (18) these may be taken together to form c_r: this will account for $[c_{r-1}, xy]$. Collecting the remaining commutators and using (18) once more we obtain the required result. □

Lemma 7.39 (Gupta and Newman [1]). Let G be a metabelian group and let a be a right n-Engel element of G belonging to G'. Then for arbitrary x and y in G

$$[a, x, {}_{n-1}y]^{n!} = 1.$$

Proof. Let i be an integer such that $0 \leqq i < n$; we will prove that

$$[a, {}_{n-i}x, {}_iy]^{f(i)} = w_{i+1}(x, y)$$

where

$$f(i) = n(n-1)\cdots(n-i+1) = n!/(n-i)!$$

and $w_{i+1}(x, y)$ is a possibly empty product of commutators of the form

$$\left.\begin{aligned} &[a, {}_jx, {}_ky]^{\pm 1} \\ \text{with}\quad & \\ &j + k \geqq n \quad\text{and}\quad k \geqq i + 1. \end{aligned}\right\} \tag{24}$$

The lemma will follow at once from this by setting $i = n - 1$ and using (18) to show that $w_n(x, y) = 1$.

Observe that our assertion is true for $i = 0$ with $w_1(x, y) = 1$ because $[a, {}_nx] = 1$. Let $i > 1$ and suppose that the assertion is true for $i - 1$. Then

$$[a, {}_{n-i+1}xy, {}_{i-1}y]^{f(i-1)}\,[a, {}_{n-i+1}x, {}_{i-1}y]^{-f(i-1)} = w_i(xy, y)\,w_i(x, y)^{-1}.$$

If we use Lemma 7.38 to expand the first term of the left side, we obtain

$$[a, {}_{n-i}x, {}_iy]^{f(i)}\,[a, {}_{n-i+1}x, {}_iy]^{f(i)} = cw_i(xy, y)\,w_i(x, y)^{-1}$$

where c is a product of commutators of type (24): notice here that $f(i) = (n - i + 1)\, f(i - 1)$. Now

$$[a, {}_{n-i+1}x, {}_iy]^{f(i)} = [[a, {}_{n-i+1}x, {}_{i-1}y]^{f(i-1)}, y]^{n-i+1}$$
$$= [w_i(x, y), y]^{n-i+1}.$$

From this it follows that

$$[a, {}_{n-i}x, {}_iy]^{f(i)} = cw_i(xy, y)\, w_i(x, y)^{-1}\, ([w_i(x, y), y]^{n-i+1})^{-1}$$
$$= w_{i+1}(x, y).$$

It remains only to verify that $w_{i+1}(x, y)$ is a product of commutators of type (24). In fact it is sufficient to check that $w_i(xy, y)\, w_i(x, y)^{-1}$ has the prescribed form. Let $[a, {}_ux, {}_vy]$ be a commutator occurring in the expression for $w_i(x, y)$ where $u + v \geqq n$ and $v \geqq i$, and let

$$z = [a, {}_uxy, {}_vy]\, [a, {}_ux, {}_vy]^{-1}.$$

Expanding by Lemma 7.38, we obtain after cancellation

$$z = [a, {}_{u-1}x, {}_{v+1}y]^u\, [a, {}_ux, {}_{v+1}y]^u\, [c', {}_vy]$$

where c' is a product of commutators of type $[a, {}_kx, {}_ly]^{\pm 1}$ with $k + l \geqq u$ and $l \geqq 2$. Hence z is a product of commutators of type (24) and therefore so is $w_i(xy, y)\, w_i(x, y)^{-1}$. ▯

Proof of Theorem 7.36. First of all let G be a metabelian n-Engel group where $n > 1$. Then Lemma 7.37 assures us that

$$\gamma_{n+1}(G)/\gamma_{n+2}(G) \text{ has exponent dividing } (n + 1)\,((n - 1)!). \tag{25}$$

Let $a \in G'$, $x \in G$ and $y \in G$. Then

$$1 = [x, {}_nya] = [[x, y]\,[x, a], {}_{n-1}ya] = [x, a, {}_{n-1}y]$$

since $[x, {}_ny] = 1$ and G' is abelian. Therefore

$$[a, x, {}_{n-1}y] = 1 \tag{26}$$

and $[a, x]$ is a right $(n - 1)$-Engel element of G.

Now define N_i to be the subgroup generated by all the commutators $[a, x, x_1, \ldots, x_i, {}_{n-i-1}y]$ where $a \in G'$ and $x, x_j, y \in G$. Then $N_0 = 1$ by (26) and

$$1 = N_0 \leqq N_1 \leqq \cdots \leqq N_{n-2} = \gamma_{n+2}(G).$$

In addition $N_i \lhd G$. Clearly $[a, x, x_1, \ldots, x_i]\, N_i$ is a right $(n - i - 1)$-Engel element of G/N_i lying in $(G/N_i)'$, so by Lemma 7.39

$$[a, x, x_1, \ldots, x_i, x_{i+1}, {}_{n-i-2}\,y]^{(n-i-1)!} \in N_i.$$

Thus N_{i+1}/N_i has finite exponent dividing $(n-i-1)!$ and consequently

$$(\gamma_{n+2}(G))^{(n-1)^*} = 1 \tag{27}$$

where

$$(n-1)^* = (1!)\,(2!)\cdots((n-1)!).$$

Combining (25) and (27) we obtain

$$(\gamma_{n+1}(G))^m = 1 \tag{28}$$

where

$$m = (n+1)\,((n-1)!)\,(n-1)^*.$$

Clearly $(n-1)^*$ divides a power of $((n-1)!)$, so the theorem is proved for G metabelian.

Now suppose that $c > 0$ and

$$G \in \mathfrak{N}_c\mathfrak{A},$$

i.e. G' is nilpotent of class at most c; also, of course, G is an n-Engel group. We will prove that

$$(\gamma_r(G))^s = 1 \tag{29}$$

where

$$r = r(n, c) = nc + 1, \quad s = s(n, c) = m^{c'}$$

and

$$c' = 1 + (c-1) + (c-1)\,(c-2) + \cdots + (c-1)\,(c-2)\cdots 2\cdot 1.$$

If $c = 1$, then G is metabelian and this has already been proved, so let $c > 1$. Let $r' = r(n, c-1)$ and $s' = s(n, c-1)$; then by induction on c applied to the group $G/\zeta(G')$, we have $(\gamma_{r'}(G))^{s'} \leqq \zeta(G')$ and therefore

$$[G', (\gamma_{r'}(G))^{s'}] = 1.$$

Let X be a normal subgroup of G contained in G'; then

$$[X, \gamma_{r'}(G)]^{s'} \leqq [X, \gamma_{r'}(G), \gamma_{r'}(G)]. \tag{30}$$

Now $r' \geqq 2$, so $\gamma_{r'}(G) \leqq G'$; thus, by repeated applications of (30), we obtain

$$[G', \gamma_{r'}(G)]^{(s')^{c-1}} \leqq \gamma_{c+1}(G') = 1. \tag{31}$$

Denote by H the holomorph of

$$\gamma_{r'}(G)/[G', \gamma_{r'}(G)]$$

by the automorphism group induced in this factor by G. Then H is a metabelian n-Engel group, and by the first part of the proof $(\gamma_{n+1}(H))^m = 1$ where $m = (n+1)\,((n-1)!)\,(n-1)^*$. Therefore

$$[\gamma_{r'}(G), \underbrace{G, \ldots, G}_{n}]^m = (\gamma_{r'+n}(G))^m \leqq [G', \gamma_{r'}(G)]. \tag{32}$$

Combining (31) and (32), we obtain

$$(\gamma_{r'+n}(G))^{(s')^{c-1}m} = 1.$$

Finally, observe that $r' + n = r(n, c)$ and $(s')^{c-1} m = s(n, c)$.

Now we assume only that G is a soluble n-Engel group with positive derived length d. We have to show that

$$(\gamma_f(G))^e = 1$$

where $f = f(n, d) = n^{d-1} + 1$ and $e = e(n, d)$ divides a power of $(n + 1)((n - 1)!)$. If $d = 1$, this is clear with $e(n, 1) = 1$, so let $d > 1$ and write $f' = f(n, d - 1)$ and $e' = e(n, d - 1)$; the latter divides a power of $(n + 1)((n - 1)!)$. By induction on d we have $(\gamma_{f'}(G'))^{e'} = 1$. Since $G/\gamma_{f'}(G')$ belongs to the class $\mathfrak{N}_{f'-1}\mathfrak{A}$, the result of the previous paragraph (equation (29)) may be applied and we obtain

$$(\gamma_{r_1}(G))^{s_1} \leqq \gamma_{f'}(G')$$

where $r_1 = r(n, f' - 1)$ and $s_1 = s(n, f' - 1)$. Hence $(\gamma_{r_1}(G))^{s_1 e'} = 1$. Finally, note that $r_1 = n(f' - 1) + 1 = n^{d-1} + 1 = f(n, d)$ and set $e(n, d) = s_1 e'$, which divides a power of $(n + 1)((n - 1)!)$.

This completes the proof of the first part of the theorem. The second part is established in an entirely analogous manner using equation (27) as the first step of the induction. ∎

7.4 Groups in which Every Subgroup is Subnormal with Bounded Index

In Section 6.1 we pointed out that the class of groups in which every subgroup is subnormal lies strictly between the class of nilpotent groups and the class of Baer groups: this relies on the example of Heineken and Mohamed [1].

As an application of Engel theory we will prove that, in contrast to the result just stated, a group in which every subgroup is subnormal with bounded subnormal index is nilpotent of bounded class. This theorem is due to Roseblade [6]. Let us begin with some definitions.

Let s be a non-negative integer and let n be a positive integer. Let

$$\mathfrak{U}_{s,n}$$

denote the class of groups in which every n-generator subgroup is subnormal with subnormal index at most s. Then

$$\mathfrak{U}_s = \bigcap_{n=1,2,\ldots} \mathfrak{U}_{s,n} \tag{33}$$

is the class of groups in which each subgroup is subnormal with subnormal index at most s: for, let G belong to $\mathfrak{U}_{s,n}$ for $n = 1, 2, \ldots$, and let $H \leqq G$; then $H^{G,s}$, the sth term of the series of successive normal closures of H in G, is the union of all $X^{G,s}$, where X is a finitely generated subgroup of H; but $X \lhd^s G$, so $X^{G,s} = X \leqq H$. Thus $H^{G,s} = H$ and $H \lhd^s G$.

If G belongs to $\mathfrak{U}_{s,1}$, then $\langle x \rangle \lhd^s G$ for all x in G; hence

$$[G, \underbrace{x, \ldots, x}_{s+1}] = 1$$

and consequently

$$\mathfrak{U}_{s,1} \leqq \mathfrak{E}_{s+1}, \tag{34}$$

which accounts for the intervention of Engel theory. Notice also the obvious inclusion

$$\mathfrak{N}_s \leqq \mathfrak{U}_s.$$

It is clear that $\mathfrak{U}_0$ is the class of trivial groups and that $\mathfrak{U}_1$ is the class of *Dedekind groups* or groups in which every subgroup is normal: it is well-known that a Dedekind group is either abelian or the direct product of a quaternion group of order 8 and a periodic abelian group without elements of order 4 (Dedekind [1] and Baer [1]: see also M. Hall [2], p. 190, Theorem 12.5.4).

Lemma 7.41. (i) Each subgroup of a group G is subnormal if and only if G is a Baer group with the generalized subnormal join property.

(ii) Each subgroup of a group G is subnormal with subnormal index at most s if and only if G is a Baer group whose finitely generated subgroups have subnormal indices at most s.

Proof. Let G be a Baer group with the generalized subnormal join property and let $H \leqq G$. If $h \in H$, then $\langle h \rangle$ *sn* G; hence H, being generated by its cyclic subgroups, is subnormal in G. The converse is obvious. The second part of the lemma is a consequence of the remark following equation (33). ▯

The principal result of this section is the next theorem.

Theorem 7.42 (Roseblade [6], Theorem 1). There exist functions f_1 and f_2 such that

$$\mathfrak{U}_{s,f_2(s)} \leqq \mathfrak{N}_{f_1(s)}$$

for each non-negative integer s.

Here and throughout this section f_i will denote a function defined on the set of non-negative integers assuming positive integral values. The functions f_1 and f_2 are defined recursively in a complicated manner and no attempt will be made to estimate them.

Corollary. A group in which each subgroup is subnormal with index at most s is nilpotent of class not exceeding $f_1(s)$.

The proof of Theorem 7.42 is broken up into a series of steps, the first being the proof of the soluble case.

Lemma 7.43. There exists a function f_3 (of two variables) such that

$$\mathfrak{U}_{s,s} \cap \mathfrak{A}^m \leqq \mathfrak{N}_{f_3(s,m)}$$

for all $s > 0$ and $m \geqq 0$.

Proof. We remind the reader that $\mathfrak{A}^m$ is the class of soluble groups with derived length at most m. Since

$$\mathfrak{U}_{1,1} = \mathfrak{U}_1 \leqq \mathfrak{N}_2,$$

we can put $f_3(1, m) = 2$, so assume that $s > 1$. Let G be a metabelian group in the class $\mathfrak{U}_{s,s}$; then $G \in \mathfrak{U}_{s,1}$, and G is an $(s+1)$-Engel group by (34). By the Gruenberg-Gupta-Newman theorem (7.36) we have

$$(\gamma_{s+2}(G))^l = 1, \tag{35}$$

where in fact

$$l = f_4(s) = (s+2)\,(s!) \prod_{i=1}^{s} (i!)\,.$$

Now G' is abelian, so (35) implies that

$$[(G')^l, \underbrace{G, \ldots, G}_{s}] = 1$$

and

$$(G')^l \leqq \zeta_s(G)\,.$$

Let $H = G/\zeta_s(G)$; then $(H')^l = 1$ and also H is metabelian and belongs to the class $\mathfrak{U}_{s,s}$. If X is a subset of H, the commutativity of H' and a simple induction on i enable us to prove that

$$\gamma_i(X^H) = \gamma_i(X)\,(\gamma H X^i); \tag{36}$$

here we are writing

$$\gamma H X^i = [H, \underbrace{X, \ldots, X}_{i}]$$

for brevity. Let $x \in H$; then $\langle x\rangle \lhd^s H$, so $\gamma H \langle x\rangle^{s+1} = 1$. It follows by (36) that $\gamma_{s+1}(x^H) = 1$ and x^H is nilpotent of class s at most. Let $\{x_1, \ldots, x_s\}$ be any set of s elements of H and let X be the subgroup that they generate. Then $X^H = x_1^H \cdots x_s^H$, so X is nilpotent of class at most s^2 by Fitting's Theorem (2.18). Now if A is an r-generator group, the ith lower central factor $\gamma_i(A)/\gamma_{i+1}(A)$ can be generated by r^i elements. Thus every subgroup of X can be generated by

$$f_5(s) = s + s^2 + \cdots + s^{s^2}$$

elements. Since X is an s-generator subgoup of H, we have $X \lhd^s H$ and by (36) this implies that

$$K = \gamma_s(X^H) = \gamma_s(X)\ (\gamma H X^s) \leqq X \cap H'.$$

Since $(H')^l = 1$, the group K has exponent dividing l; it is also abelian and can be generated by $f_5(s)$ elements. Therefore K is finite and its order is at most $l^{f_5(s)}$. Clearly there is a chief-series of H containing K. Now H is a Baer group and so is locally nilpotent, and, by a theorem of Mal'cev (Corollary 1 to Theorem 5.27), a chief-factor of a locally nilpotent group is central. Hence $K \leqq \zeta_{f_6(s)}(H)$ where

$$f_6(s) = [\log_2 (l^{f_5(s)})] = [f_5(s) \log_2 l].$$

From this and the definition of K, we conclude that every s-generator subgroup of $H/\zeta_{f_6(s)}(H)$ is nilpotent of class at most $s-1$, which obviously implies that $H/\zeta_{f_6(s)}(H)$ is nilpotent of class at most $s-1$. Hence G is nilpotent of class at most $f_7(s)$ where

$$f_7(s) = f_6(s) + s - 1 + s = f_6(s) + 2s - 1.$$

Now let G be soluble with derived length d and suppose that $G \in \mathfrak{U}_{s,s}$. If $d \leqq 2$, the result is already established if we define $f_3(s, 2) = f_7(s)$. Let $d > 2$: then G/G'' is nilpotent of class $\leqq f_3(s, 2)$ and by induction on d the group G' is nilpotent of class $\leqq f_3(s, d-1)$. Theorem 2.27 now permits us to conclude that G is nilpotent of class not exceeding $f_3(s, d)$ where

$$f_3(s, d) = \binom{f_3(s, d-1)+1}{2} f_3(s, 2) - \binom{f_3(s, d-1)}{2}. \ \square$$

In the next part of the proof we will need the following property of p-groups of automorphisms of abelian p-groups.

Lemma 7.44. Let G be an abelian p-group with finite rank r. Then the Sylow p-subgroups of Aut G are finite and have rank at most $\frac{1}{2} r(5r-1)$.

When G is finite, this result is due to P. Hall (see Roseblade [6], Lemma 5): the slight generalization given here occurs in a paper of Baer and Heineken ([1], Proposition 1.6).

Proof. G is written additively throughout the proof. Notice first that G satisfies Min and is the direct sum of r cyclic or quasicyclic groups (see Fuchs [3], p. 68). Let A be an arbitrary p-subgroup of Aut G; then A is finite, by the Corollary to Theorem 3.29.2. Denote by G_i the characteristic subgroup consisting of all g in G such that $p^i g = 0$ and let A_i be the centralizer of G_i in A. Then $A = A_0 \geqq A_1 \geqq A_2 \geqq \cdots$ and, since A is finite, we must have $A_i = A_{i+1} =$ etc. for some integer i, so that $A_i = 1$. Hence A is isomorphic with a p-group of automorphisms of G_i and we may assume that G is a finite group.

Let

$$C = C_A(G/pG).$$

Then G/pG is an elementary abelian p-group with rank r and the group A/C is isomorphic with a subgroup of the general linear group $GL(r, p)$; the Sylow p-subgroups of the latter have order $p^{\frac{1}{2}r(r-1)}$ and from this it is clear that they have rank at most $\frac{1}{2}r(r-1)$. Thus A/C may be generated by at most $\frac{1}{2}r(r-1)$ elements and therefore it is enough to show that C can be generated by

$$\frac{1}{2}r(5r-1) - \frac{1}{2}r(r-1) = 2r^2$$

elements under the hypothesis that G has exponent $p^e > p$.

Let $1 \neq \gamma \in C$; then there is largest positive integer

$$m = m(\gamma)$$

such that $m < e$ and γ centralizes $G/p^m G$. Now G can be generated by r elements, say $a_1, \ldots, a_r$; therefore we can write

$$a_i\gamma = a_i + p^{m(\gamma)} \sum_{j=1}^{r} \gamma_{ij} a_j$$

where the integers γ_{ij} are not all divisible by p. In an obvious way we can represent γ by the matrix

$$1 + p^{m(\gamma)} M(\gamma)$$

where $M(\gamma)$ is the $r \times r$ matrix whose (i, j)th element equals γ_{ij}. Clearly γ^{-1} is represented by the matrix

$$1 + \sum_{i=1}^{l(\gamma)} (-p^{m(\gamma)} M(\gamma))^i$$

where $l(\gamma)$ is the largest integer such that $l(\gamma)\, m(\gamma) < e$.

Let $\{\gamma_1, \ldots, \gamma_d\}$ be a minimal set of generators of C and define $M_i = M(\gamma_i)$, $m_i = m(\gamma_i)$ and

$$m(\gamma_1, \ldots, \gamma_d) = m_1 + \cdots + m_d.$$

Now choose $\{\gamma_1, \ldots, \gamma_d\}$ to be a minimal generating set for C whose m-value $m(\gamma_1, \ldots, \gamma_d)$ is maximal. We will assume that the m_i have been ordered so that $m_1 \leqq m_2 \leqq \cdots \leqq m_d$. If $d \leqq 2r^2$, then of course the lemma is proved, so let $d > 2r^2$.

Suppose first of all that $m_{r^2+1} = 1$, so that $m_j = 1$ for all $j \leqq r^2 + 1$. Let i be the least positive integer such that $M_1, \ldots, M_i$ are linearly dependent over the field $GF(p)$: since the vector space of all $r \times r$ matrices over any field has dimension r^2, we see that $1 < i \leqq r^2 + 1$. Then we can write

$$M_i \equiv \sum_{j=1}^{i-1} k_j M_j \bmod p$$

for certain integers k_j. Let

$$\gamma = \gamma_i^{-1} \gamma_1^{k_1} \cdots \gamma_{i-1}^{k_{i-1}}.$$

Since $1 = m_1 = \cdots = m_i$, the automorphism γ is represented by a matrix $1 + pM$ where

$$M \equiv -M_i + \sum_{j=0}^{i-1} k_j M_j \equiv 0 \bmod p.$$

This shows that γ centralizes G/p^2G and $m(\gamma) > 1$. Since the elements $\gamma_1, \ldots, \gamma_{i-1}, \gamma, \gamma_{i+1}, \ldots, \gamma_d$ generate C and since $m(\gamma) > 1 = m(\gamma_i)$, these elements constitute a minimal generating set with m-value exceeding that of the set $\{\gamma_1, \ldots, \gamma_{i-1}, \gamma_i, \gamma_{i+1}, \ldots, \gamma_d\}$.

By this contradiction $m_{r^2+1} > 1$. Observe that $r^2 < d - r^2$, so $M_{r^2+1}, \ldots, M_d$ are linearly dependent modulo p. Let t be the least integer such that $r^2 + 1 < t \leqq d$ and M_t is expressible modulo p as a linear combination of $M_{r^2+1}, \ldots, M_{t-1}$. Then for some integers k_j

$$M_t \equiv \sum_{j=r^2+1}^{t-1} k_j M_j \bmod p.$$

Let

$$\gamma = \gamma_t^{-1} \prod_{j=r^2+1}^{t-1} \gamma_j^{k_j p^{m_t - m_j}};$$

notice here that $m_j \leqq m_t$ if $j \leqq t$. A binomial expansion shows that γ is represented by the matrix $1 + p^{m_t} M$ where

$$M \equiv -M_t + \sum_{j=r^2+1}^{t-1} k_j M_j \equiv 0 \bmod p.$$

Therefore $m(\gamma) > m_t$ and the minimal generating set

$$\{\gamma_1, \ldots, \gamma_{t-1}, \gamma, \gamma_{t+1}, \ldots, \gamma_d\}$$

has a larger m-value than $\{\gamma_1, \ldots, \gamma_{t-1}, \gamma_t, \gamma_{t+1}, \ldots, \gamma_d\}$, which is our final contradiction. $\square$

Lemma 7.45. Let the group G be a product of normal r-generator abelian subgroups. Then the $(f_8(r) + 1)$th term of the lower central series of G is abelian where $f_8(r) = \frac{1}{2} r(5r - 1)$.

Proof. Clearly G is locally nilpotent, so the elements of finite order form a subgroup T of G and G/T is torsion-free. Then G/T is a product of normal free abelian subgroups, each with rank $\leqq r$. Lemma 6.37 implies that G/T is nilpotent of class at most r, so that

$$\gamma_{r+1}(G) \leqq T.$$

By hypothesis there is a normal subset $\{x_\lambda : \lambda \in \Lambda\}$ of G such that x_λ^G is an r-generator abelian group and G is the product of the x_λ^G. Let $\lambda_i \in \Lambda$ and define

$$X = [x_{\lambda_1}, \ldots, x_{\lambda_{r+1}}]^G. \tag{37}$$

Then $X \leqq T \cap (x_{\lambda_1}^G)$, so if p is any prime, the p-component P of X is a finite abelian p-group which can be generated by r elements. Let $H = G/C_G(P)$; since G is locally nilpotent, H is a finite p-group and by Lemma 7.44 H can be generated by $f_8(r)$ elements. Let $\mathrm{Frat}(H)$ be the Frattini subgroup of H: then, since $H/\mathrm{Frat}(H)$ is an elementary abelian p-group, its order cannot exceed $p^{f_8(r)}$. Also H can be generated by the elements $x_\lambda C_G(P)$, so by the Burnside Basis Theorem (see M. Hall, p. 176, Theorem 12.2.1) H can be generated by $f_8(r)$ of the elements $x_\lambda C_G(P)$. Hence H is a product of $f_8(r)$ normal abelian subgroups and so is nilpotent of class at most $f_8(r)$, by Theorem 2.18. Thus $\gamma_{f_8(r)+1}(G)$ centralizes P. Since p is arbitrary, $\gamma_{f_8(r)+1}(G)$ centralizes X and consequently $\gamma_{r+1}(G)$; for $\gamma_{r+1}(G)$ is the product of all such X by (37). Finally $r+1 \leqq f_8(r)+1$, so $\gamma_{f_8(r)+1}(G)$ is abelian. □

Lemma 7.46. Let $\mathfrak{X}_r$ be the class of all groups G such that $\gamma_r(H^G) \leqq H$ for each r-generator subgroup H of G. Then

$$\mathfrak{X}_r \leqq \mathfrak{N}_{f_9(r)}$$

for some function f_9.

Proof. Let $G \in \mathfrak{X}_r$ and let H be an r-generator subgroup of G; then $\gamma_r(H^G) \leqq H^G$. Since $H^G/\gamma_r(H^G) \in \mathfrak{N}_{r-1}$, it follows that $H \lhd^{r-1} H^G$ and $H \lhd^r G$. Hence

$$\mathfrak{X}_r \leqq \mathfrak{U}_{r,r}. \tag{38}$$

It will be sufficient if we can show that G is soluble with derived length at most $f_{10}(r)$ for a suitable function f_{10}; for then equation (38) and Lemma 7.43 will imply that G is nilpotent of class at most

$$f_9(r) = f_3(r, f_{10}(r)).$$

We define an ascending normal series

$$1 = G_0 \leqq G_1 \leqq \cdots \leqq G_\alpha \leqq \cdots$$

as follows: $G_{\alpha+1}/G_\alpha$ is the product of all the normal abelian subgroups of G/G_α. If $x \in G$, then $\gamma_r(x^G) \leq \langle x \rangle$, so x centralizes $\gamma_r(x^G)$: since the latter is normal in G, we can even say that x^G centralizes $\gamma_r(x^G)$ and that $\gamma_{r+1}(x^G) = 1$. Hence x^G is nilpotent of class at most r and consequently $x^G \leq G_r$. It follows that $G_r = G$. If we can prove that each G_{i+1}/G_i is soluble with derived length at most $f_{11}(r)$, it will follow that G is soluble with derived length at most

$$f_{10}(r) = rf_{11}(r).$$

Therefore we may assume that G is the product of its normal abelian subgroups. This implies that we can choose a set of generators $\{x_\lambda : \lambda \in \Lambda\}$ for G such that x_λ^G is abelian. Let

$$X = \langle x_{\lambda_1}, \ldots, x_{\lambda_r} \rangle$$

where $\lambda_i \in \Lambda$. Then $X^G = x_{\lambda_1}^G \cdots x_{\lambda_r}^G$ and X^G is nilpotent of class at most r. Since $G \in \mathfrak{X}_r$, we have $\gamma_r(X^G) \leq X$; now, being an r-generator nilpotent group with class $\leq r$, the subgroup X has rank $\leq r + r^2 + \cdots + r^r$ and therefore $\gamma_r(X^G)$ can be generated by at most $r + r^2 + \cdots + r^r$ elements. In addition, $\gamma_r(X^G)$ is abelian and the product of all such subgroups is $\gamma_r(G)$. Hence we deduce from Lemma 7.45 that $\gamma_k(\gamma_r(G))$ is abelian where

$$k = f_8(r + r^2 + \cdots + r^r) + 1.$$

It follows that G is soluble with derived length at most

$$f_{11}(r) = 3 + [\log_2 (k - 1)] + [\log_2 (r - 1)];$$

here we are using the Corollary to Lemma 2.21. □

Our final lemma is the basis for the inductive step in the proof of Theorem 7.42.

Lemma 7.47. Let the group G belong to the class $\mathfrak{U}_{s,r}$ and let H be a t-generator subgroup of G where $t \leq r$. Then

$$H^{G,s-j}/H^{G,s-j+1} \in \mathfrak{U}_{j,r-t}$$

provided that $0 \leq j \leq s$.

Proof. Let $H_i = H^{G,i}$. If X/H_{s-j+1} is an $(r-t)$-generator subgroup of H_{s-j}/H_{s-j+1}, we can write $X = YH_{s-j+1}$ where the subgroup Y has $r - t$ generators. Now $K = \langle H, Y \rangle$ has $t + (r - t) = r$ generators, so by hypothesis $K \lhd^s G$ and consequently $K = K_s \lhd^j K_{s-j}$. Now $K \leq H_{s-j} \leq K_{s-j}$, so $K \lhd^j H_{s-j}$. Since $X = YH_{s-j+1} = KH_{s-j+1}$, it follows that

$$X/H_{s-j+1} \lhd^j H_{s-j}/H_{s-j+1}$$

as required. □

Proof of Theorem 7.42. We have to show that

$$\mathfrak{U}_{s,f_2(s)} \leqq \mathfrak{N}_{f_1(s)}$$

for suitable functions f_1 and f_2. For $s = 0$, this is obvious if we set $f_1(0) = 0$ and $f_2(0) = 1$. If $s = 1$, this follows from the theorem of Dedekind and Baer that $\mathfrak{U}_{1,1} \leqq \mathfrak{N}_2$; for we can take $f_1(1) = 2$ and $f_2(1) = 1$.

Thus we can assume that $s > 1$ and that the theorem is true for $s - 1$. Define

$$m = (s-1)\,([\log_2 f_1(s-1)] + 1), \qquad f_{12}(s) = f_3(s, m) + 1$$

and

$$f_2(s) = f_2(s-1) + f_{12}(s).$$

It will first be shown that

$$\mathfrak{U}_{s,f_2(s)} \leqq \mathfrak{X}_{f_{12}(s)}. \tag{39}$$

Let $G \in \mathfrak{U}_{s,f_2(s)}$ and let H be a subgroup of G with $f_{12}(s)$ generators. If $0 \leqq j < s$, then by Lemma 7.47

$$H^{G,s-j}/H^{G,s-j+1} \in \mathfrak{U}_{j,f_2(s-1)},$$

since $f_2(s) - f_{12}(s) = f_2(s-1)$. Hence, by the induction hypothesis on s,

$$H^{G,s-j}/H^{G,s-j+1} \in \mathfrak{U}_{s-1,f_2(s-1)} \leqq \mathfrak{N}_{f_1(s-1)}.$$

This implies that $(H^G)^{(m)} \leqq H^{G,s} = H$ since $m = (s-1)\,([\log_2 f_1(s-1)] + 1)$. Now $f_2(s) \geqq s$, so $H^G/(H^G)^{(m)}$ is a soluble $\mathfrak{U}_{s,s}$-group; hence it is nilpotent of class at most $f_3(s, m) = f_{12}(s) - 1$, by Lemma 7.43. Thus

$$\gamma_{f_{12}(s)}(H^G) \leqq (H^G)^{(m)} \leqq H$$

and $G \in \mathfrak{X}_{f_{12}(s)}$.

Finally, if $f_1(s) = f_9(f_{12}(s))$, Lemma 7.46 shows that

$$\mathfrak{X}_{f_{12}(s)} \leqq \mathfrak{N}_{f_1(s)}. \tag{40}$$

The theorem now follows from (39) and (40). □

Chapter 8

Local Theorems and Generalized Soluble Groups

8.1 A Survey of Classes of Generalized Soluble Groups

We recall from the definition in Section 6.1 that a class of *generalized soluble groups* is a class of groups $\mathfrak{X}$ satisfying

$$\mathfrak{F} \cap \mathfrak{X} \leqq \mathfrak{S} \leqq \mathfrak{X},$$

i.e. a finite $\mathfrak{X}$-group is soluble and every soluble group is an $\mathfrak{X}$-group. Numerous classes of generalized soluble groups have arisen in previous chapters. Here several more will be introduced and we will also attempt to summarize what is known and what is still unknown about their relationships.

A. Radical Groups, SN^*-Groups, Subsoluble Groups and Hyperabelian Groups

Let us recall that a *radical group* is a group which has an ascending locally nilpotent series. This is a useful class because it contains all soluble groups and locally nilpotent groups but does not include the class of locally soluble groups, which is usually harder to deal with. A group is an SN^*, *subsoluble* or *hyperabelian* group if it has an ascending, ascending subnormal or ascending normal series with abelian factors respectively: subsoluble groups have also been called *SJ^*-groups* and hyperabelian groups *SI^*-groups*. By Lemma 2.35 every SN^*-group is a radical group, and it is clear that SN^*, subsoluble and hyperabelian are successively stronger properties.

Notice that *a group is radical, SN^*, subsoluble or hyperabelian if and only if it coincides with the limit of the upper Hirsch-Plotkin (i.e. locally nilpotent), Gruenberg, Baer or Fitting series respectively*: the first three of these results were proved in Section 2.3 (Part 1, pp. 59 and 62) while the fourth is obvious.

We recall from Lemma 7.33 that a locally nilpotent SN^*-group is a Gruenberg group. However, the locally dihedral 2-group is an example of locally nilpotent, soluble group which is not even a Baer group.

B. *SN*-Groups, *SJ*-Groups and *SI*-Groups

One of the widest classes of generalized soluble groups to have received significant attention is the class of *SN-groups* or groups with an abelian series; this is the class

$$SN = \hat{\boldsymbol{P}}\mathfrak{A}.$$

It turns out that there exist non-cyclic simple *SN*-groups—as we shall see in Section 8.4. On account of this the class of *SJ-groups*,

$$\hat{\boldsymbol{P}}_{sn}\mathfrak{A},$$

or groups which have a subnormal abelian series, is definitely smaller than *SN*. Narrower still is the class

$$\hat{\boldsymbol{P}}_{n}\mathfrak{A}$$

of *SI-groups* or groups which have a normal abelian series. A group with a descending (not necessarily normal) abelian series is an *SI*-group because its derived series terminates with the identity subgroup after a finite or infinite number of steps: such groups are called *hypoabelian*. Finally, *residually soluble groups* are clearly hypoabelian. We recall from Section 6.2 a result of Mal'cev according to which there exist hypoabelian groups of arbitrary hypoabelian (or derived) length.

Just how wide all these classes are may be judged from the fact that free groups are residually nilpotent, by the theorem of Magnus.

Notice that *every radical group is an SJ-group*: for a locally nilpotent group is a *Z*-group (by Corollary 1 to Theorem 5.27) and hence is certainly an *SJ*-group; also, a radical group has an ascending *normal* series with locally nilpotent factors, so it is an *SJ*-group.

C. $\overline{SN}$-Groups and $\overline{SI}$-Groups

An $\overline{SN}$-*group* is a group all of whose composition factors are abelian: here by a *composition factor* we mean a factor of a composition series or series of general type which has no proper refinements (see Section 1.2).

A minimal serial subgroup of an *SN*-group is easily seen to be abelian. Hence, using the fact that every group has a composition series, we conclude that *$\overline{SN}$ is the largest subclass of SN that is closed with respect to forming homomorphic images and serial subgroups.* Let

$$\hat{\boldsymbol{S}}$$

be the closure operation which consists in forming serial subgroups. Then $\hat{\boldsymbol{S}}\boldsymbol{H} \leq \boldsymbol{H}\hat{\boldsymbol{S}}$ and $\langle \hat{\boldsymbol{S}}, \boldsymbol{H} \rangle = \boldsymbol{H}\hat{\boldsymbol{S}}$. Thus

$$\overline{SN} = (SN)^{\boldsymbol{H}\hat{\boldsymbol{S}}} = (\hat{\boldsymbol{P}}\mathfrak{A})^{\boldsymbol{H}\hat{\boldsymbol{S}}}.$$

It seems unlikely that the class $\overline{SN}$ is closed under the formation of subgroups, so that

$$\overline{\overline{SN}} = (\overline{SN})^{\boldsymbol{S}} = (SN)^{\boldsymbol{HS}}$$

is a proper subclass of $\overline{SN}$, but this has not yet been settled.

We mentioned above that every radical group is an SJ-group: since the class of radical groups is $\boldsymbol{S}$ and $\boldsymbol{H}$-closed, it follows that *every radical group is an* $\overline{\overline{SN}}$*-group*. Another subclass of $\overline{\overline{SN}}$ is the class $\tilde{N}$ of groups in which all subgroups are serial; this class was discussed in Section 6.1.

A group is called an $\overline{SI}$*-group* if every chief factor is abelian. A minimal normal subgroup of an SI-group is clearly abelian and also every group has a chief series. Hence $\overline{SI}$ *is the largest subclass of* SI *that is closed with respect to forming homomorphic images*; thus

$$\overline{SI} = (SI)^{\boldsymbol{H}} = (\acute{\boldsymbol{P}}_n\mathfrak{A})^{\boldsymbol{H}}.$$

The class $\overline{SI}$ is not closed with respect to forming subgroups, by examples of Merzljakov [1] and P. Hall [14]. Thus the class

$$\overline{\overline{SI}} = (\overline{SI})^{\boldsymbol{S}} = (SI)^{\boldsymbol{HS}}$$

is properly contained in $\overline{SI}$; the class $\overline{\overline{SI}}$ was first studied in a paper of O. J. Schmidt [5].

It is obvious that every hyperabelian group is an $\overline{\overline{SI}}$-group and that every $\overline{\overline{SI}}$-group is an $\overline{\overline{SN}}$-group. Also, the chief factors of a locally soluble group are abelian (Corollary 1, Theorem 5.27), so *every locally soluble group is an* $\overline{\overline{SI}}$*-group*.

Diagram of classes of groups

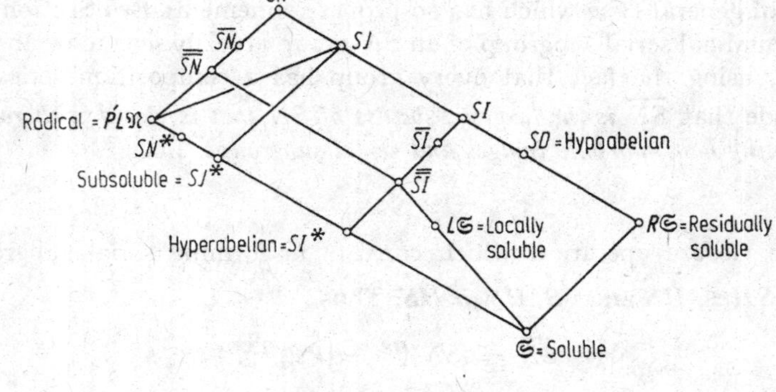

The following questions are oustanding.

(i) Is there an $\overline{SI}$-group which is not an $\overline{SN}$-group?

(ii) Does a subgroup of an $\overline{SN}$-group have to be an $\overline{SN}$-group? This is equivalent to asking whether $\overline{SN} = \overline{\overline{SN}}$.

Theorem 8.11. Apart from the two unsettled questions mentioned above, there are no inclusions between the fifteen classes of groups other than those indicated in the diagram.

Proof

(i) Neither hyperabelian nor locally soluble implies hypoabelian. This is shown by one of McLain's characteristically simple groups; these are perfect Fitting groups.

(ii) Hyperabelian does not imply locally soluble. Let M be the McLain group $M(Q, F)$ where Q is the set of rational numbers in their natural order and F is a countable field. By applying a construction of P. Hall it is possible to obtain a group $G = \langle t, u \rangle$ where u is an element of the cartesian product C of isomorphic copies M_i, $i = 0, \pm 1, \pm 2, \ldots$, of M and t is the so-called "shift operator" on C, such that $(u^G)'$ equals the direct product of the M_i'. It is not difficult to see that G is hyperabelian; of course G cannot be soluble since $M = M'$. For further details see P. Hall [11] (pp. 342—343) and [13] (p. 539).

(iii) Subsoluble does not imply SI. This construction too is due to P. Hall ([11], pp. 350—351); see also Čan Van Hao [2]. Let Q_2 be the set of all rational numbers of the form $m2^n$ (where m and n are integers) in their natural order and let F be any field. Notice that Q_2 is order-isomorphic with Q, the set of all rational numbers in their natural order; hence

$$M = M(Q_2, F) \simeq M(Q, F).$$

Let ξ and η be the automorphisms of M defined by the r ules

$$(1 + ae_{\lambda\mu})^{\xi} = 1 + ae_{2\lambda\, 2\mu} \quad \text{and} \quad (1 + ae_{\lambda\mu})^{\eta} = 1 + ae_{\lambda+1\, \mu+1}$$

where $\lambda, \mu \in Q_2$ and $a \in F$. Here the notation of Section 6.2 (p. 14) is employed. Now let G be the holomorph of M by $H = \langle \xi, \eta \rangle$. Since

$$\eta^{\xi} = \eta^2,$$

H is a metabelian group. M is a Baer group, so G is subsoluble by Lemma 2.35. We will prove that M is a minimal normal subgroup of G: since M is not abelian, this will show that G is not an SI-group. Suppose that N is a non-trivial normal subgroup of G contained in M. By Theorem 6.21, the definition of η and the commutator properties of the

$1 + ae_{\lambda\mu}$, the subgroup N contains $1 + ae_{0_2 n}$ for some integer n and all $a \in F$. Applying $\xi^{-n}\eta^l$, we find that $1 + ae_{ll+1} \in N$ for every integer l. Thus N contains $1 + ae_{lm}$ for all integers l and m; hence N contains all $1 + ae_{\alpha\beta}$, $(\alpha, \beta \in Q_2)$, and $N = M$. (It is easy to show that G is in fact monolithic with monolith M).

Observe that H satisfies Max-n by Theorem 5.34; therefore *G satisfies Max-n*. If F is a finite field, G is also finitely generated.

(iv) SN^* does not imply subsoluble. For example, consider the wreath power $W = Wr\, C_p^{Z^-}$ where C_p is a cyclic group of prime order p in its regular representation and Z^- is the set of negative integers in their natural order. W is a Gruenberg group and hence is an SN^*-group, but the Baer radical of W is 1 by Theorem 6.23.

(v) Radical does not imply SN^*, by the Kargapolov-Kovács-Neumann example: see Theorem 6.27.

(vi) $\overline{\overline{SN}}$ does not imply SJ. For there exist non-cyclic simple groups in the class $\overline{\overline{SN}}$ by the Corollary to Theorem 8.44.

(vii) Locally soluble does not imply radical. This is shown by McLain's example of an insoluble locally soluble group satisfying Max-n (Part 1, pp. 167—168); this group has trivial Hirsch-Plotkin radical.

(viii) Hypoabelian does not imply residually soluble since there are hypoabelian groups of arbitrary hypoabelian length; see p. 17.

(ix) Residually soluble does not imply $\overline{SN}$ or $\overline{SI}$: this is shown by any non-cyclic free group.

(x) $\overline{SI}$ does not imply $\overline{\overline{SN}}$. This can be seen from the group G consisting of all matrices

$$\begin{pmatrix} \alpha & \beta \\ \gamma & \delta \end{pmatrix}$$

where the p-adic integers α, β, γ and δ satisfy

$$\alpha \equiv \delta \equiv 1 \bmod p, \quad \beta \equiv \gamma \equiv 0 \bmod p$$

and

$$\alpha\delta - \beta\gamma = 1.$$

In [14] P. Hall has shown that G has no minimal normal subgroups and that if N is a non-trivial normal subgroup of G, then G/N is a finite p-group. It follows that every chief factor of G is central; thus G is a $\overline{Z}$-group and in particular an $\overline{SI}$-group. However, by a theorem of Brenner [1], the matrices

$$\begin{pmatrix} 1 & p \\ 0 & 1 \end{pmatrix} \text{ and } \begin{pmatrix} 1 & 0 \\ p & 1 \end{pmatrix}$$

generate a free group. Hence G is not an $\overline{\overline{SN}}$-group. A similar example may be found in the paper [1] by Merzljakov: see also Robertson [1]. □

We conclude our survey by considering three classes of generalized soluble groups which have been introduced in recent years.

Normally Supplemented Groups, Normalizer-Rich Groups and Residually Commutable Groups

We will say that a group G is *normally supplemented* if, whenever M is a non-normal maximal subgroup of a subgroup H, there exists a proper normal *supplement* of M in H, by which we mean a proper normal subgroup N of H such that

$$H = MN.$$

This class of groups has been studied by Baer [49] (p. 392) and Ayoub [2]. It is clearly closed with respect to forming subgroups.

Following Baer ([49], p. 358), we will call a group G *normalizer-rich (normalisator-reich)* if each proper self-normalizing subgroup H normalizes some *incomparable subgroup* K, i.e. a subgroup such that

$$H \nleq K \quad \text{and} \quad K \nleq H.$$

It is almost immediate from the definitions that *a group whose subgroups are all normalizer-rich is normally supplemented.*

Finally, we will say that a group G is *residually commutable*, if, given a pair of non-trivial elements a and b, there exists a normal subgroup N of G which contains $[a, b]$ but does not contain *both* a and b: this is clearly equivalent to the condition

$$\text{either } a \notin [a, b]^G \text{ or } b \notin [a, b]^G \text{ for all } a \neq 1 \text{ and } b \neq 1. \tag{1}$$

This property was introduced by Ayoub in [2].

The class of residually commutable groups is clearly $\boldsymbol{S}$ and $\boldsymbol{R}$-closed and equation (1) implies that it is also $\boldsymbol{L}$-closed. We remark that it is easy to show that *every SI-group is residually commutable* (Ayoub [2]).

Let M be a minimal normal subgroup of a residually commutable group G and suppose there exist $a, b \in M$ such that $[a, b] \neq 1$. Then there is an $N \lhd G$ such that N contains $[a, b]$ but not both a and b. Since $[a, b] \in M \cap N$, it follows that $M \leqq N$. However, this implies that $a \in N$ and $b \in N$. By this contradiction *a minimal normal subgroup of a residually commutable group is abelian.* From the $\boldsymbol{L}$-closure of the class of residually commutable groups we infer that a chief factor of a locally soluble group is abelian and a simple locally soluble group is cyclic of prime order: *thus we have another simple proof of these important theorems of Mal'cev* (see also Corollary 1 to Theorem 5.27).

Theorem 8.12. If G is a residually commutable group, then every subgroup of G is normalizer-rich.

Proof. Since a subgroup of a residually commutable group is also residually commutable, it suffices to prove that a proper self-normalizing subgroup H of G normalizes some incomparable subgroup of G. Suppose that this is not the case. Let $K = H^G$ and note that $H < K$ since $H = N_G(H) < G$. Let $b \in K \setminus H$; then $b \notin N_G(H)$. Replacing b by b^{-1} if necessary, we can find an $a \in H$ such that $a^b \notin H$. Then $c = [a, b] = a^{-1}a^b \notin H$. Now there exists a normal subgroup N of G which contains c but not both a and b. Since $c \notin H$, we see that $N \nleqq H$; consequently $H \leqq N$ and therefore $K \leqq N$. But this implies that a and b belong to N, which is not the case. □

We can therefore state that SI, *residually commutable, normalizer-rich subgroups and normally supplemented are successively weaker properties.*

The next theorem completes the proof that these lead to classes of generalized soluble groups.

Theorem 8.13 (Baer [49], Hilfssatz 3.6, Ayoub [2]). The normally supplemented groups which satisfy Min are precisely the soluble Černikov groups.

Proof. Let G be a normally-supplemented group satisfying Min. Supposing G not to be a soluble Černikov group, we may replace it by a group (also denoted by G) which is minimal with respect to not being a soluble Černikov group. If G is not finitely generated, each finitely generated subgroup of G is soluble and G is locally soluble. However, we know that a locally soluble group satisfying Min *is* a soluble Černikov group (by Theorem 3.12 and Corollary 1 to Theorem 5.27). Therefore G is finitely generated and, like all finitely generated groups, it possesses at least one maximal subgroup M. Now M is a soluble Černikov group, so M is not normal in G—for if it were, then G/M would be cyclic of prime order and G would be soluble. Since G is normally supplemented, there is a proper normal subgroup N of G such that $G = MN$. Now N is soluble and so is G/N because $G/N \simeq M/M \cap N$; therefore G is soluble and we know that this is impossible. □

It should be pointed out that the normalizer-rich groups do not form a class of generalized soluble groups because a finite normalizer-rich group may not be soluble, as the direct product of an alternating group of degree 5 and a cyclic group of order 2 shows.

Theorem 8.14. The class of normally supplemented groups is $\hat{\boldsymbol{P}}_n$-closed, that is to say, a group is normally supplemented if it has a normal series with normally supplemented factors. The class of normalizer-rich groups is likewise $\hat{\boldsymbol{P}}_n$-closed.

Proof. Let the group G have a normal series $\{\Lambda_\sigma, V_\sigma : \sigma \in \Sigma\}$ whose factors are normally supplemented and suppose that there is, nevertheless, a non-normal maximal subgroup M of G which has no proper normal supplement in G. If N is a normal subgroup of G, then either $N < M$ or $N = G$. Now let $g \in G \setminus M$ and suppose that $g \in \Lambda_\sigma \setminus V_\sigma$. Then $\Lambda_\sigma \nleqq M$, so $\Lambda_\sigma = G$; also $V_\sigma \neq G$ and $V_\sigma \leqq M$. Hence M/V_σ is a non-normal maximal subgroup of the normally supplemented group G/V_σ. Thus M/V_σ has a proper normal supplement in G/V_σ, from which we infer that M has a proper normal supplement in G. Since the class of normally supplemented groups is $\boldsymbol{S}$-closed, this contradiction shows that G is normally supplemented.

Now let $\{\Lambda_\sigma, V_\sigma : \sigma \in \Sigma\}$ be a normal series in a group G with normalizer-rich factors. Suppose that $H = N_G(H) < G$ and H normalizes no incomparable subgroup of G. There exist $x \in G \setminus H$ and $y \in H$ such that $y^x \notin H$. Suppose that $y \in \Lambda_\sigma \setminus V_\sigma$. Then $y^x \in \Lambda_\sigma$ and $y^x \notin H$ imply that $\Lambda_\sigma \nleqq H$, so $H < \Lambda_\sigma$; also $y \in H$ and $y \notin V_\sigma$ imply that $H \nleqq V_\sigma$, so $V_\sigma < H$. Thus H/V_σ, as a proper self-normalizing subgroup of the normalizer-rich group Λ_σ/V_σ, must normalize an incomparable subgroup of Λ_σ/V_σ, which implies that H normalizes an incomparable subgroup of G. ▯

With the aid of Theorem 1.36 we deduce

Corollary. The class of normally supplemented groups and the class of groups all of whose subgroups are normalizer-rich are residual classes.

We remark that the class of residually commutable groups, on the other hand, is not even closed under forming extensions. For, if G is the group of part (iii) of the proof of Theorem 8.11, it is clear that $G \in (\boldsymbol{L}\mathfrak{N})\,\mathfrak{S}$, so G is an extension of one residually commutable group by another, yet G is not residually commutable. Hence by Theorem 8.14 the class of groups whose subgroups are normalizer-rich properly contains the class of residually commutable groups.

We recall that a subgroup H is *almost-normal* in a group G if there exists a normal subgroup N of G such that

$$G = HN \quad \text{and} \quad H \cap N \lhd G.$$

Let us consider the following property $\mathscr{P}$: *if M is maximal in a subgroup H of a group G, then M is almost normal in H.* Obviously every group with $\mathscr{P}$ is normally supplemented, but $\mathscr{P}$ is a much more stringent requirement since it is inherited by homomorphic images. It is straightforward to show that *every hyperabelian group has $\mathscr{P}$*. By Theorem 8.13 a finite group which has $\mathscr{P}$ is soluble (Ore [1], p. 451, Theorem 9; Kuroš [9], Vol. 2, p. 181).

Properties related to but weaker than $\mathscr{P}$ are studied by Baer in [48]. We mention also that a full discussion of normalizer-rich groups may be found in Baer [49]: see also Amberg [2]. Some generalizations of solubility akin to residual commutability are in a papers of Durbin [4] and Newell [2].

The following problems are unsolved.

(i) Does there exist a residually commutable group that is not an SI-group?

(ii) Does there exist a normally supplemented group that is not normalizer-rich?

Generalizations of Supersolubility

We mention briefly some classes of generalized supersoluble groups. Firstly there is the class of *hypercyclic groups* or groups with an ascending normal series with cyclic factors. More generally, there is the class of groups having a normal series with cyclic factors

$$\hat{P}_n\mathfrak{C}.$$

Since the automorphism group of a cyclic group is abelian, we see that *the derived subgroup of a hypercyclic group is a hypercentral group* and *the derived subgroup of a group with a normal cyclic series is a Z-group*. By Theorem 3.18 a hypercyclic group is locally supersoluble, *so the properties locally hypercyclic and locally supersoluble are identical*. Other generalizations of supersolubility can be found in papers of Baer [30], [44], Durbin [4] and Newell [2]. Conditions for wreath products to be supersoluble or hypercyclic or to have a normal cyclic series have been obtained by Durbin [1] and Scott and Sonneborn [1].

The Effect of Finiteness Conditions

By the Corollary to Theorem 3.16, an SN-group satisfying Min is a soluble Černikov group and by Theorem 8.13 a normally supplemented group satisfying Min is a soluble Černikov group. Thus all the classes of generalized soluble groups which we have considered coincide with the class of soluble groups when Min is imposed. The effect of Min-n is less drastic. However it is obvious that a hypoabelian group which satisfies Min-n is soluble. In addition there is the following result.

Theorem 8.15. A residually commutable group which satisfies Min-n is hyperabelian.

Proof. Let G be residually commutable and suppose that G satisfies Min-n but is not hyperabelian. By Baer's criterion (Theorem 2.15) there exist two sequences of elements of G

$$x_1, x_2, \ldots \quad \text{and} \quad y_1, y_2, \ldots$$

such that

$$1 \neq x_{i+1} = [x_i, y_i, x_i]$$

for each positive integer i. Let $N_1 = G$ and suppose that we have constructed a normal subgroup N_i of G containing x_i. Now, since each $x_j \neq 1$, we see that $[x_i, y_i] \neq 1$. Since G is residually commutable, there is a normal subgroup N of G such that $x_{i+1} = [x_i, y_i, x_i] \in N$ but N does not contain both x_i and $[x_i, y_i]$. Set $N_{i+1} = N_i \cap N$; then $x_{i+1} \in N_{i+1}$ and $N_i > N_{i+1}$ since either x_i or $[x_i, y_i]$ belongs to $N_i \backslash N$. But this construction evidently produces an infinite descending chain of normal subgroups, $G = N_1 > N_2 > \cdots$, which is impossible by Min-n. □

Corollary (Ayoub [2]). A residually commutable group is finite if and only if it is finitely generated and satisfies Min-n.

Proof. We need only show that a finitely generated hyperabelian group G which satisfies Min-n is finite. If this is not the case, then, by Lemma 6.17, we may suppose that every proper factor group of G is finite. Now G has a non-trivial normal abelian subgroup A, so G/A is finite and A is finitely generated. Since A is infinite, $A > A^2 > A^4 > \cdots$, which contradicts Min-n. □

On the other hand, there exist non-cyclic simple $\overline{\overline{SN}}$-groups and these are certainly insoluble and satisfy Max-n and Min-n. It is unknown if a subsoluble group satisfying Min-n need be hyperabelian.

The effect of imposing maximal conditions is a good deal less certain. For example, it is not known if a residually soluble group satisfying Max need be soluble. On the other hand, a radical group with either of the properties Max-ab or Max-sn is polycyclic, by Theorems 3.31 and 5.46: the corresponding statements for the minimal conditions are also true. Clearly a hyperabelian group which satisfies Max-n is soluble. However, the properties hyperabelian, subsoluble, SN^* and radical remain distinct under Max-n: to see this we require Theorem 8.11 (part (iii) of proof), Corollary, Theorem 6.25 and also the following: apply the construction of Theorem 6.27 with Mc Lain's group $M(Q, GF(p))$ for each H_α: take the Z *th* standard wreath power W of the result and form $G = \langle W, a \rangle$, where a is the shift operator automorphism of W (see p. 25). Then G is radical with Max-n, but is not SN^*.

We have already seen that there are insoluble locally soluble groups which satisfy Max-n (Part 1, pp. 167—168.). Further interesting questions are whether a locally soluble group satisfying either Max-ab or Max-sn is polycyclic: it is known that a locally soluble group satisfying Min-ab or Min-sn is a Černikov group (Theorem 3.45 and Theorem 3.12).

A discussion of the effect on generalized solubility of maximal and minimal conditions on subnormal or ascendant subgroups may be found in Baer [50] and Robinson [9] (Theorems E and E*): see also Section 5.4 (v).

Radical Groups with Finite Rank

Theorem 8.16. Let G be a radical group whose Hirsch-Plotkin radical has finite rank r. Then G has a normal subgroup N such that N' is hypercentral and $|G:N|$ is finite and does not exceed a number depending only on r.

Notice that G itself need not have finite rank—consider for example the holomorph of the additive group of rational numbers.

During the proof we will need a property of the Hirsch-Plotkin radical of a radical group. If H is a set of operators on a group G, a series $\{\Lambda_\sigma, V_\sigma: \sigma \in \Sigma\}$ in G is called *H-stable** if for each σ in Σ

$$[\Lambda_\sigma, H] \leq V_\sigma,$$

i.e. $[x, \theta] = x^{-1}x^\theta \in V_\sigma$ for all $x \in \Lambda_\sigma$ and $\theta \in H$. For example, a G-stable series is simply a central series of G. The result referred to above is

Lemma 8.17. Let R be the Hirsch-Plotkin radical of a radical group G and let $N \lhd G$. Then R has an N-stable ascending series if and only if N is hypercentral.

Proof. If N is hypercentral, then $N \leq R$ and $[R, N] \leq N$; thus the upper central series of N with R adjoined is an N-stable ascending series of R. Conversely, let R have such a series: in order to prove that N is hypercentral it is sufficient to show that $N \leq R$. Suppose that $N \nleq R$; then by Lemma 2.16 the subgroup NR/R contains a non-trivial normal locally nilpotent subgroup L/R. By Dedekind's modular law $L = L \cap (NR) = MR$ where $M = L \cap N$. If X is a finitely generated subgroup of M, then $X/X \cap R$ is nilpotent, while R—and hence $X \cap R$—has an X-stable ascending series. Hence X is hypercentral. It follows that M is locally nilpotent. Hence $M \leq R$ and $L = MR = R$, which is a contradiction. □

Proof of Theorem 8.16. The Hirsch-Plotkin radical R of the radical group G is assumed to have finite rank r. Then we see from Corollary 1 to Theorem 6.36 that R is hypercentral. We now refine the upper central series of R to an ascending G-admissible series whose factors are either elementary abelian or torsion-free; by further refinement we may assume

* For an account of the theory of stability groups see Hall and Hartley [1] and Hartley [2], [4].

that the action of G is irreducible on a finite factor and rationally irreducible on an infinite factor. Thus each factor gives rise to a representation of G as an irreducible linear group of degree r. Bearing in mind that a radical linear group is soluble (Corollary to Theorem 3.23), we can apply Mal'cev's theorem on soluble linear groups (Theorem 3.21): we conclude that there is an integer $m = m(r)$ such that if $N = G^m$, then R has an N'-stable ascending series. Then N' is hypercentral by Lemma 8.17. Also, G/N is a periodic radical group and hence is locally finite (Section 2.3, p. 59). Now *if H is a finite group with rank r and exponent e, the order of H cannot exceed some $d = d(r, e)$*: to prove this, first estimate the order of a Sylow subgroup of H using Lemma 2.19.1 and then observe that the order of H is divisible by at most $[\log_2 e]$ primes. It follows that G/N has finite order not exceeding $d(r, m)$ and the theorem is proved. □

Radical and Locally Soluble Groups with Finite Abelian Subgroup Rank

We mention next some results which are related to Mal'cev's theory of locally nilpotent groups with finite abelian subgroup rank (Section 6.3). Recently Baer and Heineken [1] have shown that *if G is a radical group with finite abelian subgroup rank, then G is hyperabelian and each abelian section of G has boundedly finite p-rank for $p = 0$ or a prime.* These authors also show that *a radical group whose abelian subgroups have finite rank itself has finite rank*: this generalizes previous work of Kargapolov [9] who proved the theorem for soluble groups. Similar results are to be found in two papers of Čarin [10], [11]. See also Theorem 10.35 below.

The investigation of locally soluble groups with finite abelian subgroup rank is beset with difficulties. For example, Merzljakov [2] * has constructed a locally soluble group whose abelian subgroups have finite but unbounded rank: this group cannot therefore have finite rank. In the same paper Merzljakov proves that *a locally soluble group whose abelian subgroups are all of boundedly finite rank itself has finite rank.* Also Gorčakov [4] has shown that *a periodic locally soluble group whose abelian subgroups are all of finite rank itself has finite rank.*

In conclusion we note

Theorem 8.18 (Baer [44], Theorem F). A locally supersoluble group G with finite abelian subgroup rank is hypercyclic.

Proof. Since the derived subgroup of a supersoluble group is nilpotent, G' is locally nilpotent. By Corollary 1 to Theorem 6.36, there is an ascending G-admissible central series in G' whose factors are either finite

* See also Algebra i Logika **8**, 686–690 (1969) = Algebra and Logic **8**, 391–393 (1969), for a correction.

elementary abelian p-groups or torsion-free groups of finite rank; moreover, we may suppose that the action of G is irreducible on finite factors and rationally irreducible on infinite factors. Since each factor is central with respect to G', the group G induces an abelian group of automorphisms in every factor.

A factor of our series gives rise to an irreducible representation of G by an abelian group G^* of linear transformations of a finite dimensional vector space V. Let $x \in G^*$; since G is locally supersoluble, there is a non-zero characteristic vector for x, say v. If $y \in G^*$, then vy is also a characteristic vector for x, in view of the commutativity of G^*. But the vy generate V since G^* acts irreducibly on V, so x is scalar. Therefore G^* is scalar and V has dimension 1, which shows that G is hypercyclic. $\square$

Normal Products of Generalized Soluble Groups

It is noteworthy that the product of two normal locally soluble subgroups need not be locally soluble—as was first pointed out by P. Hall. This is, of course, in marked contrast to the behaviour of both soluble groups and locally nilpotent groups (cf. the Hirsch-Plotkin Theorem).

We will now describe Hall's construction (unpublished; see also Baumslag, Kovács and Neumann [1]). The basis for this is

Lemma 8.19. Let H be a normal subgroup of a group G and suppose that G/H is an infinite cyclic group. Then G can be embedded in a group G^* which is a normal product of two isomorphic copies of the standard wreath product of H with an infinite cyclic group.

Proof. Since G/H is infinite cyclic, there is an element x such that

$$G = \langle x, H \rangle \quad \text{and} \quad \langle x \rangle \cap H = 1.$$

Let D be the group of all restricted sequences

$$a = (a_n)_{n \in Z}$$

where $a_n \in H$ and Z is the set of all integers. Two automorphisms ξ and η of D are defined by the rules

$$(a^{\xi})_n = a_n^x \quad \text{and} \quad (a^{\eta})_n = a_{n+1}.$$

Clearly $\xi\eta = \eta\xi$ and $X = \langle \xi, \eta \rangle$ is abelian. Notice that both η and $\xi\eta$ have infinite order. Let G^* be the holomorph of D by X. If H_0 is the subgroup of all a in H such that $a_n = 1$ for $n \neq 0$, then

$$G^* = \langle \xi, \eta, H_0 \rangle. \tag{2}$$

Evidently $G \simeq \langle \xi, H_0 \rangle$.

We define $M = \langle \eta, D \rangle$ and $N = \langle \xi\eta, D \rangle$. Since X is abelian, $(G^*)' \leq D \leq M \cap N$, which shows that $M \lhd G^*$ and $N \lhd G^*$. Also $G^* = MN$. Now $M \simeq H \wr \langle \eta \rangle$ by definition of η, and the mappings $\eta \to \xi\eta$ and $a \to a'$ determine an isomorphism of M with N if a' is defined by $(a')_n = (a_n)^{x^{-n}}$. □

Theorem 8.19.1. (P. Hall)

(i) There exists a finitely generated group which is the product of two normal locally soluble, hyperabelian subgroups but which is not hyperabelian.

(ii) There exists a finitely generated group which is the product of two normal locally soluble subgroups but which is not an SI-group or a radical group.

Proof. Let H be the McLain group $M(Z, GF(p))$ where p is a prime and Z is the set of integers in their natural order; let x be the automorphism of H defined by

$$1 + e_{ij} \to 1 + e_{i+1\,j+1}$$

(see Section 6.2). If G is the holomorph of H by $\langle x \rangle$, then G/H is infinite cyclic; thus by Lemma 8.19 the group G can be embedded in a group G^* which is a normal product of two isomorphic copies of $H \wr \langle x \rangle$. One verifies without difficulty that the classes of locally soluble groups and hyperabelian groups are closed with respect to forming finite wreath products; thus G^* is the product of two normal locally soluble, hyperabelian subgroups. Now H can be generated by all the conjugates $(1 + e_{01})^{x^i}$, so it follows from equation (2) that G^* is a 3-generator group. However G, and therefore G^*, is not hyperabelian, as we shall now show. Let N be a non-trivial normal subgroup of G; if $H \cap N = 1$, then $[H, N] \leq H \cap N = 1$: however $[1 + e_{rs}, x^i] = 1$ implies that $i = 0$; therefore $N \leq H$ and $N = 1$. It follows that $H \cap N \neq 1$, and by Theorem 6.21 the subgroup N contains an element $1 + e_{ij}$. Since $N \lhd G$, the element $(1 + e_{ij})^{x^{j-i}} = 1 + e_{j\,2j-i}$ belongs to N. But $[1 + e_{ij}, 1 + e_{j\,2j-i}] = 1 + e_{i\,2j-i}$, so N is not abelian and therefore G cannot be hyperabelian.

To prove the second part of the theorem take H to be the wreath power

$$Wr\, C_\infty^Z$$

where C_∞ is a cyclic group of infinite order and Z is the set of integers in their natural order, and denote by x the automorphism of H which permutes the generating subgroups J_n according to the rule

$$(J_n)^x = J_{n+1},$$

(see p. 25). Let G be the holomorph of H by $\langle x \rangle$. Since G/H is infinite cyclic, we can embed G in a group G^* which is a normal product of two copies of $H \wr \langle x \rangle$. Since H is locally soluble, so is $H \wr \langle x \rangle$. Equation (2) and the definition of x show that G^* is a 3-generator group. But G—and therefore G^*—is not an SI-group or a radical group: for by Theorem 6.25 the subgroup H' is a minimal normal subgroup of G and H' is not locally nilpotent. ▯

Corollary. The following classes of groups are not closed with respect to forming products of pairs of normal subgroups: locally soluble groups, hyperabelian groups, locally hyperabelian groups, locally subsoluble groups, locally SN^*-groups, locally radical groups, SI, $\overline{SI}$, $\overline{\overline{SI}}$.

Normal Products of Free Groups

Theorem 8.19.2 (Hall and Hartley [1], Lemma 13). An arbitrary group G can be embedded in a group G^* which is the product of two normal free subgroups.

Proof. Let F be the free group on a set $\{x(g): g \in G\}$. If $g \in G$, then the mapping $x(g_1) \to x(g_1^g)$ extends to an automorphism of F and this enables us to represent G faithfully as a group of automorphisms of F. Let G^* be the semi-direct product of F by G which is thus determined.

If $g \in G$, we define

$$y(g) = gx(g);$$

we will write

$$\overline{F} = \langle y(g): g \in G \rangle.$$

Now $F \lhd G^*$ and also $\overline{F} \lhd G$ since $(y(g))^{g_1} = y(g^{g_1})$; clearly $G^* = F\overline{F}$, so it is enough to prove that the $y(g)$ freely generate $\overline{F}$.

Let w be an element of $\overline{F}$ and write

$$w = y(g_1)^{\varepsilon_1} \cdots y(g_r)^{\varepsilon_r}$$

where $\varepsilon_i = \pm 1$ and $\varepsilon_i = \varepsilon_{i+1}$ whenever $g_i = g_{i+1}$. Now g and $x(g)$ commute and this enables us to prove by induction on r that

$$w = g_1^{\varepsilon_1} \cdots g_r^{\varepsilon_r} x(g_1^{h_1})^{\varepsilon_1} \cdots x(g_r^{h_r})^{\varepsilon_r}$$

where

$$h_r = 1 \text{ and } h_i = g_{i+1}^{\varepsilon_{i+1}} \cdots g_r^{\varepsilon_r} \text{ if } 1 \leqq i < r.$$

Suppose that $r \geqq 1$ but $w = 1$. Since $F \cap G = 1$,

$$x(g_1^{h_1})^{\varepsilon_1} \cdots x(g_r^{h_r})^{\varepsilon_r} = 1. \tag{3}$$

If $g_i^{h_i} = g_{i+1}^{h_{i+1}}$, then the equation $h_i = g_{i+1}^{\varepsilon_{i+1}} h_{i+1}$ leads to $g_i = g_{i+1}$ and hence to $\varepsilon_i = \varepsilon_{i+1}$. Thus (3) represents a non-trivial relation between the free generators $x(g)$ of F. By this contradiction $w \neq 1$ and the $y(g)$ freely generate $\overline{F}$. ▯

Corollary 1. None of the following classes of groups is closed with respect to forming products of pairs of normal subgroups; SN, SJ, SI, hypoabelian groups, residually soluble groups, residually nilpotent groups.

For by Magnus's theorem free groups are residually nilpotent. Of the classes of generalized soluble groups in the diagram (other than $\mathfrak{S}$ itself) only the classes $\overline{SN}$, $\overline{\overline{SN}}$, radical groups, SN^* and subsoluble groups are $\boldsymbol{N}_0$-closed: notice that $\overline{SN}$ and $\overline{\overline{SN}}$ are $\boldsymbol{P}$-closed and $\boldsymbol{H}$-closed, so they are $\boldsymbol{N}_0$-closed by the relation $\boldsymbol{N}_0 \leqq \boldsymbol{PH}$. In fact all five classes are even $\boldsymbol{N}$-closed, either for obvious reasons or, in the case of $\overline{SN}$ and $\overline{\overline{SN}}$, by $\boldsymbol{L}$-closure.

The reasoning behind Corollary 1 can be more sharply expressed by the following relation between closure operations $\boldsymbol{S}$, $\boldsymbol{F}$ and $\boldsymbol{N}_0$: here $\boldsymbol{F}$ is the closure operation that consists in forming free products.

Corollary 2 (cf. Göbel [1], Satz 3.8). $\boldsymbol{SN}_0\boldsymbol{SF} = \boldsymbol{U}$, the universal closure operation.

Proof. To prove this we need only show that if $\mathfrak{X}$ is any class other than the class of trivial groups, then $\boldsymbol{SF}\mathfrak{X}$ contains all free groups. Let $1 \neq G \in \mathfrak{X}$: the free product $G_1 * G_2$, where $G_1 \simeq G \simeq G_2$, has an infinite cyclic subgroup which therefore belongs to $\boldsymbol{SF}\mathfrak{X}$. Since $\boldsymbol{FS} \leqq \boldsymbol{SF}$, every free group belongs $\boldsymbol{FSF}\mathfrak{X} = \boldsymbol{SF}\mathfrak{X}$. ▯

8.2 Local Theorems of Mal'cev and Applications

A theorem asserting that a class of groups is $\boldsymbol{L}$-closed will be called a *local theorem (of Mal'cev's type)* and such a class is called a *local class*. Our object here is to establish local theorems for classes of generalized soluble or nilpotent groups such as SN, SI and Z. A general method of proving local theorems was given by Mal'cev in 1941 utilizing a theorem in logic (Mal'cev [2]). Subsequently, other methods have been published by Kuroš and Černikov ([1], § 6), Kuroš ([9], § 55), McLain [6], Mal'cev [10] and Cleave [1]. The approach which we employ is a variation on McLain's method and its formulation is due to P. Hall.

Obvious examples of local classes are: any variety, the class of periodic groups and the class of torsion-free groups. It has already been shown that *the classes* $\overline{Z}$, $\overline{SI}$, $\tilde{N}$ *and* U *are local classes* (Theorem 5.27, Theorem 5.38 and its sequel, Theorem 6.18). Also, it is easy to prove that *for each i the class* $\mathfrak{B}_i$ *is a local class* (see Part 1, p. 173 for the definition). We mention without proof that Mal'cev has proved that *the class of linear groups of degree n is a local class* (Mal'cev [1], Theorem IV).

Local Systems and Mal'cev Functions

Let S be a non-empty set; a *local system*

$$\mathscr{L}$$

on S is a collection of subsets of S such that each finite subset of S lies within some member of $\mathscr{L}$. Thus the class $\boldsymbol{L}\mathfrak{X}$ is just the class of groups which have a local system consisting of $\mathfrak{X}$-subgroups. (Some authors use the term "local system" in a narrower sense, cf. Kuroš [9], § 55).

We select a fixed local system $\mathscr{L}$ on S and denote by

$$\Sigma$$

the set of all finite decompositions of $\mathscr{L}$ of the form

$$\sigma\colon \mathscr{L} = \mathscr{L}_1^{(\sigma)} \cup \cdots \cup \mathscr{L}_{r_\sigma}^{(\sigma)}$$

where the $\mathscr{L}_i^{(\sigma)}$ are distinct non-empty subsets of $\mathscr{L}$ but need not be local systems.

The following basic result is really a special case of the theorem of Steenrod that the projective limit of a system of non-empty compact sets is non-empty (Steenrod [1], Theorem 2.1, Lefschetz [1], p. 32): see also McLain [6].

Lemma 8.21. Let P be the set of all positive integers and let Σ be the set of all finite decompositions of a local system $\mathscr{L}$ on a set S. Then there exists a function $f\colon \Sigma \to P$ such that if Φ is a non-empty finite subset of Σ,

$$\mathscr{M}_\Phi = \bigcap_{\sigma\in\Phi} \mathscr{L}_{f(\sigma)}^{(\sigma)}$$

is a local system on S.

Proof. Consider the set of all functions $f\colon T \to P$ where T is a subset of Σ; we shall call f *admissible* if

$$\mathscr{M}_\Phi(f) = \bigcap_{\sigma\in\Phi} \mathscr{L}_{f(\sigma)}^{(\sigma)}$$

is a local system on S for all non-empty finite subsets Φ of T. Let σ be the trivial decomposition of $\mathscr{L}$

$$\mathscr{L} = \mathscr{L}_1^{(\sigma)}$$

and let $f\colon \{\sigma\} \to P$ be defined by $f(\sigma) = 1$. Then f is an admissible function and the set of admissible functions is not empty.

Let $f\colon T \to P$ and $f_1\colon T_1 \to P$ be two functions where T and T_1 are subsets of Σ. We define $f \leqq f_1$ to mean that $T \subseteq T_1$ and $f(\sigma) = f_1(\sigma)$ for all $\sigma \in T$. Clearly the relation $\leqq$ is a partial ordering of functions. Next let $\{f_\alpha\colon \alpha \in A\}$ be a chain of admissible functions and denote by T_α the domain of f_α: let $T = \bigcup_{\alpha \in A} T_\alpha$ and let $f\colon T \to P$ be defined by the rule $f(\sigma) = f_\alpha(\sigma)$ if $\sigma \in T_\alpha$. Since the f_α form a chain, the function f is well-defined. If Φ is a finite subset of T, then $\Phi \leqq T_\alpha$ for some $\alpha \in A$ and $f(\sigma) = f_\alpha(\sigma)$ if $\sigma \in \Phi$; thus f is admissible. We may therefore apply Zorn's Lemma and conclude that there is a maximal admissible function, say

$$f\colon T \to P.$$

If $T = \Sigma$, the lemma is established, so let us assume that this is not the case. Let $\sigma \in \Sigma \backslash T$. Then f cannot be extended to $T \cup \{\sigma\}$ and remain admissible. Now to extend f to $T \cup \{\sigma\}$ is to assign to f an integer $f(\sigma)$ satisfying $1 \leqq f(\sigma) \leqq r_\sigma$. Consequently, to each integer i such that $1 \leqq i \leqq r_\sigma$ there corresponds a finite subset Φ_i of T such that

$$\mathscr{M}_{\Phi_i}(f) \cap \mathscr{L}_i^{(\sigma)} \tag{4}$$

is *not* a local system on S. Hence there is a finite subset F_i of S which is not a subset of any member of (4). Let

$$\Phi = \bigcup_{i=1}^{r_\sigma} \Phi_i \quad \text{and} \quad F = \bigcup_{i=1}^{r_\sigma} F_i.$$

Since Φ and F are finite sets and $f\colon T \to P$ is admissible, there is a subset H belonging to $\mathscr{M}_\Phi(f)$ such that $F \subseteq H$. Now $H \in \mathscr{L}$, so $H \in \mathscr{L}_i^{(\sigma)}$ for some i satisfying $1 \leqq i \leqq r_\sigma$. Also $\Phi_i \leqq \Phi$; therefore

$$\mathscr{M}_\Phi(f) \leqq \mathscr{M}_{\Phi_i}(f)$$

and consequently H belongs to $\mathscr{M}_{\Phi_i}(f) \cap \mathscr{L}_i^{(\sigma)}$. But $F_i \subseteq F \subseteq H$, so F_i is contained in a member of $\mathscr{M}_{\Phi_i}(f) \cap \mathscr{L}_i^{(\sigma)}$, in contradiction to our choice of F_i. ▯

Corollary. If $\sigma \in \Sigma$, at least one of the $\mathscr{L}_i^{(\sigma)}$ is a local system on S.

We will call such a function f a *Mal'cev function* for the local system $\mathscr{L}$. In what follows the notation

$$S^{[n]}$$

will be used for the nth cartesian power of S.

Lemma 8.22. Let $\mathscr{L}$ be a local system on a set S, let F be a finite set and let n be a positive integer. Suppose that for each $H \in \mathscr{L}$ there is given a function $\alpha_H: H^{[n]} \to F$. Then there is a function $\alpha: S^{[n]} \to F$ such that for every finite subset $\{x_1, \ldots, x_m\}$ of $S^{[n]}$ there is an $H \in \mathscr{L}$ such that $x_i \in H^{[n]}$ and $\alpha(x_i) = \alpha_H(x_i)$ for $i = 1, \ldots, m$.

Proof. We may assume that $F = \{1, 2, \ldots, r\}$. For each x in $S^{[n]}$ define a finite decomposition $\sigma(x)$ of $\mathscr{L}$ as follows:

$$\sigma(x): \mathscr{L} = \mathscr{L}_1^{(x)} \cup \cdots \cup \mathscr{L}_{r+1}^{(x)}$$

where $H \in \mathscr{L}_{r+1}^{(x)}$ if $x \notin H^{[n]}$ and $H \in \mathscr{L}_i^{(x)}$ $(1 \leqq i \leqq r)$ if $x \in H^{[n]}$ and $\alpha_H(x) = i$. Notice that $\mathscr{L}_{r+1}^{(x)}$ is *not* a local system on S: for if it were and $x = (s_1, \ldots, s_n)$, then $X = \{s_1, \ldots, s_n\}$, as a finite subset of S, would have to be contained in some $H \in \mathscr{L}_{r+1}^{(x)}$, which is impossible because $x \notin H^{[n]}$ for any such H.

By Lemma 8.21 there exists a Mal'cev function $f: \Sigma \to P$ for $\mathscr{L}$: here Σ is the set of all finite decompositions of $\mathscr{L}$ and P is the set of all positive integers. A function $\alpha: S^{[n]} \to F$ is defined by the rule

$$\alpha(x) = f(\sigma(x)), \quad (x \in S^{[n]}).$$

Since $\mathscr{L}_{r+1}^{(x)}$ is not a local system and $\mathscr{L}_{f(\sigma(x))}^{\sigma(x)}$ is, it follows that

$$1 \leqq f(\sigma(x)) \leqq r.$$

and $\alpha(x) \in F$. Let $\{x_1, \ldots, x_m\}$ be a finite subset of $S^{[n]}$ and let

$$\Phi = \{\sigma(x_1), \ldots, \sigma(x_m)\}.$$

Since

$$\mathscr{M}_\Phi = \bigcap_{\sigma \in \Phi} \mathscr{L}_{f(\sigma)}^{(\sigma)}$$

is a local system on S, there is an $H \in \mathscr{M}_\Phi$ such that $\{x_1, \ldots, x_m\} \subseteq H^{[n]}$. Now $H \in \mathscr{L}_{f(\sigma(x_i))}^{\sigma(x_i)}$, so $\alpha_H(x_i) = f(\sigma(x_i)) = \alpha(x_i)$ for each i, as required. ∎

It will be convenient to write

$$\alpha = \lim_{H \xrightarrow{(f)} S} \alpha_H.$$

Application to Local Theorems

We must now explain how these results may be applied to prove local theorems. Suppose that $\mathfrak{X}$ is a class of groups such that a group H belongs to $\mathfrak{X}$ if and only if there exists a collection of functions $\{\alpha_{H,i}: i \in I\}$ where $\alpha_{H,i}: H^{[n_i]} \to F_i$ with n_i a positive integer and F_i a finite set, which satisfy certain conditions. Let $G \in L\mathfrak{X}$ and let $\mathscr{L}$ be a local system on G consisting of $\mathfrak{X}$-subgroups. If f is a Mal'cev function for $\mathscr{L}$, we define

$$\alpha_{G,i} = \lim_{\substack{(f) \\ H \longrightarrow G}} \alpha_{H,i} \quad (i \in I)$$

in accordance with Lemma 8.22. If the conditions on the functions $\alpha_{H,i}$ are *hereditary*, i.e. are inherited by the $\alpha_{G,i}$, then $G \in \mathfrak{X}$ and $\mathfrak{X}$ is a local class. We shall now describe some situations when these techniques can be applied.

Applications

a) Local classes defined by series

Theorem 8.23. Let $\mathfrak{B}$ be a variety. Then the class $\hat{P}\mathfrak{B}$ of group with a $\mathfrak{B}$-series and the class $\hat{P}_n\mathfrak{B}$ of groups with a normal $\mathfrak{B}$-series are local classes.

Taking $\mathfrak{B}$ to be the class of abelian groups here, we obtain

Corollary (Mal'cev [2]; see also Černikov [7]). The classes SN and $\grave{S}I$ are local classes.

In order to apply Lemma 8.22 we need to obtain a functional representation of a series (see Mal'cev [10]).

Let G be any group and let

$$\{\Lambda_\sigma, V_\sigma : \sigma \in \Sigma\}$$

be a series in G. The series determines a binary relation on G

$$\prec$$

defined as follows: $x \prec y$ means that either $x = 1$ or $x \neq 1$ and $\sigma(x) \leqq \sigma(y)$. Recall here that $\sigma(x)$ is the unique element of Σ such that $x \in \Lambda_{\sigma(x)} \setminus V_{\sigma(x)}$. The following properties of $\prec$ are easily verified:

(i) $x \prec y$ and $y \prec z$ imply that $x \prec z$,
(ii) either $x \prec y$ or $y \prec x$ (possibly both),
(iii) $x \prec 1$ implies that $x = 1$,
(iv) $x \prec y$ and $z \prec y$ imply that $xz^{-1} \prec y$,
(v) $y \prec x^y$ implies that $y \prec x$.

Of these (iv) expresses the fact that $\Lambda_{\sigma(y)}$ is a subgroup and (v) that $V_{\sigma(y)} \lhd \Lambda_{\sigma(y)}$. If the given series is normal, i.e. each term is normal in G, then in addition we have

(vi) $x^y \prec x$ for all x and y.

If each factor of the series belongs to a variety $\mathfrak{B}$, then:

(vii) given $x_i \prec y$, $(i = 1, \ldots, n)$ and θ a word in n variables determining $\mathfrak{B}$, it follows that

$$y \not\prec \theta(x_1, \ldots, x_n).$$

Conversely, we will show that if $\prec$ is a binary relation on G satisfying (i)—(v), then it determines a series in G. Let us define

$$x \sim y$$

to mean that both $x \prec y$ and $y \prec x$ are valid. Then $\sim$ is an equivalence relation on G by (i) and (ii). Let

$$\Sigma$$

be the set of all $\sim$-equivalence classes on G other than the obvious one $\{1\}$—notice that $x \sim 1$ implies that $x = 1$ by (iii). A linear ordering

$$<$$

of Σ is defined as follows: if $\sigma, \tau \in \Sigma$, then $\sigma < \tau$ means that $\sigma \neq \tau$ and there exist $x \in \sigma$ and $y \in \tau$ such that $x \prec y$. By (i) the relation $<$ is well-defined and by (i) and (ii) it is a linear ordering of Σ.

We are now ready to define the series that corresponds to $\prec$. If $\sigma \in \Sigma$, let

$$\Lambda_\sigma = \{x \colon x \in G, x \prec y \text{ for some } y \in \sigma\}$$

and

$$V_\sigma = \bigcup_{\tau < \sigma} \Lambda_\tau .$$

If $\sigma \leqq \tau$, then it is easy to see that $\Lambda_\sigma \subseteq \Lambda_\tau$. By (iv) we see that Λ_σ is a subgroup and, since the Λ_σ form a chain, V_σ is a subgroup. From (v) we deduce that $V_\sigma \lhd \Lambda_\sigma$ and it it now easy to verify that $\{\Lambda_\sigma, V_\sigma \colon \sigma \in \Sigma\}$ is a series in G. Furthermore, if the relation $\prec$ also satisfies. (vi), ((vii)), then the series is a normal series (a $\mathfrak{B}$-series). Evidently we have obtained a one-one correspondence between series in G and binary operations $\prec$ on G satisfying the conditions (i) to (v).

Suppose that $\mathscr{L}$ is a local system of subgroups on a group G and that each $H \in \mathscr{L}$ possesses a series determined by a binary operation $\prec_H$ on H. A function $\alpha_H \colon H^{[2]} \to \{0, 1\}$ is defined as follows

$$\alpha_H(x, y) = 1 \quad \text{if} \quad x \prec y$$

and

$$\alpha_H(x, y) = 0 \quad \text{otherwise}.$$

The properties (i) to (vii) may be stated purely as properties of the function α_H and it is easy to see that in this form they are hereditary. Thus, if f is a Mal'cev function for $\mathscr{L}$, the function

$$\alpha = \lim_{\substack{(f) \\ H \longrightarrow G}} \alpha_H$$

satisfies those of the properties (i) to (vii) which are satisfied by the α_H. Thus we obtain a series in G of the same kind as those in the subgroups H. This completes the proof of Theorem 8.23.

Theorem 8.24. If $\mathfrak{V}$ is any variety, the class of groups which have a $\mathfrak{V}$-marginal series is a local class.

Proof. We recall that a *$\mathfrak{V}$-marginal series* $\{\Lambda_\sigma, V_\sigma: \sigma \in \Sigma\}$ in a group G is a normal series such that Λ_σ/V_σ is $\mathfrak{V}$-marginal in G/V_σ for all $\sigma \in \Sigma$. If $\prec$ is the binary relation on G which determines the series, $\mathfrak{V}$-marginality is equivalent to the following property of $\prec$:

(viii) if θ is a word in n variables determining $\mathfrak{V}$ and if $x_i^{-1} y_i \prec z \neq 1$ for $i = 1, \ldots, n$, then

$$z \nprec \theta(x_1, \ldots, x_n)^{-1}\, \theta(y_1, \ldots, y_n).$$

That this property is hereditary is immediate. ▯

Taking $\mathfrak{V}$ to be the class of abelian groups, we obtain

Corollary (Mal'cev [2]). The class Z is a local class.

Further local theorems may be derived with the aid of

Lemma 8.25. Let $\boldsymbol{A}$ and $\boldsymbol{B}$ be closure operations satisfying $\boldsymbol{BA} \leq \boldsymbol{AB}$ and let $\boldsymbol{B}$ be unary. Then $\boldsymbol{A}(\mathfrak{X}^{\boldsymbol{B}}) \leq (\boldsymbol{A}\mathfrak{X})^{\boldsymbol{B}}$ for an arbitrary class of groups $\mathfrak{X}$.

Proof. Obviously $\boldsymbol{A}(\mathfrak{X}^{\boldsymbol{B}}) \leq \boldsymbol{A}\mathfrak{X}$ and since $\boldsymbol{BA} \leq \boldsymbol{AB}$,

$$\boldsymbol{B}(\boldsymbol{A}(\mathfrak{X}^{\boldsymbol{B}})) \leq \boldsymbol{AB}(\mathfrak{X}^{\boldsymbol{B}}) = \boldsymbol{A}(\mathfrak{X}^{\boldsymbol{B}}).$$

Hence $\boldsymbol{A}(\mathfrak{X}^{B})$ is contained in $(\boldsymbol{A}\mathfrak{X})^{\boldsymbol{B}}$, since the latter is by definition the largest $\boldsymbol{B}$-closed subclass of $\boldsymbol{A}\mathfrak{X}$. ▯

Theorem 8.26 (Mal'cev [2], [10]). The classes

$$\overline{SN}, \overline{\overline{SN}}, \overline{SI}, \overline{\overline{SI}}, \overline{Z}, \overline{\overline{Z}}$$

are local classes.*

* That $\overline{\overline{SI}}$ is a local class was also proved by O. J. Schmidt ([5], Theorem 3).

Proof. Recall that $\overline{SN} = (SN)^{\boldsymbol{H}\hat{\boldsymbol{S}}}$, $\overline{\overline{SN}} = (SN)^{\boldsymbol{HS}}$, $\overline{SI} = (SI)^{\boldsymbol{H}}$, $\overline{\overline{SI}} = (SI)^{\boldsymbol{HS}}$, $\overline{Z} = Z^{\boldsymbol{H}}$ and $\overline{\overline{Z}} = Z^{\boldsymbol{HS}}$. Now apply Lemma 8.25 with $\mathfrak{X} = SN$, SI or Z (all of which are $\boldsymbol{L}$-closed by Theorems 8.23 and 8.24), $\boldsymbol{A} = \boldsymbol{L}$ and $\boldsymbol{B} = \boldsymbol{H}$, $\boldsymbol{H}\hat{\boldsymbol{S}}$ or $\boldsymbol{HS}$. ▯

We have proved that six of the classes of generalized soluble groups in the diagram on p. 80 are local classes: obviously there is a seventh local class, the class of locally soluble groups. We now assert that *none of the remaining eight classes of groups in the diagram is a local class.* To see this observe that any $\boldsymbol{L}$-closed class of generalized soluble groups must contain all the locally soluble groups: thus we can rule out at once the classes of radical, SN^*, subsoluble, hyperabelian, soluble, residually soluble and hypoabelian groups, in view of Theorem 8.11. The remaining class SJ is not a local class since there exist non-cyclic simple groups belonging to the class $(\boldsymbol{LP})^{\omega+1}\mathfrak{A}$—see Theorem 8.45 and its corollary.

b) Criteria for Non-Simplicity of Locally Finite Groups

We consider next a somewhat different application of the technique embodied in Lemma 8.22.

Let G be a group and let $\mathscr{L}$ be a local system of subgroups on G. In each H in $\mathscr{L}$ we select a normal series determined by a function α_H as described in a). If f is a Mal'cev function for $\mathscr{L}$ and

$$\alpha = \lim_{H \xrightarrow{(f)} G} \alpha_H ,$$

then α determines a normal series in G. If G is simple, this must, of course, reduce to the trivial series $1 \lhd G$. Hence $\alpha(x, y) = 1$ unless $x \neq 1$ and $y = 1$. Now let X be a finite non-empty subset of G. By Lemma 8.22 there exists an $H \in \mathscr{L}$ such that $X \subseteq H$ and α and α_H coincide on $X^{[2]}$; hence $\alpha_H(x, y) = 1$ for all x and y in X unless $x \neq 1$ and $y = 1$. If $\{\Lambda_\sigma, V_\sigma : \sigma \in \Sigma\}$ is the prescribed normal series in H, it follows that

$$X \backslash \{1\} \subseteq \Lambda_\sigma \backslash V_\sigma \tag{5}$$

for some $\sigma \in \Sigma$, by definition of the function α_H.

On the basis of these remarks we will prove the following.

Theorem 8.27. Let G be an infinite locally finite group and suppose that if H is a finite subgroup of G, the index in H of its soluble radical cannot surpass a certain integer d. Then G is not a simple group.

Proof. Suppose that G is a simple group. Let $\mathscr{L}$ be the set of all finite subgroups of G: this is certainly a local system on G. If H is a finite subgroup of G, we form a normal series in H by adjoining H to the derived series of its soluble radical. Let X be an arbitrary finite subgroup of G.

By our previous discussion, particularly equation (5), there is an $H \in \mathscr{L}$ such that $X \leqq H$ and $X \backslash \{1\} \leqq L \backslash M$ where L/M is a factor of the normal series in H. Hence $X \leqq L$ and $X \cap M = 1$, so that $X \simeq XM/M \leqq L/M$. Therefore X is either abelian or finite of order $\leqq d$.

Let x and y be any two elements of G. Since G is infinite, there exists a countably infinite subgroup K containing x and y. Now K, being locally finite, is the union of a chain of finite subgroups only finitely many of which can have order $\leqq d$. Hence K is abelian and $[x, y] = 1$. Therefore G is abelian. But G is also simple, so it is finite, which is not the case. □

Corollary (cf. Baer [34], Satz 2.1). Let G be a locally finite group satisfying Min-*sn*. Then G is a Černikov group if and only if each subnormal composition factor has the property that each finite subgroup has its soluble radical of bounded index.

Proof. A subnormal composition factor of a Černikov group is easily seen to have boundedly finite order, so the condition is necessary. Conversely, if G satisfies the condition, each subnormal composition factor of G is finite by the theorem. It now follows from the Corollary to Theorem 5.46 that G is a Černikov group. □

For comparison with Theorem 8.27 we will prove another criterion for non-simplicity, which does not depend on the theory of local systems developed in Lemma 8.21 and 8.22.

If $\mathscr{L}$ is a local system of subgroups on a group G, then $\mathscr{L}$ is *ascendantly linked* if, given H and K in $\mathscr{L}$ such that $H \leqq K$, it follows invariably that H *asc* K. Every group G has ascendantly linked local systems, for example $\{G\}$ and $\{1, G\}$.

Theorem 8.28. Let G be a group which is not finitely generated and let $\mathscr{L}$ be an ascendantly linked local system on G. Suppose that there is in $\mathscr{L}$ a finitely generated subgroup H such that either the Hirsch-Plotkin radical of H is non-trivial or H contains a minimal subnormal subgroup. Then G is not simple.

Proof. Let us assume that G is a simple group. Suppose that the Hirsch-Plotkin radical of H is non-trivial and let $1 \neq x \in \varrho_{L\mathfrak{N}}(H)$. Denote by X an arbitrary finitely generated subgroup of G. Then $\langle H, X\rangle$ is finitely generated and therefore lies in some member of $\mathscr{L}$, say K. Now H *asc* K, so $\varrho_{L\mathfrak{N}}(H)$ *asc* K and therefore $\varrho_{L\mathfrak{N}}(H) \leqq \varrho_{L\mathfrak{N}}(K)$ by Theorem 2.31. Thus x^K, and therefore x^X, is locally nilpotent. It follows that x^G is locally nilpotent and $\varrho_{L\mathfrak{N}}(G) \neq 1$. But G is simple, so $\varrho_{L\mathfrak{N}}(G) = G$ and G, being a simple locally nilpotent group, is cyclic of prime order. This is impossible because G is not finitely generated.

It follows that H must contain a non-abelian minimal subnormal subgroup M. Once again let X be a finitely generated subgroup of G and let K be a member of $\mathscr{L}$ containing $\langle H, X\rangle$, so that M *asc* K. From the Corollary to Theorem 5.45 we obtain $M \lhd^2 K$, so $M \lhd^2 \langle M, X\rangle$, for all X. Therefore $M \lhd^2 G$. Since G is simple, this must mean that $G = M = H$, and that G is finitely generated, contrary to hypothesis. $\square$

Corollary. Let G be a non-finitely generated group which has an ascendantly linked local system at least one of whose members is finite and non-trivial. Then G is not a simple group.

A special case of this corollary is in a paper of Plotkin ([11], Theorem 3.14).

c) Sylow Subgroups of FC-Groups

As a final application we will prove an extension of Sylow's Theorem to FC-groups. First we recall some terminology. If p is a prime, a *Sylow p-subgroup* of a possibly infinite group is a maximal p-subgroup; such subgroups always exist by Zorn's Lemma. In a finite group a maximal p-subgroup is just a p-subgroup with order equal to the highest power of p dividing the group order, by Sylow's Theorem. There are well-known examples which show that in general two Sylow p-subgroups of group need not even have the same cardinality: see for example Kuroš [9], Vol. 2, pp. 269—270.

Several other counterexamples to standard Sylow theorems are contained in the paper [1] of Kovács, Neumann and de Vries.

In the other direction there is a result of Dietzmann, Kuroš and Uzkov [1] which asserts that *if a Sylow p-subgroup P of a group G has only finitely many conjugates in G, then every Sylow p-subgroup of G is conjugate to P.*

Following Gol'berg [2], we call an automorphism α of a group G *locally inner* if, given a finite subset F of G, there exists an element $g = g(F)$ in G such that $x^\alpha = x^g$ for all $x \in F$. If H and K are subgroups of a group G and $H^\alpha = K$ where α is a locally inner automorphism of G, then H and K are said to be *locally conjugate.* Obviously locally conjugate subgroups are isomorphic.

Theorem 8.29. (B. H. Neumann [17]). Any two Sylow p-subgroups of an FC-group are locally conjugate.

For periodic FC-groups this theorem is due to Gol'berg [2]—a slightly weaker version is in an earlier paper of Baer [8] (Theorem 4.4).

Proof. If X is a finite subset of G, then X^G is finitely generated since G is an FC-group. Hence the set

$$\mathscr{L}$$

of all normal finitely generated subgroups of G is a local system on G.

Let P be a Sylow p-subgroup of G and let $H \in \mathscr{L}$. By Theorem 4.32 and its second corollary, the elements of finite order in H form a finite subgroup T and clearly $T \lhd G$. Evidently $P \cap H = P \cap T$. Let $L = PT$ and set $C = \mathrm{Core}_L\, P$. Since $|L: P| = |T: P \cap T|$, the group L/C is finite. Clearly P/C is a Sylow p-subgroup of L/C, so $|L: P|$ is prime to p and consequently $P \cap T$ is a Sylow p-subgroup of T. Therefore $P \cap H$ is a Sylow p-subgroup of H. If P_1 is another Sylow p-subgroup of G, then by Sylow's Theorem (applied to the finite group T)

$$P_1 \cap H = (P \cap H)^{g(H)}$$

for some $g(H)$ in H. From now on P and P_1 will be two fixed Sylow p-subgroups.

We define a function $\alpha_H: H^{[2]} \to \{0, 1\}$ as follows:

$$\alpha_H(x, x^{g(H)}) = 1 \text{ and } \alpha_H(x, y) = 0 \text{ if } y \neq x^{g(H)}.$$

Choose a Mal'cev function f for $\mathscr{L}$ and let

$$\alpha = \lim_{H \xrightarrow{(f)} G} \alpha_H.$$

Now let $x \in G$. Since G is an FC-group, the set $\{x^G\}$ of all conjugates of x in G is finite and hence is contained in some $H \in \mathscr{L}$ such that α and α_H coincide on the set $\{x^G\}^{[2]}$: here we are using Lemma 8.22. Hence $\alpha(x, x^{g(H)}) = \alpha_H(x, x^{g(H)}) = 1$. Suppose that $\alpha(x, y) = 1 = \alpha(x, y')$ for some y and y' in G. Then $\alpha_H(x, y) = 1 = \alpha_H(x, y')$ for a suitable H in $\mathscr{L}$ and therefore $y = x^{g(H)} = y'$. It follows that to each x in G there corresponds a unique y such that $\alpha(x, y) = 1$.

Now define

$$x^\theta = y$$

where $\alpha(x, y) = 1$. Then one verifies at once that on a finite subset of G the mapping θ coincides with an inner automorphism of G. Hence θ is a locally inner automorphism of G. Also $P^\theta \leqq P_1$, so $P^\theta = P_1$ by maximality of P. ▯

Corollary (B. H. Neumann [17].) If each element of order a power of a prime p has only a finite number of conjugates in a group G, then the Sylow p-subgroups of G are isomorphic.

Proof. All the elements with order a power of p lie in F, the FC-centre of G, and F is a FC-group. The result now follows directly from the theorem. ▯

On the other hand, the hypotheses of the corollary do not imply that the Sylow p-subgroups of G are locally conjugate or even conjugate under an automorphism of G. This is shown by the following example (B. H. Neumann [17]).

The alternating group A_4 of degree 4 can be generated by elements a and b subject to defining relations

$$a^3 = b^2 = (ab)^3 = 1.$$

The mappings

$$a \to a^{-1} \quad \text{and} \quad b \to b \tag{6}$$

determine an automorphism of A_4 of order 2. Let $\{H_1, H_2, \ldots\}$ be a countably infinite set of isomorphic copies of A_4 and let D be their direct product. t is the automorphism of D which induces in each H_i the automorphism corresponding to (6). Let G be the holomorph of D by $\langle t \rangle$. Then D is the FC-centre of G and

$$S = \langle a_i : i = 1, 2, \ldots \rangle \quad \text{and} \quad \bar{S} = \langle a_i b_i : i = 1, 2, \ldots \rangle$$

are Sylow 3-subgroups of G lying inside D. Suppose $S^\alpha = \bar{S}$ for some $\alpha \in \operatorname{Aut} G$. Then $t' = t^\alpha$ must transform each $a_i b_i$ into its inverse; however there is no such element t' in G.

In general the Sylow p-subgroups of an FC-group need not be conjugate. For example, if G is the direct product of a countable infinity of symmetric groups of degree 3, then G has uncountably many Sylow 2-subgroups and these cannot all be conjugate because G is countable. Indeed Kargapolov [1] has proved that *the Sylow p-subgroups of a periodic FC-group are conjugate if and only if they are finite in number*: this had previously been established for countable groups by Baer ([8], Theorem 4.5).

Much attention has been devoted to the extension of other classical theorems of finite group theory to infinite groups: for example, Baer [8] and Gol'berg [2] have established P. Hall's theory of Sylow bases of finite soluble groups for locally soluble, periodic FC-groups. The more recent theory of Carter and Gaschütz has also been extended. The following papers have appeared in recent years: Baer [53], Platonov [1], Stonehewer [1]—[4], Tomkinson [1]—[4], Tyskevič [1], Wehrfritz [1], [2], [8]. The reader may also refer to the set of lectures notes by Kegel ([8], Chapter 2) and to Kuroš [13], Vol. 3, § 85.

8.3 Countably Recognizable Classes and Local Theorems of Baer

If $\mathfrak{c}$ is a cardinal number and $\mathfrak{X}$ a class of groups, we recall that

$$L_{\mathfrak{c}}\mathfrak{X}$$

is the class of groups G such that every subset of cardinality $\leq \mathfrak{c}$ is contained in an $\mathfrak{X}$-subgroup of G. Here we are concerned with classes of groups that are closed with respect to the closure operation

$$L_{\aleph_0}.$$

Such classes will be called *countably recognizable*; (cf. Baer [32], p. 345). A theorem asserting that a class of groups is countably recognizable may be called a *local theorem of Baer's type.* Observe that $\boldsymbol{L}_{\aleph_0} \leq \boldsymbol{L}$, so $\boldsymbol{L}_{\aleph_0}$-closure is weaker than $\boldsymbol{L}$-closure; in fact we will find that many interesting classes of groups which are not $\boldsymbol{L}$-closed are $\boldsymbol{L}_{\aleph_0}$-closed.

Obviously the intersection of any set of countably recognizable classes is also countably recognizable. However the corresponding result for unions is false—see Baer [32], p. 346—unless we restrict ourselves to countable unions.

Lemma 8.31 (Baer [32], Satz 1.1). The union of a countable set of countably recognizable classes is itself countably recognizable.

Proof. Let $\mathfrak{X}$ be the union of a countable set of countably recognizable classes $\mathfrak{X}_1, \mathfrak{X}_2, \ldots$ and suppose that the group G does not belong to $\mathfrak{X}$. Then for each positive integer i we have $G \notin \mathfrak{X}_i = \boldsymbol{L}_{\aleph_0}\mathfrak{X}_i$, so there is a countable subset S_i of G which does not lie in any $\mathfrak{X}_i$-subgroup of G. Let S denote the union of the S_i; then S is countable and is not contained in any $\mathfrak{X}$-subgroup of G. Hence $G \notin \boldsymbol{L}_{\aleph_0}\mathfrak{X}$. It follows that $\mathfrak{X}$ is countably recognizable. ▯

From this we draw a more or less obvious conclusion.

Corollary. The class of soluble groups and the class of nilpotent groups are countably recognizable.

Theorem 8.32. Let f be a group theoretical functional such that if X is a subgroup of a group G and $H \in f(G)$, then $H \cap X \in f(X)$. Then the properties Max-f and Min-f are countably recognizable.

Proof. Let G be a group which does not satisfy Max-f: then there is a *countably* infinite, ascending chain $H_1 < H_2 < \cdots$ in $f(G)$. Choose x_i to be any element of $H_{i+1} \backslash H_i$. Let X be a subgroup of G which contains every x_i. Since $x_i \in (X \cap H_{i+1}) \backslash H_i$,

$$X \cap H_1 < X \cap H_2 < \cdots$$

is an infinite ascending chain in $f(X)$. Therefore X does not satisfy Max-f and $G \notin \boldsymbol{L}_{\aleph_0}$(Max-$f$). It follows that Max-$f$ is countably recognizable: for Min-f the proof is similar. ▯

Corollary The properties Max, Max-*ab*, Max-*sn*, Max-*n* and Min, Min-*ab*, Min-*sn*, Min-*n* are countably recognizable.

The countable recognizability of Max and Min was proved by Baer ([32], Lemma 1.4). Since a group with the property Max is countable, we can even assert that *any class of groups satisfying Max is countably recognizable*. For example, *the class of polycyclic-by-finite groups is countably recognizable.*

Lemma 8.33. Let $\circ$ be a relation between elements and subgroups of a group such that (i) $a \circ H$ implies that $a \in H$, (ii) $a \in K \leqq H$ and $a \circ H$ imply that $a \circ K$ and (iii) if $a \in H$ and $a \circ K$ for every countable subgroup of K of H which contains a, then $a \circ H$.

Then if G is a non-trivial group and if each non-trivial countable subgroup K of G contains a non-trivial element $a(K)$ such that $a(K) \circ K$, there is a non-trivial element a such that $a \circ G$.

Proof. Suppose that G contains no non-trivial element a such that $a \circ G$. Then each non-trivial element a is contained in a countable subgroup $S(a)$ such that $a \circ S(a)$ is false. Now define a countable ascending chain $K_1 \leqq K_2 \leqq \cdots$ as follows: let K_1 be any non-trivial countable subgroup of G and let

$$K_{i+1} = \langle K_i, S(a) : 1 \neq a \in K_i \rangle.$$

Clearly each K_i is countable, so the union K of the K_i is also countable. By hypothesis K contains a non-trivial element a such that $a \circ K$. Now $a \in K_i$ for some i, so $S(a) \leqq K_{i+1} \leqq K$ and hence $a \circ S(a)$, which is false. ▯

If $\mathfrak{X}$ and $\mathfrak{Y}$ are classes of groups, let

$$(\mathfrak{X}, \mathfrak{Y})$$

be the class of all groups G such that if H is a non-trivial homomorphic image of G, then H has a non-trivial normal subgroup N such that

$$N \in \mathfrak{X} \quad \text{and} \quad \operatorname{Aut}_H N \in \mathfrak{Y}.$$

Theorem 8.34 (Baer [32], Satz 4.4). Let $\mathfrak{X}$ and $\mathfrak{Y}$ be countably recognizable, $\boldsymbol{S}$ and $\boldsymbol{H}$-closed classes. Then $(\mathfrak{X}, \mathfrak{Y})$ is countably recognizable.

Proof. Let G be any group and let a be an element and H a subgroup of G. We define $a \circ H$ to mean that $a \in H$, $a^H \in \mathfrak{X}$ and $\operatorname{Aut}_H (a^H) \in \mathfrak{Y}$. We verify that the relation $\circ$ satisfies conditions (i)—(iii) of Lemma 8.33. Assume the validity of $a \circ H$ and let $a \in K \leqq H$; then $a^K \in \boldsymbol{S}\mathfrak{X} = \mathfrak{X}$ and $\operatorname{Aut}_K (a^K)$ is isomorphic with a section of $\operatorname{Aut}_H(a^H)$ and hence belongs to $\mathfrak{Y}$: thus $a \circ K$ is true. Assume next that $a \circ K$ is valid for every countable subgroup K of G that contains a. Now every countable subset of a^G lies in a subgroup of the form a^K where K is countable and contains a; by hypothesis $a^K \in \mathfrak{X}$. Since $\mathfrak{X}$ is countably recognizable, it follows that $a^G \in \mathfrak{X}$. A countable subset of $\operatorname{Aut}_G (a^G)$ lies in a subgroup $\operatorname{Aut}_K (a^G)$ where K is a countable subgroup of G containing a. Thus to prove that $\operatorname{Aut}_G (a^G) \in \mathfrak{Y}$, it suffices to show that $\operatorname{Aut}_K (a^G) \in \mathfrak{Y}$. Let $k \in K$ and assume that $k \notin C_G(a^G)$; then we can find a $g(k) \in G$ such that k does not commute with $a^{g(k)}$. Let X be the subgroup generated by K and $g(k)$ for all k in $K \backslash (C_G(a^G))$. Then X is countable and no element in K lying outside $C_K(a^G)$ can centralize a^X. Hence $C_K(a^X) \leq C_K(a^G)$; but the opposite inclusion is obviously valid, so

$$C_K(a^X) = C_K(a^G)$$

and therefore

$$\mathrm{Aut}_K(a^G) \simeq \mathrm{Aut}_K(a^X) \leq \mathrm{Aut}_X(a^X) \in \mathfrak{Y},$$

in view of $K \leq X$ and the validity of $a \circ X$.

Observe that a group G belongs to $(\mathfrak{X}, \mathfrak{Y}))$ if and only if each non-trivial homomorphic image H of G contains an element a such that $a \circ H$. Also, $(\mathfrak{X}, \mathfrak{Y}))$—and hence $\boldsymbol{L}_{\aleph_0}(\mathfrak{X}, \mathfrak{Y}))$—is $\boldsymbol{S}$ and $\boldsymbol{H}$-closed: $\boldsymbol{S}$-closure follows from the characterization of $(\mathfrak{X}, \mathfrak{Y}))$ by ascending normal series provided by Lemma 1.22. The desired result can now be read off from Lemma 8.33. □

Corollary 1. Let $\mathfrak{X}$ be a countably recognizable $\boldsymbol{S}$ and $\boldsymbol{H}$-closed class. Then the class of hyper-$\mathfrak{X}$ groups is countably recognizable.

This follows directly from the theorem by setting $\mathfrak{Y} = \mathfrak{O}$, the class of all groups. For example, the property hyperabelian is countably recognizable (see also the Corollary to Theorem 2.15), and so is the property hyperfinite. Since Min is countably recognizable and the hyperfinite groups that satisfy Min are just the Černikov groups (Corollary 2 to Theorem 5.21), we deduce that *the class of Černikov groups is countably recognizable* (Baer [34], Zusatz 2.5).

Corollary 2. The class of hypercentral groups and the class of FC-hypercentral groups are countably recognizable.

To prove these, set $\mathfrak{X} = \mathfrak{O}$ and let $\mathfrak{Y} = \mathfrak{I}$ or $\mathfrak{F}$ in Theorem 8.34.

The range of applicability of Lemma 8.33 is extended by the following result.

Lemma 8.35 (cf. Baer [32], Lemma 2.3). Let $\mathfrak{X}$ be a countably recognizable $\boldsymbol{S}$-closed class of groups and let H be a subgroup of a group G. Then the following statements are equivalent.

(i) H is contained in a subnormal $\mathfrak{X}$-subgroup of G.

(ii) If K is a countable subset of G, then H is contained in a subnormal $\mathfrak{X}$-subgroup of $\langle H, K\rangle$.

Proof. That (i) implies (ii) is clear from the $\boldsymbol{S}$-closure of $\mathfrak{X}$. Let us assume therefore the validity of (ii). Suppose that H does not lie within any subnormal $\mathfrak{X}$-subgroup of G. If $H^{G,i}$ denotes the ith normal closure of H in G, then $H^{G,i} \notin \mathfrak{X}$ for each integer $i \geq 0$. Since $\mathfrak{X}$ is countably recognizable, there exists a countable subset S_i of $H^{G,i}$ such that S_i is not contained in any $\mathfrak{X}$-subgroup of $H^{G,i}$. If $x \in S_i$, then $x \in H^{T(i,x),i}$ for some finite subset $T(i, x)$ of G. Let

$$T = \bigcup_{\substack{x \in S_i \\ i=0,1,2,\dots}} T(i, x),$$

a countable set. By hypothesis H is contained in a subnormal $\mathfrak{X}$-subgroup of $U = \langle H, T \rangle$ and, by the $\boldsymbol{S}$-closure of $\mathfrak{X}$, this means that $H^{U,i} \in \mathfrak{X}$ for some integer i. But $S_i \subseteq H^{U,i}$, although S_i was chosen not to be contained in any $\mathfrak{X}$-subgroup of $H^{G,i}$. By this contradiction (ii) implies (i). ▯

Corollary. If $\mathfrak{X}$ is a countably recognizable $\boldsymbol{S}$-closed class, the property "every element is contained in a subnormal $\mathfrak{X}$-subgroup" is countably recognizable.

Taking $\mathfrak{X} = \mathfrak{A}$, we find that *the class of Baer groups is countably recognizable.*

Theorem 8.36. Let $\mathfrak{X}$ be a countably recognizable, $\boldsymbol{S}$ and $\boldsymbol{H}$-closed class of groups. Then the radical class generated by $\mathfrak{X}$ is countably recognizable.

Proof. Theorem 1.33 shows that Rad $\mathfrak{X} = \acute{\boldsymbol{P}}_{sn}\mathfrak{X}$. If a is an element and H is a subgroup of a group G, let $a \circ H$ mean that a belongs to a subnormal $\mathfrak{X}$-subgroup of H. By Lemma 8.35 the relation $\circ$ satisfies the hypotheses of Lemma 8.33. The theorem now follows from Lemma 8.33, the $\boldsymbol{H}$-closure of Rad $\mathfrak{X}$ and Lemma 1.23. ▯

Taking $\mathfrak{X} = \mathfrak{A}$, we obtain

Corollary (Baer [32], p. 358). The class of subsoluble groups is countably recognizable.

However, neither the class of Gruenberg groups nor the class of SN^*-groups is countably recognizable in view of the existence of locally finite p-groups with trivial Gruenberg radical (Theorem 6.27).

Theorem 8.37 (Baer [32], Satz 2.6). The class of groups which satisfy the normalizer condition is countably recognizable.

Proof. Assume that every countable subgroup of a group G is an N-group but that G nevertheless possesses a proper self-normalizing subgroup H. Let X be a countable subgroup of G such that

$$1 < X \cap H < X;$$

such an X exists because $1 < H < G$. Let $x \in X \setminus (X \cap H)$; then x cannot normalize H, so there is an element $x^* \in H$ such that $(x^*)^x \notin H$ or $(x^*)^{x^{-1}} \notin H$. Let X^* be the subgroup generated by X and all x^* for $x \in X \setminus (X \cap H)$. Then X^* is countable. Also $X \leqq X^* \nleqq H$, so

$$1 < X^* \cap H < X^*.$$

Now define $X_1 = X$ and $X_{i+1} = (X_i)^*$. Then $X_1 \leqq X_2 \leqq \cdots$ is a countable chain of countable subgroups, so its union U is countable. Hence U is an N-group. Since $U \nleqq H$, we have $U \cap H < U$, and there is an element $y \in U \setminus (U \cap H)$ such that $(U \cap H)^y = U \cap H$. Now $y \in X_i$ for some positive integer i, so $y \in X_i \setminus (X_i \cap H)$ and therefore

$$y^* \in X_{i+1} \cap H \leqq U \cap H.$$

But $(y^*)^y$ and $(y^*)^{y^{-1}}$ do not both belong to H, so y cannot normalize $U \cap H$, and we have a contradiction. ▯

Theorem 8.38 The following group theoretical properties are countably recognizable.

(i) Each subgroup has all its maximal subgroups of finite index (Phillips [2], Theorem 4.3).

(ii) Each subgroup has all its minimal normal subgroups finite.

Proof. (i) Suppose that the theorem is false. Then there exists a group G such that every countable subgroup of G has all its maximal subgroups of finite index but G has a maximal subgroup M of infinite index. Hence there exists a countable infinity of distinct cosets $x_1M, x_2M, \ldots$ Let

$$K_1 = \langle x_1, x_2, \ldots \rangle$$

and observe that $|K_1M : M|$ is infinite. Let $a \in G \setminus M$, so that $G = \langle a, M \rangle$; if $x \in G$, then it follows that

$$x \in \langle a, S(a, x) \rangle \tag{7}$$

where $S(a, x)$ is a finite subset of M. We now define a countable chain of subgroups $K_1 \leqq K_2 \leqq \cdots$ by the rule

$$K_{i+1} = \langle K_i, S(a, x) : a \in K_i \setminus M, x \in K_i \rangle.$$

Clearly each K_i is countable, so the union U of the K_i is countable. We note that $|UM : M|$ is infinite since $K_1 \leqq U$. Also

$$|U : U \cap M| = |UM : M|,$$

so $U \cap M$ cannot be maximal in U. It follows that there exist elements $a \in U \setminus M$ and $x \in U \setminus \langle a, U \cap M \rangle$. Now for some integer i the elements a and x belong to K_i. But by construction $S(a, x) \subseteq K_{i+1} \cap M \leqq U \cap M$, and by (7) we obtain the contradiction $x \in \langle a, U \cap M \rangle$. The proof of the second part is similar. ▯

Corollary. The property "each subgroup has all its chieffactors finite" is countably recognizable.

Finally, we mention without proof the following result: *if* $\mathfrak{B}$ *is a variety, the property* "$\mathfrak{B}$*-marginal-by-finite*" *is countably recognizable*: here a group G is said to be $\mathfrak{B}$-marginal-by-finite if the $\mathfrak{B}$-marginal subgroup of G has finite index. In particular *the property central-by-finite is countably recognizable* (Baer [32], p. 345). Further countably recognizable classes are determined in the paper of Baer [32].

8.4 Simple Generalized Soluble Groups

In this section we consider which of the various classes of generalized soluble groups can contain non-abelian simple groups. We recall the fundamental result of Mal'cev that a non-abelian simple group cannot be locally soluble (Corollary 1 to Theorem 5.27: see also Ayoub's proof on p. 83). In contrast to this, P. Hall [13] has constructed examples of non-abelian simple $\overline{\overline{SN}}$-groups. Hall's groups possess ascending series of arbitrary limit ordinal type and are therefore non-strictly simple: the existence of non-strictly simple groups has also been established by Levič in a later paper [1].

We begin with a generalization of Mal'cev's theorem.

Theorem 8.41 (Hartley and Stonehewer [1]). Let $\mathfrak{X}$ be a class of groups which is $\boldsymbol{L}$, $\boldsymbol{R}$ and $\boldsymbol{S}$-closed. Then any simple group which belongs to the class $(\boldsymbol{L}\hat{\boldsymbol{P}}_n)^\omega\mathfrak{X}$ must belong to $\mathfrak{X}$.

The proof depends on

Lemma 8.42. Let $\mathfrak{X}$ be a class of groups which is $\boldsymbol{L}$, $\boldsymbol{R}$ and $\boldsymbol{S}$-closed and let G be a group belonging to the class $\boldsymbol{L}\hat{\boldsymbol{P}}_n\mathfrak{X}$. Then each minimal normal subgroup of G belongs to $\mathfrak{X}$.

Proof. Let N be a minimal normal subgroup of G and suppose that $N \notin \mathfrak{X}$. Since $\mathfrak{X} = \boldsymbol{L}\mathfrak{X}$, the subgroup N contains a finitely generated subgroup X such that $X \notin \mathfrak{X}$. Let S be the $\mathfrak{X}$-residual of X; then $X/S \in \mathfrak{X}$ and $S \neq 1$ by the $\boldsymbol{R}$-closure of $\mathfrak{X}$. By minimality of N we have $N = S^G$; consequently $X \leqq S^Y$ for some finitely generated subgroup Y containing X. Since $G \in \boldsymbol{L}\hat{\boldsymbol{P}}_n\mathfrak{X}$ and $\mathfrak{X} = \boldsymbol{S}\mathfrak{X}$, we conclude that $Y \in \hat{\boldsymbol{P}}_n\mathfrak{X}$. Hence there is a normal $\mathfrak{X}$-series in Y, say $\{\Lambda_\sigma, V_\sigma \colon \sigma \in \Sigma\}$. Now let $\{x_1, \ldots, x_n\}$ be a set of non-trivial generators of X. Since $X \leqq Y$, there is a $\sigma_i \in \Sigma$ such that $x_i \in \Lambda_{\sigma_i} \backslash V_{\sigma_i}$. If σ is the last of the elements $\sigma_1, \ldots, \sigma_n$ in the ordering of Σ, then $X \leqq \Lambda_\sigma$ and $X \nleqq V_\sigma$. Now

$$X/X \cap V_\sigma \simeq XV_\sigma/V_\sigma \leqq \Lambda_\sigma/V_\sigma \in \mathfrak{X}.$$

In view of $\mathfrak{X} = \boldsymbol{S}\mathfrak{X}$, we conclude that $X/X \cap V_\sigma \in \mathfrak{X}$ and $S \leq X \cap V_\sigma$. Now $V_\sigma \lhd Y$, so

$$X \leq S^Y \leq (X \cap V_\sigma)^Y \leq V_\sigma,$$

which is a contradiction. □

For a generalization of this lemma see Newell [2].

Proof of Theorem 8.41. Let G be a simple group belonging to the class $(\boldsymbol{L}\hat{\boldsymbol{P}}_n)^\omega\mathfrak{X}$. Then $G \in (\boldsymbol{L}\hat{\boldsymbol{P}}_n)^i\mathfrak{X}$ for some integer i and we can assume that $i > 0$ since otherwise $G \in \mathfrak{X}$. Let

$$\mathfrak{Y} = (\boldsymbol{L}\hat{\boldsymbol{P}}_n)^{i-1}\mathfrak{X},$$

so that $G \in \boldsymbol{L}\hat{\boldsymbol{P}}_n\mathfrak{Y}$. Clearly $\mathfrak{Y}$ is $\boldsymbol{L}$-closed: the $\boldsymbol{S}$-closure of $\mathfrak{Y}$ follows from the obvious relations $\boldsymbol{SL} \leq \boldsymbol{LS}$ and $\boldsymbol{S}\hat{\boldsymbol{P}}_n \leq \hat{\boldsymbol{P}}_n\boldsymbol{S}$, while the $\boldsymbol{R}$-closure of $\mathfrak{Y}$ follows from the easily established

$$\boldsymbol{RL} \leq \boldsymbol{LRS} \quad \text{and} \quad \boldsymbol{R}\hat{\boldsymbol{P}}_n \leq \hat{\boldsymbol{P}}_n\boldsymbol{S}_n;$$

see Lemma 1.37. Since G is simple, Lemma 8.42 may be applied and we conclude that $G \in \mathfrak{Y}$. By induction on i it follows that $G \in \mathfrak{X}$. □

Taking $\mathfrak{X} = \mathfrak{A}$, we obtain the generalization of Mal'cev's theorem referred to above.

Corollary 1. The class $(\boldsymbol{L}\hat{\boldsymbol{P}}_n)^\omega\mathfrak{A}$ contains no non-abelian simple groups.

A subnormal composition factor of a group which belongs to the class $(\boldsymbol{L}\hat{\boldsymbol{P}}_n)^\omega\mathfrak{A}$ is therefore abelian (of prime order). Hence by the Corollary to Theorem 5.46 we obtain

Corollary 2. A group belonging to the class $(\boldsymbol{L}\hat{\boldsymbol{P}}_n)^\omega\mathfrak{A}$ satisfies Min-*sn* if and only if it is a soluble Černikov group.

Since $\boldsymbol{L}\acute{\boldsymbol{P}}_n\boldsymbol{L}\mathfrak{N}$ is the class of locally radical groups and

$$\boldsymbol{L}\acute{\boldsymbol{P}}_n\boldsymbol{L}\mathfrak{N} \leq (\boldsymbol{L}\hat{\boldsymbol{P}}_n)^2\mathfrak{A},$$

these corollaries apply in particular to the class of locally radical groups.

Non-Strictly Simple Groups

P. Hall's construction for non-strictly simple groups depends upon properties of wreath powers, especially Theorem 6.25. The following technical result is also required.

Lemma 8.43. Let G be a group and let λ be a non-zero limit ordinal. Assume that for each ordinal $\alpha < \lambda$ there exist subgroups H_α and L_α of G with the following properties.

(i) $H_\alpha < H_{\alpha+1}$,

(ii) $H_\mu = \bigcup_{\alpha<\mu} H_\alpha$ for each limit ordinal $\mu < \lambda$,

(iii) $G = \bigcup_{\alpha<\mu} H_\alpha$,

(iv) $H_\alpha \cap L_\alpha = 1$,

(v) L_α is normalized by H_α,

(vi) $H_\alpha L_\alpha$ is subnormal but not normal in $H_{\alpha+1}$.

Then G' has an ascending series with order type λ. Also, G' is simple if the following additional conditions are satisfied. Let $K_\alpha = H_\alpha^{H_{\alpha+1}}$.

(vii) Each normal subgroup of K'_α is normal in K_α.

(viii) Each non-trivial normal subgroup of $H_{\alpha+1}$ contains K'_α.

Proof. By (vi) $L_\alpha \leqq H_{\alpha+1}$, so $L_\alpha \leqq H_\beta$ and L_β is normalized by L_α for all $\alpha < \beta$, by (i) and (v). It follows that the L_β permute and each $L_{\beta_1} \cdots L_{\beta_n}$ is a subgroup. We define

$$P_\alpha = \langle L_\beta : \alpha \leqq \beta < \lambda \rangle. \tag{8}$$

Let $1 \neq x \in P_\alpha$ and suppose that $x \in H_\alpha$: now we can write $x = x_1 x_2 \cdots x_n$ where $1 \neq x_i \in L_{\beta_i}$ and $\alpha \leqq \beta_1 < \beta_2 < \cdots < \beta_n < \lambda$. Hence

$$x_n = x_{n-1}^{-1} \cdots x_2^{-1} x_1^{-1} x \in H_{\beta_n} \cap L_{\beta_n} = 1$$

by (iv). This contradiction shows that

$$H_\alpha \cap P_\alpha = 1. \tag{9}$$

From (v) we see that H_α normalizes P_α, so

$$Q_\alpha = H_\alpha P_\alpha$$

is a subgroup. Now $H_\alpha L_\alpha \; sn \; H_{\alpha+1}$ by (vi) and $P_{\alpha+1} \lhd H_{\alpha+1} P_{\alpha+1} = Q_{\alpha+1}$, so $Q_\alpha = H_\alpha L_\alpha P_{\alpha+1} \; sn \; Q_{\alpha+1}$.

Let μ be a limit ordinal such that $\mu < \lambda$. If $\alpha < \mu$, then

$$Q_\alpha = H_\alpha P_\alpha \leqq H_\mu P_\alpha;$$

if $\alpha \leqq \beta < \mu$, then $L_\beta \leqq H_{\beta+1} \leqq H_\mu$, so that $H_\mu P_\alpha \leqq H_\mu P_\mu = Q_\mu$ by (8). Therefore $Q_\alpha \leqq Q_\mu$. On the other hand, if $x \in Q_\mu$, then

$$x \in H_\alpha P_\mu \leqq H_\alpha P_\alpha = Q_\alpha$$

for some $\alpha < \mu$. Hence

$$Q_\mu = \bigcup_{\alpha<\mu} Q_\alpha.$$

Since G is the union of the H_α, it is also the union of the Q_α. Now define

$$R_\alpha = G' \cap Q_\alpha.$$

Then $R_\alpha \; sn \; R_{\alpha+1}$ and, if μ is a limit ordinal, R_μ is the union of the R_α for $\alpha < \mu$. In addition G' is the union of the R_α.

Next we will show that $R_\alpha < R_{\alpha+1}$. Since $P_\alpha = L_\alpha P_{\alpha+1}$, equation (9) and Dedekind's modular law yield

$$H_{\alpha+1} \cap Q_\alpha = H_{\alpha+1} \cap (H_\alpha L_\alpha P_{\alpha+1}) = H_\alpha L_\alpha (H_{\alpha+1} \cap P_{\alpha+1}) = H_\alpha L_\alpha. \quad (10)$$

By (vi) we have $(H_{\alpha+1})' \not\leq H_\alpha L_\alpha$, so $H'_{\alpha+1} \not\leq Q_\alpha$ by equation (10). Therefore $R_\alpha < R_{\alpha+1}$.

Since $R_\alpha \; sn \; R_{\alpha+1}$, we may refine the chain $\{R_\alpha : \alpha < \lambda\}$ to an ascending series in G', which will have order type equal to λ since λ is a limit ordinal.

Now let us assume that conditions (vii) and (viii) are also satisfied, and suppose that N is a non-trivial normal subgroup of G'. Since $H_\alpha \leqq K_\alpha \leqq H_{\alpha+1}$, it follows via (iii) that G is the union of the K_α and G' is the union of the K'_α. Now $N \cap K'_\alpha \lhd K'_\alpha$, so $N \cap K'_\alpha \lhd K_\alpha$ by (vii). Therefore N, being the union of the $N \cap K'_\alpha$, is normal in G. There exists a first ordinal $\alpha_0 < \lambda$ such that $N \cap H_{\alpha_0} \neq 1$. If $\alpha \geqq \alpha_0$, then $1 \neq N \cap H_{\alpha+1} \lhd H_{\alpha+1}$, so $K'_\alpha \leqq N$, by (viii), and $N = G'$. Hence G' is simple. □

The Construction

In order to exploit Lemma 8.43 we have to show how the hypothetical situation of that result can be realized. It is at this point that wreath powers become relevant.

Let H be any non-trivial group and let Z be the set of integers in their natural order. We form the wreath power

$$W = Wr \, H^Z,$$

taking H in its regular representation. Let a be the automorphism of W induced by the order-automorphism of Z in which $n \to n+1$. Let the holomorph of W by $\langle a \rangle$ be denoted by

$$G = f(H).$$

Then, according to Theorem 6.25, the group G is monolithic with monolith W'.

There is one further property of the group W that we will need: *each normal subgroup of W' is normal in W.* To prove this let $1 \neq N \lhd W'$ and let $\xi \in N$. Let η be any element of W. Then we choose a bisection $Z = Z_1 \cup Z_2$ such that ξ and η both belong to $J^{(1)}$: our notation here is that adopted throughout the discussion of wreath products in Section 6.2. Let $n \in Z_2$. Since $H_n \neq 1$, there is an element $\zeta \in H_n$ such that $1_n\zeta \neq 1_n$. Denoting the corresponding element of J_n also by ζ, we observe that $\zeta^{-1}J^{(1)}\zeta$ is a direct factor of the base group of $W = W^{(1)} \wr W^{(2)}$ and is distinct from $J^{(1)}$. Hence η^ζ commutes with both ξ and η and therefore

$$[\eta^{-1}, \xi] = [\eta^{-1}\eta^\zeta, \xi] = [\eta, \zeta, \xi]. \qquad (11)$$

Since $[\eta, \zeta] \in W'$ and $\xi \in N \lhd W'$, we can deduce from (11) that $[\eta^{-1}, \xi] \in N$. Thus $\xi^{\eta^{-1}} \in N$ for all $\eta \in W$ and therefore $N \lhd W$.

Turning now to the situation of Lemma 8.43, let us suppose that H is any non-trivial group and let

$$H_0 = H.$$

Suppose that a group H_α has been defined—α being an ordinal—and let

$$H_{\alpha+1} = f(H_\alpha).$$

We identify H_α with the canonical image in

$$W = Wr\, H_\alpha^Z$$

corresponding to 0. Let Z_1 and Z_2 denote respectively the sets of negative and non-negative integers. Then the bisection $Z = Z_1 \cup Z_2$ leads to a decomposition

$$W = W^{(1)} \wr W^{(2)} \qquad (12)$$

in the usual way. Now let L_α be the base group of the wreath product (12). Clearly $H_\alpha < H_{\alpha+1}$: also $H_\alpha \cap L_\alpha = 1$ and L_α is normalized by H_α since $H_\alpha \leqq J^{(2)}$. We can write $J^{(2)} = H_\alpha \wr P$ where $P = Wr\, H^{Z^+}$ and Z^+ is the set of positive integers; let B be the base group of the wreath product $J^{(2)}$. Then, in view of our identification, H_α is a direct factor of B. Hence $H_\alpha \lhd B \lhd J^{(2)}$; also $L_\alpha \lhd W$, so $H_\alpha L_\alpha \lhd BL_\alpha \lhd J^{(2)}L_\alpha = W$ and $H_\alpha L_\alpha$ *sn* W. Also $W \lhd H_{\alpha+1}$, so $H_\alpha L_\alpha$ *sn* $H_{\alpha+1}$. On the other hand, $H_\alpha \neq J^{(2)}$; thus

$$K_\alpha = H_\alpha^{H_{\alpha+1}} = W \nleqq H_\alpha L_\alpha$$

and $H_\alpha L_\alpha$ is not normal in $H_{\alpha+1}$. In addition W' is the monolith of $H_{\alpha+1}$ by Theorem 6.25. We verified above that all normal subgroups of W' are normal in W. Thus $H_{\alpha+1}$ and L_α have the properties of Lemma 8.43. Finally, define H_μ to be the union of all H_α for $\alpha < \mu$, where μ is a limit ordinal. Let λ be a limit ordinal > 0 and write $G = H_\lambda$. Then G' is a simple group which has ascending series of order type λ. We will write

$$G' = g(H, \lambda).$$

Theorem 8.44 (P. Hall [13], Theorem A.) Let $\mathfrak{X}$ be a class of groups which is closed with respect to forming semi-direct products, derived subgroups and unions of ascending chains. Assume moreover that $\mathfrak{X}$ contains an infinite cyclic group. If $H \in \mathfrak{X}$ and λ is a limit ordinal > 0, the group $g(H, \lambda)$ is a simple $\mathfrak{X}$-group with an ascending series of order type λ.

Proof. By the argument which precedes the theorem it is sufficient to prove that $g(H, \lambda) \in \mathfrak{X}$. Since $\mathfrak{X}$ is closed with respect to forming derived subgroups and unions of ascending chains, we need only show that $f(X) \in \mathfrak{X}$ follows from $X \in \mathfrak{X}$. Now the closure of $\mathfrak{X}$ with respect to forming semi-direct products and unions of ascending chains implies that a wreath product of a pair—and hence any finite set—of $\mathfrak{X}$-groups is an $\mathfrak{X}$-group. It follows that $W = Wr\, X^{\mathbf{Z}} \in \mathfrak{X}$. Finally $f(X)$ is a semi-direct product of W by an infinite cyclic group, so $f(X) \in \mathfrak{X}$. $\square$

Corollary. There exist non-cyclic simple $\overline{\overline{SN}}$-groups which have an ascending series of order-type λ where λ is an arbitrary limit ordinal > 0.

Proof. We have only to show that the class $\overline{\overline{SN}}$ satisfies the conditions of Theorem 8.44. It is clear that $\overline{\overline{SN}}$ is $\mathbf{S}$ and $\mathbf{P}$-closed and contains all infinite cyclic groups. Moreover $\overline{\overline{SN}}$ is $\mathbf{L}$-closed by Theorem 8.26, so it is certainly closed with respect to forming unions of ascending chains. $\square$

Of all the classes of generalized soluble groups in the diagram in Section 8.1, only SN, $\overline{SN}$ and $\overline{\overline{SN}}$ contain non-cyclic simple groups; for a simple SJ-group is obviously cyclic.

Theorem 8.45 (P. Hall [13], Theorem C). Let A be an arbitrary non-trivial abelian group. Then $g(A, \omega)$ is a non-cyclic simple group belonging to the class $(\mathbf{LP})^{\omega+1}\mathfrak{A}$.

Proof. We have only to prove that $g(A, \omega) \in (\mathbf{LP})^{\omega+1}\mathfrak{A}$. Let $\mathfrak{X}$ be any class of groups, let $X \in \mathfrak{X}$ and $Y \in \mathbf{PLP}\mathfrak{X}$: a direct power of X belongs to the class $\mathbf{LP}\mathfrak{X}$, so any wreath product $X \wr Y$ belongs to

$$(\mathbf{LP}\mathfrak{X})\,(\mathbf{PLP}\mathfrak{X}) \leqq \mathbf{P}(\mathbf{PLP}\mathfrak{X}) = \mathbf{PLP}\mathfrak{X}.$$

Hence any finite wreath product of $\mathfrak{X}$-groups belongs to $\mathbf{PLP}\mathfrak{X}$ and therefore $Wr\, X^{\mathbf{Z}} \in \mathbf{LPLP}\mathfrak{X}$ if $X \in \mathfrak{X}$. Hence $X \in \mathfrak{X}$ and $\mathfrak{A} \leqq \mathfrak{X}$ imply that

$$f(X) \in \mathbf{PLPLP}\mathfrak{X} \leqq (\mathbf{LP})^3\mathfrak{X}. \tag{13}$$

Now define $A_0 = A$ and $A_{i+1} = f(A_i)$ where i is an integer $\geqq 0$; we identify A_i with the appropriate subgroup of A_{i+1}. Then (13) shows that

$$A_i \in (\mathbf{LP})^{3i}\,\mathfrak{A} \leqq (\mathbf{LP})^{\omega}\,\mathfrak{A}.$$

Therefore

$$g(A, \omega) = \Big(\bigcup_{i=0,1,2,\ldots} A_i\Big)' \in \boldsymbol{SL}(\boldsymbol{LP})^{\omega}\,\mathfrak{A} \leqq (\boldsymbol{LP})^{\omega+1}\,\mathfrak{A},$$

as required. ▯

Theorem 8.45 and the theorem of Hartley and Stonehewer (Theorem 8.41) show that *the first ordinal* α *for which the class* $(\boldsymbol{L}\hat{\boldsymbol{P}}_n)^{\alpha}\mathfrak{A}$ *contains a non-abelian simple group is* $\omega + 1$.

We will draw one further conclusion from Theorem 8.45.

Corollary. The class SJ is not a local class.

Proof. For suppose that SJ is $\boldsymbol{L}$-closed. Since SJ is obviously $\boldsymbol{P}$-closed, it follows that $(\boldsymbol{LP})^{\omega+1}\mathfrak{A} \leqq SJ$ and SJ contains a non-abelian simple group, which is plainly impossible. ▯

We mention without proof that $g(H, \lambda)$ *is orderable if* H *is orderable* (P. Hall [13], Theorem B). The first example of a simple orderable group was given by Chehata [1]—see also Dlab [4]. Since every orderable group is a SN-group (Iwasawa [2]) a simple orderable group cannot be absolutely simple. It is unknown if there exist strictly simple groups which are not absolutely simple: the group $g(H, \lambda)$ is, of course, not strictly simple and it is apparently uncertain whether Chehata's group is strictly simple.

In conclusion we remark that the possible order types of an abelian series in a non-abelian simple SN-group are studied in the papers [13] of P. Hall (Theorem E) and [1] of Hartley and Stonehewer.

Chapter 9

Residually Finite Groups

Our object in this chapter is to discover groups which are residually finite: in particular we shall find that free groups and free products of residually finite groups are residually finite and that several significant types of soluble groups share this property. It soon emerges that groups of many disparate kinds are residually finite, so that the residually finite groups form a rather complex class. For a further account of the present state of the theory and for an estimate of its significance for other branches of mathematics the reader may consult the survey paper of Magnus [3].

9.1 The Residual Finiteness of Free Groups and Free Products

Theorem 9.11 (Iwasawa [1]). A free group is residually a finite p-group for all primes p.

Corollary 1 (Magnus [1], p. 269). A free group is residually nilpotent.

Other proofs of Magnus' theorem have been given by Fuchs-Rabinovič ([1], p. 205) and Witt [1]; see also Kuroš [9], vol. 2, p. 38.

Proof of Theorem 9.11 (P. Hall: see Hall and Hartley [1], pp. 27—28). Let F be a free group and let X be a set of free generators for F. If $1 \neq f \in F$, we can write

$$f = x_{i_1}^{m_1} \cdots x_{i_r}^{m_r}$$

where $x_{i_u} \in X$, $m_u \neq 0$ and the positive integers i_u satisfy $i_u \neq i_{u+1}$. Let $q = \max\{i_1, \ldots, i_r\}$. To show that F is residually a finite p-group it is sufficient to find a finite p-group $G = \langle g_1, \ldots, g_q\rangle$ such that

$$g = g_{i_1}^{m_1} \cdots g_{i_r}^{m_r} \neq 1.$$

For then the mappings $x_{i_u} \to g_{i_u}$ and $x_j \to 1$ $(j \neq i_1, \ldots, i_r)$ extend to a homomorphism of F onto G whose kernel does not contain f.

Let n be a positive integer such that p^n does not divide $m_1 \cdots m_r$ and let e_{uv} be the $(r+1)\times(r+1)$ matrix over the ring $R = Z/p^nZ$ of integers modulo p^n such that the (u, v)th coefficient is 1 and all others are 0. Define

$$g_j = \prod_{i_u = j} (1 + e_{u\,u+1})$$

for $j = 1, \ldots, q$. If none of the i_u equals j, then of course we mean that $g_j = 1$. Since $i_u \neq i_{u+1}$, each pair of factors of g_j commute, so the order of the factors is immaterial. In addition

$$g_{i_j}^{m_j} = 1 + m_j \sum_{i_u = i_j} e_{u\,u+1}$$

and the coefficient of $e_{1\,r+1}$ in the expression for g is $m_1 \cdots m_r$, which is not zero. Hence $g \neq 1$. Now $G = \langle g_1, \ldots, g_q \rangle$ is a subgroup of $U = U(r+1, R)$, the group of all $(r+1) \times (r+1)$ upper unitriangular matrices over R. Clearly U is a finite p-group of order

$$p^{\frac{1}{2} nr(r-1)},$$

so the theorem is proved. ▯

For alternative proofs of Iwasawa's theorem see M. Hall [1] (§ 5), Hall and Hartley [1] (p. 26), Neuwirth [1] and Takahasi [1]. We record a second corollary of Iwasawa's theorem.

Corollary 2 (Fuchs-Rabinovič [2]). A locally free group cannot be simple.

Proof. A free group is residually nilpotent and therefore is certainly a Z-group. By the local theorem for Z-groups (Corollary to Theorem 8.24), a locally free group is a Z-group. Finally, a simple Z-group is obviously cyclic of prime order. ▯

However a locally free group need not be residually finite, as we see from the additive group of rational numbers. In fact *there exists a perfect countably infinite, locally free group G such that $G = G^n$ for every positive integer n* (a group with the latter property is said to be *semi-radicable*).

Such a group can be constructed in the following manner. For each positive integer i we choose a free group F_i of countably infinite rank and a positive integer n_i such that every positive integer occurs infinitely often in the sequence $n_1, n_2, \ldots$ Since $F_i \simeq F_{i+1} \simeq (F'_{i+1})^{n_i}$, there is a monomorphism $\theta_i \colon F_i \to F_{i+1}$ mapping F_i onto $(F'_{i+1})^{n_i}$. Let G be the direct limit determined by the F_i and θ_i. With the appropriate identification, $F_i = (F'_{i-1})^{n_i}$ and G is the union of the chain $F_1 < F_2 < \ldots$ Let $x \in G$ and suppose that $x \notin G' \cap G^n$ where $n > 0$. Now $x \in F_i$ for some i and $n = n_j$ for some $j \geqq i$ by construction. Hence

$$x \in F_i \leqq F_j = (F'_{j+1})^{n_j} \leqq G' \cap G^n.$$

By this contradiction $G = G' = G^n$. Obviously G is locally free.

Recently Katz and Magnus, generalizing an earlier theorem of Peluso [1], have established the following results. *Let* $\mathfrak{X}$ *be the class of all trivial groups and finite alternating groups and let* $\mathfrak{Y}_r$ *be the class of all trivial groups and projective special linear groups of degree 2 over a Galois field of* p^r *elements. Then every free group is residually an* $\mathfrak{X}$*-group and a* $\mathfrak{Y}_r$*-group* (Katz and Magnus [1], Theorems 1 and 2). The first of these results has the following consequence: *the intersection of all subgroups of prime index in a free group is the identity subgroup.*

Automorphism Groups of Residually Finite Groups

In 1963 Baumslag and Smirnov proved that the automorphism group of a finitely generated residually finite group is again residually finite. The proof is remarkably simple and allows us to establish a somewhat more general result.

Theorem 9.12. If the intersection of all the characteristic subgroups of finite index in a group G is trivial, then Aut G is residually finite.

Proof. Let $1 \neq \alpha \in \text{Aut}\, G$; then there exists an element g in G such that $g^\alpha \neq g$. Hence $h = g^{-1}g^\alpha \neq 1$. By hypothesis there is a characteristic subgroup N with finite index in G such that $h \notin N$. Let Γ be the normal subgroup of Aut G formed by all automorphisms of G which induce the trivial automorphism in G/N. Then $\alpha \notin \Gamma$ and $(\text{Aut}\, G)/\Gamma$ is finite. Hence Aut G is residually finite. ▯

Since a finitely generated group G has only finitely many subgroups of each finite index, a subgroup of finite index in G contains a characteristic subgroup of finite index. Hence

Corollary (Baumslag [19], Smirnov [8]). The automorphism group of a finitely generated residually finite group is residually finite.

The class of residually finite groups is not closed under forming extensions: indeed

$$\mathfrak{F}(\boldsymbol{R}\mathfrak{F}) \nleq \boldsymbol{R}\mathfrak{F};$$

for example, let G be the central product of a countably infinite set of quaternion groups of order 8; then $G \in \mathfrak{F}(\boldsymbol{R}\mathfrak{F})$, but G is not residually finite because every non-trivial normal subgroup of G contains the centre. However, Mal'cev [9] has proved a weaker result.

Lemma 9.13. Let G be the semi-direct product of a group N by a group X and assume that the intersection of all the G-admissible subgroups of N with finite index is trivial and that X is residually finite. Then G is residually finite.

Proof. Let $1 \neq g \in G$. We have to find a normal subgroup with finite index in G to which g does not belong. Clearly we may assume that $g \in N$. Then there is a G-admissible subgroup M with finite index in N such that $g \notin M$. Since $M \lhd G$, we may suppose N to be finite. Now set $C = C_X(N)$; then X/C is finite and $C \lhd XN = G$, so G/C is finite. But $C \leqq X$, so we obtain $g \notin C$ as required. □

Using Theorem 9.12 and Lemma 9.13, we obtain

Corollary. If the intersection of all the characteristic subgroups of finite index in G is trivial, then the holomorph of G is residually finite. In particular this is the case if G is a finitely generated residually finite group (Smirnov [8]).

Residual Finiteness and Linear Groups

We have already mentioned (in Chapter 1) the theorem of Mal'cev that a finitely generated linear group (over a field) is residually a finite linear group of the same degree (Mal'cev [1], Theorem VII and VIII). Some extensions of this are known: *a linear group G over a finitely generated integral domain R has a subgroup of finite index which is residually a finite p-group where p is the characteristic of R if this is non-zero; otherwise p can be any prime with at most a finite number of exceptions*. (Merzljakov [4], Platonov [5], Wehrfritz [10], Proposition 1.6): thus, in particular, G is residually a finite π-group for some finite set of primes π. See also Tokarenko [4].

Some Residual Properties of Free Products

Let $\mathfrak{X}$ be a class of groups. We will call $\mathfrak{X}$ a *root class* (cf. Gruenberg [2], p. 33) if

(i) a subgroup of an $\mathfrak{X}$-group is an $\mathfrak{X}$-group,

(ii) given $H \lhd K \lhd G$ where $G/K \in \mathfrak{X}$ and $K/H \in \mathfrak{X}$, there exists an $L \lhd G$ such that $L \leqq H$ and $G/L \in \mathfrak{X}$.

Property (ii) is a strong form of $\boldsymbol{P}$-closure. For example, soluble groups and finite π-groups form root classes (where π is any set of primes). Our aim is to prove

Theorem 9.14 (Gruenberg [2], Theorem 4.1). Let $\mathfrak{X}$ be a root class other than the class of trivial groups. Then every free product of residually $\mathfrak{X}$-groups is a residually $\mathfrak{X}$-group if and only if every free group is a residually $\mathfrak{X}$-group.

Iwasawa's theorem allows us to deduce the following.

Corollary. Let π be any set of primes. Then the class of groups which are residually finite-π and the class of residually soluble groups are closed with respect to forming free products.

Proof of Theorem 9.14. First of all assume that all free products of residually $\mathfrak{X}$-groups are residually $\mathfrak{X}$-groups and let H and K be two non-trivial $\mathfrak{X}$-groups. Then $H * K \in \boldsymbol{R}\mathfrak{X}$. Now $H * K$ has an element of infinite order and $\boldsymbol{SR}\mathfrak{X} \leq \boldsymbol{RS}\mathfrak{X} = \boldsymbol{R}\mathfrak{X}$, so $\boldsymbol{R}\mathfrak{X}$ contains an infinite cyclic group. By hypothesis this implies that $\boldsymbol{R}\mathfrak{X}$ contains all free groups.

Conversely let us assume that every free group is a residually $\mathfrak{X}$-group. Consider the free product

$$G = \underset{\lambda \in \Lambda}{Fr}\, G_\lambda$$

where $G_\lambda \in \boldsymbol{R}\mathfrak{X}$ for each $\lambda \in \Lambda$. Let $1 \neq g \in G$; then g is a product of elements from a finite number of the G_λ, say $G_{\lambda_1}, \ldots, G_{\lambda_n}$. The mappings $x \to x$, $(x \in G_{\lambda_i})$, and $x \to 1$, $(x \in G_\lambda,\ \lambda \neq \lambda_1, \ldots, \lambda_n)$, extend to an endomorphism of G with image

$$F = \underset{i=1}{\overset{n}{Fr}}\, F_i$$

where $F_i = G_{\lambda_i}$, and under this endomorphism g is mapped onto itself. Therefore it suffices to prove that g can be excluded from a normal subgroup of F whose factor group belongs to $\mathfrak{X}$.

g is a product of elements from the F_i: let us denote the finite set of elements in this product by S and put $S_i = S \cap F_i$. The set of non-identity elements of the form $s_2^{-1}s_1$ where s_1 and s_2 belong to S_i is finite, so there is a normal subgroup N_i of F_i such that $F_i/N_i \in \mathfrak{X}$ and none of the $s_2^{-1}s_1$ belong to N_i. Thus different elements of S_i lie in different cosets of N_i, and there exists a transversal T_i to N_i in F_i containing the subset $S_i \cup \{1\}$. Let

$$H = \underset{i=1}{\overset{n}{Fr}}\, (F_i/N_i).$$

The natural homomorphisms of the F_i onto the F_i/N_i extend to a homomorphism ϕ of F onto H. Suppose that $g \in K = \operatorname{Ker} \phi$ and that

$$g = g_{i_1} \cdots g_{i_k}$$

where $1 \neq g_{i_j} \in F_{i_j}$ and $i_j \neq i_{j+1}$. Since

$$g^\phi = (g_{i_1}N_{i_1}) \cdots (g_{i_k}N_{i_k})$$

is trivial, we deduce that $g_{i_j} \in N_{i_j}$ for each j. But also $g_{i_j} \in S \cap F_{i_j} = S_{i_j} \subseteq T_{i_j}$; since T_{i_j} is a transversal containing 1, it follows that $g_{i_j} = 1$ for all j and $g = 1$. This is not the case, so $g \notin K$.

Since $H \simeq F/K$, it suffices to show that $H \in \boldsymbol{R}\mathfrak{X}$. Let L be the kernel of the canonical homomorphism of H onto

$$D = \mathop{Dr}_{i=1}^{n} (F_i/N_i).$$

Since $F_i/N_i \in \mathfrak{X} = \boldsymbol{D}_0\mathfrak{X}$, it follows that $H/L \in \mathfrak{X}$. Clearly L intersects each free factor F_i/N_i trivially, so by the fundamental theorem of Kuroš on subgroups of free products (Kuroš [1], p. 651; for a short proof see MacLane [1]), L is a free group. By hypothesis $L \in \boldsymbol{R}\mathfrak{X}$. Let $1 \neq h \in H$; we have to find a normal subgroup not containing h whose factor group in H is an $\mathfrak{X}$-group. Evidently we may assume that $h \in L$. Then there exists an $M \lhd L$ such that $h \notin M$ and $L/M \in \mathfrak{X}$. Finally, since $\mathfrak{X}$ is a root class, there is an $N \lhd H$ such that $N \leq M$ and $H/N \in \mathfrak{X}$, and it is clear that $h \notin N$. □

Another proof of Gruenberg's theorem has been given by MacBeath [1]. We remark that the theorem is more generally valid, namely for *regular products* in the sense of Golovin (see Kuroš [13], Vol. 3, § 78 or Magnus, Karass and Solitar [1], § 6.4); this generalization is also due to Gruenberg ([2], Theorem 6.2).

On the other hand, the generalized free product (or free product with an amalgamated subgroup) of residually finite groups need not be residually finite: on this question see Baumslag [10], [15], Dyer [1] and Stebe [1].

Wreath products of residually finite groups have been studied by Gruenberg who showed that *the standard wreath product $G \wr H$ is residually finite if and only if both G and H are residually finite and either G is abelian or H is finite* (Gruenberg [2], Theorem 3.2).

Finally we mention that Mal'cev has obtained conditions for a free product to be residually nilpotent, particularly when the factors of the free product are nilpotent (Mal'cev [5], § 3): some related questions are discussed in papers of Baumslag [17] and Dey [3]. On the problem of the residual nilpotence of wreath products see Hartley [3].

9.2 Radicable and Semi-Radicable Groups.

Radicable Hypercentral Groups

We discuss now group theoretical properties which are quite opposite to residual finiteness.

If π is a set of primes, a multiplicative group G is said to be *π-radicable** if to each element g of G and each positive π-number n there corres-

* Some authors use the term *complete* instead of radicable.

ponds an element g_1 of G such that

$$g_1^n = g.$$

(For additive groups the term "π-divisible" would be used.) Obviously this is equivalent to the requirement that every element of G be a pth power for all $p \in \pi$.

A group G is said to be *semi-π-radicable* if $G = G^n$ for every positive π-number n. Clearly *a π-radicable group is semi-π-radicable.* When π is the set of all primes, we speak of *radicable groups* and *semi-radicable groups.*

A related class is the class of *$\mathfrak{F}_\pi$-perfect groups*

$$\mathfrak{F}_\pi^{-\boldsymbol{H}};$$

this consists of groups which have no proper normal subgroups of index a finite π-number. Obviously *every semi-π-radicable group is $\mathfrak{F}_\pi$-perfect.* We remark that it is easy to show that the class of semi-π-radicable groups and the class of $\mathfrak{F}_\pi$-perfect groups are closed with respect to forming homomorphic images, extensions and arbitrary joins: in addition both are local classes. In particular *the class of semi-π-radicable groups and the class of $\mathfrak{F}_\pi$-perfect groups are radical classes,* by Theorem 1.32.

Observe that in the Kuroš correspondence the class of $\mathfrak{F}_\pi$-perfect groups corresponds to the class $\boldsymbol{P}_n(\mathfrak{F}_\pi)$ of hypo-(finite-π) groups and the class of semi-π-radicable groups to the class of groups that are hypo-(of finite π-exponent)—see Section 1.3. Also, the $\mathfrak{F}_\pi^{-\boldsymbol{H}}$-radical of an arbitrary group coincides with the iterated $\mathfrak{F}_\pi$-residual; however the trivality of the $\mathfrak{F}_\pi^{-\boldsymbol{H}}$-radical does not in general imply that a group is residually a finite π-group; some situations where this implication is valid are discussed in Section 9.3 and Section 4.4 (ii).

Lemma 9.21. Let π be a non-empty set of primes.

(i) The π-radicable groups form a proper subclass of the class of semi-π-radicable groups.

(ii) The semi-π-radicable groups form a proper subclass of the class of $\mathfrak{F}_\pi$-perfect groups if π contains a prime $p \geqq 4381$.

Proof. Let $p \in \pi$ and consider the standard wreath product

$$W = H \wr K$$

where H and K are groups of type p^∞. Since W is generated by p^∞-groups, it is semi-radicable. However we will show that W is not p-radicable, so it certainly cannot be π-radicable.

Let $h \in H$ and $k \in K$ be elements of order p and let b be the element of the base group B defined by $b_1 = h$ and $b_x = 1$ if $1 \neq x \in K$. Suppose that kb is a pth power and $kb = (yc)^p$ where $y \in K$ and $c \in B$. Then

$$kb = (yc)^p = y^p c^{y^{p-1}+y^{p-2}+\cdots+y+1},$$

so that $k = y^p$ and

$$b = c^{y^{p-1}+y^{p-2}+\cdots+y+1}. \tag{1}$$

Transforming (1) successively by y^{pi}, $i = 0, 1, \ldots, p-1$, and forming the product of the resulting equations, we obtain

$$b^{1+y^p+\cdots+y^{p^2-p}} = c^{1+y+\cdots+y^{p^2-1}} = d,$$

say, since B is abelian. Hence

$$d^{y-1} = c^{y^{p^2}-1} = 1,$$

since y has order p^2. However the component of d^{y-1} corresponding to y^{p^2-p+1} is equal to h, and we have a contradiction. (For a discussion of π-radicability in wreath products see Baumslag [5].)

Let π contain a prime $p \geqq 4381$. By the work of Kostrikin, Novikov and Adjan there exists a finitely generated infinite simple group of exponent p—see Section 5.1. Evidently G is $\mathfrak{F}$-perfect but is not semi-π-radicable. (On the other hand it is easy to prove that "semi-p-radicable" and "$\mathfrak{F}_p$-perfect" are the same if $p = 2$ or 3.) □

Lemma 9.22. Let G be a group and let π be a set of primes. Assume that G has no non-trivial homomorphic image which is perfect and has finite exponent equal to a π-number. Then G is semi-π-radicable if and only if G/G' is semi-π-radicable.

Proof. Let n be a positive π-number and suppose that $G > G^n$. By hypothesis G/G^n is not perfect, so $G > G'G^n$ and G/G' is not semi-π-radicable. The converse is clear. □

Corollary (Gluškov [2]). Let G be a soluble group. Then G is semi-π-radicable if and only if G/G' is semi-π-radicable.

π-Radicable Hypercentral Groups

It is clear the "π-radicable" and "semi-π-radicable" are identical for abelian groups: in view of the structure of abelian groups of finite exponent (Kaplansky [1], Theorem 6, p. 17), we can add "$\mathfrak{F}_\pi$-perfect" to this list of properties identical for abelian groups. It is a fact that the three properties also coincide for hypercentral groups.

The following theorem is essentially a synthesis of work of Černikov: the relevant papers are Černikov [13] (Theorem 10) and [10] (Theorems 4 and 10): see also Baumslag [4] (Theorem 14.1 and Corollary 14.4).

Theorem 9.23. Let G be a hypercentral group, let π be a set of primes and let G/G' be π-radicable. Then

(i) G is π-radicable,

(ii) the torsion-subgroup of G is π-radicable and its π-component is contained in the centre of G,

(iii) each term of the upper central series of G is π-radicable.

Corollary 1. For hypercentral groups the properties "π-radicable", "semi-π-radicable" and "$\mathfrak{F}_\pi$-perfect" are identical.

Corollary 2. A periodic hypercentral group is π-radicable if and only if it is the direct product of a π'-group and groups of type p^∞ where $p \in \pi$. In particular, a radicable periodic hypercentral group is abelian.

Proof of Theorem 9.23. We may assume that G is non-abelian and that π is not empty. We shall write

$$Z_\alpha = \zeta_\alpha(G),$$

so that $Z_1 < Z_2$. Let $z \in Z_2 \backslash Z_1$ and observe that the mapping $x \to [x, z]$ is a homomorphism of G into Z_1 with non-trivial image. By hypothesis G/G' is π-radicable, and therefore so is every abelian homomorphic image of G. We conclude that Z_1 contains a non-trivial π-radicable subgroup. By repeated use of this argument we are able to construct an ascending central series of G with π-radicable factors,

$$\{G_\beta : 0 \leqq \beta \leqq \alpha\}$$

say.

Let g be an element of G and suppose that there exists a positive π-number m such that g is not the mth power of any element of G. Certainly $g \neq 1$ and hence $g \in G_{\alpha_1+1} \backslash G_{\alpha_1}$ for some $\alpha_1 < \alpha$. Now $G_{\alpha_1+1}/G_{\alpha_1}$ is π-radicable, so

$$g = g_1^m h_1 \tag{2}$$

where $g_1 \in G_{\alpha_1+1}$ and $h_1 \in G_{\alpha_1}$. Since $h_1 \neq 1$, we have $h_1 \in G_{\alpha_2+1} \backslash G_{\alpha_2}$ where $\alpha_2 < \alpha_1$. Again we can write

$$h_1 = g_2^m \bar{h}_2 \tag{3}$$

where $g_2 \in G_{\alpha_2+1}$ and $\bar{h}_2 \in G_{\alpha_2}$. Since $G_{\alpha_2+1}/G_{\alpha_2}$ lies in the centre of G/G_{α_2}, we have $(g_1 g_2)^m \equiv g_1^m g_2^m \bmod G_{\alpha_2}$. Therefore from (2) and (3) it follows that

$$g = (g_1 g_2)^m h_2$$

for some $h_2 \in G_{\alpha_2}$. Here $h_2 \neq 1$ and $h_2 \in G_{\alpha_3+1} \backslash G_{\alpha_3}$ for some $\alpha_3 < \alpha_2$. This procedure cannot terminate, so it leads to an infinite decreasing chain of ordinals $\alpha_1 > \alpha_2 > \alpha_3 > \cdots$. We conclude that G is π-radicable.

Next let T be the torsion-subgroup of G and let P and Q denote respectively the π and π'-components of T. Obviously Q is π-radicable; since $T = P \times Q$, the π-radicability of T will follow from that of P. Let $x \in P$ and let m be a positive π-number; since G is π-radicable, $x = y^m$ for some $y \in G$. Now G/P contains no elements other than 1 with order a π-number; consequently $y \in P$ and P is π-radicable.

Suppose that $P \nleqq Z_1$; Lemma 2.16 shows that $(PZ_1) \cap Z_2 > Z_1$ and by Dedekind's modular law this means that $P \cap Z_2 \nleqq Z_1$. Let $x \in (P \cap Z_2) \backslash Z_1$; then $x^n = 1$ for some positive π-number n. Hence if g is an arbitrary element of G,

$$1 = [g, x^n] = [g, x]^n = [g^n, x] \tag{4}$$

since $[g, x] \in Z_1$. But $G = G^n$, so $x \in Z_1$, a contradiction.

We show next that G/Z_1 is π-free. If this is not the case, then, by Lemma 2.16 again, Z_2/Z_1 contains an element xZ_1 whose order m is a positive π-number. Then $x^m \in Z_1$ and as in (4) we find that $[g^m, x] = 1$ for all g in G; thus $x \in Z_1$. It follows that G/Z_1 is π-free, and Theorem 2.25 shows that the same is true of $Z_{\alpha+1}/Z_\alpha$ for each $\alpha > 0$. Hence G/Z_α is π-free for each $\alpha > 0$. The π-radicability of G now implies that of Z_α. □

The Structure of Radicable Hypercentral Groups

Černikov has used Theorem 9.23 to give a fairly good description of the structure of radicable hypercentral groups.

Theorem 9.24. (Černikov [13], Theorem 8). G is a radicable hypercentral group if and only if there is a well-ordered set of subgroups $\{X_\beta : 0 \leqq \beta < \alpha\}$ such that

(i) X_0 is a direct product of quasicyclic groups and lies in the centre of G,

(ii) if $0 < \beta < \alpha$, then X_β is isomorphic with the additive group of rational numbers,

(iii) if $G_0 = 1$ and $G_\beta = \langle X_\gamma : 0 \leqq \gamma < \beta \rangle$, then $G_\beta \lhd G$, $G_\beta \cap X_\beta = 1$ and $G_\alpha = G$.

Proof. Let G be a radicable hypercentral group and define $X_0 = G_1$ to be the torsion-subgroup of G. Then X_0 is a radicable subgroup of $\zeta(G)$ by Theorem 9.23. The upper central factors of G/G_1 are torsion-free radicable groups by the same theorem. Thus by refinement we can find an ascending central series.

$$1 = G_0 < G_1 < G_2 < \cdots < G_\lambda = G$$

where $G_{\beta+1}/G_\beta$ is isomorphic with the additive group of rational numbers if $\beta > 0$. Here, of course, we are appealing to the structure theorem for radicable abelian groups (Fuchs [3], Theorem 19.1, p. 64).

Let $x \in G_{\beta+1} \backslash G_\beta$ where $\beta > 0$. Now $G_{\beta+1}$ is $\mathfrak{F}$-perfect, so it is radicable by Theorem 9.23. Consequently there exists a sequence of elements $x = x_1, x_2, x_3, \ldots$ such that $x_i^i = x_{i-1}$ for each integer $i > 1$. Define

$$X_\beta = \langle x_i : i = 1, 2, \ldots \rangle;$$

$G_{\beta+1}/G_\beta$ is torsion-free, so $G_\beta \cap X_\beta = 1$. Hence X_β is isomorphic with a subgroup of the additive group of rationals. But X_β is clearly radicable; hence X_β is isomorphic with the additive group of rationals. Therefore $G_{\beta+1} = G_\beta X_\beta$ and $G_\beta = \langle X_\gamma : 0 \leqq \gamma < \beta \rangle$.

Conversely let G have such a well-ordered set of subgroups. Then G is evidently $\mathfrak{F}$-perfect. Since the automorphism group of the additive group of rational numbers is a direct product of cyclic groups, G centralizes each $G_{\beta+1}/G_\beta$, so G is hypercentral. Therefore G is radicable by Theorem 9.23. □

We mention without proof a theorem of Gluškov [1]: *in a torsion-free radicable hypercentral group the normalizer of a radicable subgroup is also radicable.* For further results the reader should consult Černikov [10], [13] and Bačurin [1].

Radicable Groups in General

Radicable groups can have a very complicated structure, as the following theorem indicates.

Theorem 9.25 (B. H. Neumann [4]). An arbitrary group G can be embedded in a radicable group G^*. If G is infinite, then G and G^* have the same cardinality.

Proof. Let $C_n = \langle a \rangle$ be a cyclic group of finite order n and consider the standard wreath product $H = G \wr C_n$. Then G is isomorphic with the diagonal subgroup D of the base group B. Let $g \in G$ and let $b \in B$ be defined by the rule $b_1 = g$ and $b_x = 1$ if $1 \neq x \in C_n$. Now

$$(ab)^n = a^n b^{a^{n-1}+\cdots+a+1} = b^{a^{n-1}+\cdots+a+1} = (g, \ldots, g) \in D.$$

Thus we can embed G in H in such a way that each element of G is the nth power of an element of H.

Let $p_1, p_2, \ldots$ be the set of primes so arranged that each prime occurs infinitely often in the sequence. We define

$$G_0 = G \quad \text{and} \quad G_{i+1} = G_i \wr C_{p_{i+1}}$$

for $i = 0, 1, \ldots$, and identify G_i with the diagonal subgroup of the base group of G_{i+1}. Then

$$G = G_0 < G_1 < G_2 < \cdots.$$

Let G^* be the union of the G_i. If $g \in G^*$ and p is any prime, there is a positive integer i such that $g \in G_i$ and $p = p_i$; hence g is the pth power of some element of G_{i+1} and G is radicable. Clearly $|G| = |G^*|$ if G is infinite. □

It is easy to show by the methods of the previous proof that *if the group G is hyperabelian, subsoluble, SN^*, radical, locally soluble or locally finite-π, then G can be embedded in a radicable group G^* with the same property*: similar results are in a paper of Kargapolov, Merzljakov and Remeslennikov [2].

On a deeper level there is the well-known result of Mal'cev that *a torsion-free nilpotent or locally nilpotent group G can be embedded in a radicable torsion-free nilpotent or locally nilpotent group G^* which is determined by G up to isomorphism* (Mal'cev [4]). Other proofs of Mal'cev's theorem are due to Baumslag ([8]), P. Hall ([6], § 6), Lazard ([3], pp. 180—182) and Šmel'kin [1]. Generalizations of Mal'cev's theorem have been given by Kargapolov in [6] and [8]; see also Wiegold [5].

For further results about radicable groups and questions concerning the existence and uniqueness of roots of an element of a group the reader is referred to the extensive paper [4] of Baumslag, and also to Kuroš [13] (Vol. 3, § 86), Baumslag [3], [5] and Houang Ki [1].

9.3 Residual Finiteness and Groups with Finite Abelian Section Rank

A group G is said to have *finite abelian section rank* if every abelian section of G has finite p-rank for $p = 0$ or a prime. It is easy to see that a soluble group G has finite abelian section rank if and only if there exists a series of finite length

$$1 = G_0 \lhd G_1 \lhd \cdots \lhd G_n = G$$

such that each G_{i+1}/G_i is an abelian group with finite p-rank for $p = 0$ or a prime. We will occasionally write

$$\mathfrak{S}_0$$

for the class of soluble groups with finite abelian section rank.

We propose to derive criteria for an $\mathfrak{S}_0$-group—and more generally for a hyperabelian group with finite abelian section rank—to be residually finite. It should be kept in mind that Baer and Heineken [1] have shown that *the radical groups with finite abelian subgroup rank are just the hyperabelian groups with finite abelian section rank.*

The first criterion was obtained by Robinson ([7], Theorem B) for $\mathfrak{S}_0$-groups and later extended by Baer and Heineken ([1]) to hyperabelian groups with finite abelian section rank.

Theorem 9.31. If G is a hyperabelian group with finite abelian section rank, the $\mathfrak{F}$-residual and the $\mathfrak{F}$-perfect radical coincide and form a radicable nilpotent subgroup. Thus G is residually finite if and only if there exist no non-trivial normal radicable subgroups.

Corollary 1. The class of residually finite, hyperabelian groups with finite abelian section rank is closed with respect to forming extensions.

In proving Corollary 1 we need the result that a hyperabelian group with finite abelian section rank possesses an ascending *characteristic* abelian series: this is because of the structure of the factors of the upper nilpotent (or Fitting) series of G (see Corollary 1 to Theorem 6.36).

Corollary 2 (Hirsch [6]). Every polycyclic group is residually finite.

Corollary 3. For hyperabelian groups with finite abelian section rank, the properties $\mathfrak{F}$-perfect, semi-radicable and radicable coincide (and imply nilpotence).

These three properties coincide for two other classes of groups, namely the hypercentral groups (Corollary 1, Theorem 9.23) and the CL-groups (Section 4.4, Part 1, p. 138).

We shall precede the proof of Theorem 9.31 with three lemmas.

Lemma 9.32 (Robinson [7], Lemma 2.21: Merzljakov [6]). Let A be an abelian group and let π be a set of primes. Assume that A has finite p-rank for each $p \in \pi$. Then A is residually a finite π-group if and only if it is π-reduced.

(Here we term a group *π-reduced* if it contains no non-trivial semi-π-radicable subgroups: when π is the set of all primes, we employ the term *reduced*).

Proof. Obviously A is π-reduced if it is residually a finite-π-group. Conversely, let A be π-reduced and define

$$I = \bigcap A^m$$

where m runs over the set of all positive π-numbers. Since A/A^m is a direct product of cyclic groups (Kaplansky [1], Theorem 6, p. 17), we need only show that $I = 1$, and this will certainly be the case if I is π-radicable. Let $a \in I$ and $p \in \pi$. Since A is π-reduced and has finite p-rank, the p-component of A is finite: let p^i be its order. Then we can write

$$a = a_1^p = a_2^{p^2} = \cdots$$

where $a_i \in A^{p^i}$. Now $(a_1^{-1}a_2^p)^p = 1$ and A^{p^i} has no elements of order p; therefore $a_1 = a_2^p$ and for similar reasons

$$a_1 = a_2^p = a_3^{p^2} = \cdots.$$

Hence $a_1 \in A^{p^i}$ for each $i \geqq 0$. Let m be any positive π-number and write $m = p^i n$ where p does not divide n. Now A^{p^i}/A^m has finite exponent dividing n and therefore cannot contain an element of order p. But $a_1 \in A^{p^i}$ and $a_1^p = a \in I \leqq A^m$; hence $a_1 \in A^m$ and $a_1 \in I$. It follows that I is p-radicable for all $p \in \pi$, so that I is π-radicable. ▯

We remark that a reduced abelian p-group may easily fail to be residually finite. For example, the abelian group with generators $x_1, x_2, \ldots$ and relations

$$x_1^p = 1 \quad \text{and} \quad x_{i+1}^{p^i} = x_1, \quad (i = 1, 2, \ldots),$$

has $\langle x_1 \rangle$ as its finite residual. The hypothesis of finite rank in Lemma 9.32 is therefore essential.

Lemma 9.33 (Robinson [7], Lemma 2.31). Let N be a normal subgroup of a group G and suppose that every subnormal factor of G with finite exponent is finite. Assume also that G/N has an abelian series of finite length whose factors have finite abelian subgroup rank and reduced torsion-subgroups. If x is an element of N which does not belong to the finite residual of N, then x does not belong to the finite residual of G. Thus

$$\varrho_{\mathfrak{F}}^*(N) = \varrho_{\mathfrak{F}}^*(G) \cap N.$$

Proof. We can refine the given series in G/N by inserting the torsion-subgroup in each factor to obtain an abelian series of finite length,

$$N = G_0 \lhd G_1 \lhd \cdots \lhd G_n = G,$$

such that G_{i+1}/G_i is either a direct product of finite p-groups for various primes p or a torsion-free abelian group (of finite rank). If $n = 0$, then $N = G$ and the result is obvious, so let $n > 0$ and put

$$M = G_{n-1}.$$

By induction on n the element x does not belong to M^m for some positive integer m. By hypothesis M/M^m is finite. Let $C = C_G(M/M^m)$; then $C \lhd G$ and G/C is finite. We may therefore suppose that $x \in C$. Since G/M is abelian,

$$[C', C] \leqq [M, C] \leqq M^m,$$

so C/M^m is nilpotent of class at most 2.

Suppose first of all that G/M is periodic. Then $\bar{C} = C/M^m$ has finite Sylow subgroups; since it is also nilpotent, $\bar{C}$ is the direct product of its Sylow subgroups. Let π be the finite set of primes which divide the order of xM^m and write $\bar{C} = \bar{C}_\pi \times \bar{C}_{\pi'}$ where $\bar{C}_\pi$ and $\bar{C}_{\pi'}$ are respectively the π and π'-components of $\bar{C}$. Now $xM^m \notin \bar{C}_{\pi'}$ and $|\bar{C}: \bar{C}_{\pi'}| = |\bar{C}_\pi|$, which is finite. Since G/C is finite, we obtain a normal subgroup of finite index in G which does not contain x.

Now let G/M be torsion-free and define

$$L = C^{m^2} M^m.$$

Then $L \lhd G$ and

$$|G: L| = |G: C|\,|C: L| \leqq |G: C|\,|C: C^{m^2}|,$$

which is finite. Suppose that $x \in L$, so that $x \equiv y \bmod M^m$ for some $y \in C^{m^2}$. Now the identity

$$(ab)^d = a^d b^d [b, a]^{\binom{d}{2}}$$

is valid in any nilpotent group of class at most 2. Since $m \,\Big|\, \binom{m^2}{2}$ and $C' \leqq M$, it follows that $y \equiv z^{m^2} \bmod M^m$ for some $z \in C$, and

$$x \equiv z^{m^2} \bmod M^m. \tag{5}$$

But $x \in N \leqq M$, so $z^{m^2} \in M$. Since G/M is torsion-free, we conclude that $z \in M$; thus $x \in M^m$ by (5). This contradiction completes the proof of the lemma. □

The following is a generalization of a theorem of Mal'cev ([7], Theorem 3).

Lemma 9.34. Let G be a radical group whose abelian sections have finite 0-rank and let T be the maximal normal periodic subgroup of G. Then G/T has a characteristic abelian series of finite length whose factors have finite torsion-subgroups.

Proof. We shall assume that $T = 1$. Let H be the Hirsch-Plotkin radical of G. Then H is torsion-free, so by Theorem 6.36 it is nilpotent. Let C be the intersection of the centralizers in G of the upper central factors of H: these factors are torsion-free abelian groups of finite rank by Theorem 2.25. The corollary to the theorem of Zassenhaus (3.23) assures us that G/C is soluble. Now G is a radical group and H has a C-stable series of finite length, so $C \leqq H$ by Lemma 8.17. It follows that G/H, and hence G, is soluble.

If G is abelian, there is nothing to prove. Let G have derived length $d > 1$ and put $A = G^{(d-1)}$. Then A is a torsion-free abelian group of finite rank. Denote by S/A the maximal normal periodic subgroup of G/A: by induction on d, the group G/S has a series of the required type. Let $C = C_S(A)$; then S/C is isomorphic with a periodic group of matrices over the field of rational numbers, so it is finite by a classical theorem of Schur (see Part 1, p. 85)—here we have the additional hypothesis that S/C is soluble, so we could alternatively appeal to Lemma 5.29.1 and conclude that all subnormal abelian subgroups of S/C are finite; the finiteness of S/C is then just a short step away. Now $A \leq \zeta(C)$; thus $C/\zeta(C)$ is periodic and consequently locally finite. The Corollary to Theorem 4.12 implies that C' is locally finite. Hence $C' = 1$ and C is a torsion-free abelian group of finite rank. Since S/C is finite and A, S and C are all characteristic in G, the result follows. ▯

Proof of Theorem 9.31. (a) G is a hyperabelian group with finite abelian section rank. Let R be the $\mathfrak{F}$-perfect radical of G. By hypothesis there exists an ascending normal abelian series $\{G_\alpha\}$ in G such that $G_{\alpha+1}/G_\alpha$ is either finite or torsion-free abelian of finite rank and in the second case $\mathrm{Aut}_G(G_{\alpha+1}/G_\alpha)$ is rationally irreducible. If $G_{\alpha+1}/G_\alpha$ is finite, it is centralized by R. Otherwise R is represented through its action on $G_{\alpha+1}/G_\alpha$ by an irreducible linear group over the field of rational numbers. Lemma 5.29.1 shows that in this case R must also centralize $G_{\alpha+1}/G_\alpha$. It follows that R is hypercentral. By Theorem 9.23 the group R is radicable and its torsion-subgroup is contained in its centre. Finally, R is nilpotent by Lemma 6.37 (applied to a maximal normal abelian subgroup N of R).

It remains to show that G/R is residually finite. From now on we shall assume that $R = 1$, so that G has no non-trivial normal $\mathfrak{F}$-perfect subgroups: indeed G has no non-trivial $\mathfrak{F}$-perfect subgroups at all, for the $\mathfrak{F}$-perfect radical contains *all* $\mathfrak{F}$-perfect subgroups since $\mathfrak{F}^{-H}$ is closed with respect to forming arbitrary joins.

(b) *The periodic case.* Let G be periodic. Then G is locally finite by Theorem 1.45. Each Sylow subgroup satisfies Min-*ab* and hence is a Černikov group by Theorem 3.32. In view of the absence of quasicyclic subgroups, each Sylow subgroup of G is finite. Moreover the latter property holds in every factor group of G, by the Corollary to Lemma 3.46.

Let $1 \neq x \in G$ and choose a normal subgroup N which is maximal subject to $x \notin N$. We shall assume that $N = 1$ and prove that G is finite, which will establish the theorem in this case. Each non-trivial normal subgroup of G contains x. Hence, if π is the finite set of primes which divide the order of x, there are no normal π'-subgroups of G except 1. It follows that F, the Fitting subgroup of G, is a π-group. Each Sylow subgroup of F is finite, so F is finite. Since G is hyperabelian, $C_G(F) \leq F$ by Lemma 2.17. We deduce that G/F, and hence G, is finite.

(c) *The soluble case.* Let G be a soluble group: we may assume that the derived length of G is $d > 1$, in view of Lemma 9.32. Choose a maximal normal abelian subgroup A_1 containing $G^{(d-1)}$: this exists by Zorn's Lemma. Suppose that G/A_1 has a normal abelian subgroup which is not reduced; then G/A_1 has a non-trivial normal radicable abelian subgroup R/A_1. Let m be any positive integer. R/A_1 induces a group of automorphisms in the finite group A_1/A_1^m. But R/A_1 has no proper subgroups of finite index, so R must centralize A_1/A_1^m. Since A_1 is residually finite, it follows that

$$[A_1, R] \leq \bigcap_{m=1,2,\ldots} A_1^m = 1.$$

Hence $A_1 \leq \zeta(R)$. If $x \in R$, the mapping $yA_1 \to [y, x]$ is a homomorphism of R/A_1 into A_1; thus the image $[R, x]$ is a radicable subgroup of A_1. Hence $[R, x] = 1$, proving that R is abelian and $A_1 = R$ by maximality of A_1.

We have shown that G/A_1 has each of its normal abelian subgroups reduced: it also has derived length $d - 1$. By repeated applications of this argument we obtain a normal abelian series

$$1 = A_0 < A_1 < \cdots < A_d = G$$

in which each A_{i+1}/A_i is reduced. By induction on d we may suppose that $N = A_{d-1}$ is residually finite. Lemma 9.32 shows that G/N is residually finite. The residual finiteness of G now follows easily from Lemma 9.33.

(d) *The final step.* We assume that G is merely hyperabelian. Let $1 \neq x \in G$ and denote by T the maximal normal periodic subgroup of G. By (b) T is residually finite. Now if $H \leq G$, then H/H^m is finite: for its Sylow subgroups, and therefore its Fitting subgroup, are finite. Hence the intersection of all the characteristic subgroups with finite index in T is trivial, and by Theorem 9.12 the group Aut T is residually finite. Let $C = C_G(T)$; then G/C is residually finite and we can evidently assume that

$$x \in C.$$

Now G/T is soluble by Lemma 9.34, and if d is the derived length of G/T, then $[C^{(d)}, C] = 1$ and C is soluble. Hence C is residually finite by (c). It follows that $x \notin C^m$ for some positive integer m. Now C/C^m is finite and G/C is residually finite, so G/C^m can have no $\mathfrak{F}$-perfect subgroups except 1. Clearly we can pass to G/C^m and assume that C is finite. Therefore x is contained in the periodic radical T of G. We know from (b) that T is residually finite. Hence $x \notin T^n$ for some integer n and T/T^n is finite. Finally Lemmas 9.33 and 9.34 together imply that x does not belong to some normal subgroup with finite index in G. The proof is now complete. ▯

We will derive a second, more powerful criterion for residual finiteness. Two preliminary lemmas are required.

Lemma 9.35. (Robinson [7], Lemma 2.32; see also Smirnov [1]). Let G be a group and let π be a set of primes. If the centre of G is π-reduced, then $\zeta_{i+1}(G)/\zeta_i(G)$ is π-reduced for each non-negative integer i.

Proof. It is sufficient to show that $\zeta_2(G)/\zeta_1(G)$ is π-reduced. Let $R/\zeta_1(G)$ be a π-radicable subgroup of $\zeta_2(G)/\zeta_1(G)$. If $g \in G$, the map $a\zeta_1(G) \rightarrow [a, g]$ is a homomorphism of $R/\zeta_1(G)$ into $\zeta_1(G)$, so the image $[R, g]$ is π-radicable. Therefore $[R, g] = 1$ and $R = \zeta_1(G)$. ▯

Lemma 9.36 (cf. Robinson [7], Lemma 2.32). Let B be the Baer radical of a group G and assume that the centre of B is π-reduced where π is a set of primes. Then every normal abelian subgroup of G which has finite abelian subgroup rank is π-reduced.

Proof. Suppose that the lemma is false: then there exists a non-trivial normal π-radicable abelian subgroup R which has finite abelian subgroup rank. Clearly $R \leqq B$. By hypothesis $\zeta_1(B)$ is π-reduced and we know from Lemma 9.35 that $\zeta_{i+1}(B)/\zeta_i(B)$ is also π-reduced for each integer $i \geqq 0$. Now it is easy to see that the property "π-reduced" is **P**-closed; therefore $\zeta_i(B)$ is π-reduced for each integer i.

Let X be a finitely generated subgroup of B; then X is nilpotent and subnormal in B, so

$$\gamma B X^n = [B, \underbrace{X, \ldots, X}_{n}] = 1 \tag{6}$$

for some integer n. Let p be any prime and denote by P the p-component of R. If P_0 is the maximal radicable subgroup of P, then P/P_0 is finite since P has finite rank. It is straightforward to deduce from (6) that if P_0 has rank $r(p)$, then $\gamma P X^{n(p)} = 1$ where

$$n(p) = r(p) + |P: P_0|;$$

the argument here is the same as in the proof of Theorem 6.35. Since $n(p)$ is independent of X, we conclude that $\gamma P B^{n(p)} = 1$ and $P \leqq \zeta_{n(p)}(B)$. However, as we have indicated, $\zeta_{n(p)}(B)$ is π-reduced, and it is easy to see that the π-radicability of R is inherited by P (whether or not $p \in \pi$). Thus $P = 1$ and R is torsion-free. If r_0 is the 0-rank of R, then $R \leqq \zeta_{r_0}(B)$ by Lemma 6.37. Thus $R = 1$. ▯

In any group the centre of the Gruenberg radical is contained in the centre of the Baer radical. However, Lemma 9.36 (and Theorem 9.37 below) would not be true if we weakened the hypothesis to "the centre of the Gruenberg radical is π-reduced". For example, the locally dihedral 2-group is a Gruenberg group with centre of order 2, yet it possesses a normal 2^∞-subgroup.

The second criterion for residual finiteness mentioned above runs as follows.

Theorem 9.37 (cf. Robinson [7], Theorem A). Let G be a hyperabelian group with finite abelian section rank. Then G is residually finite if and only if the centre of its Baer radical is reduced.

Proof. Let R be the finite residual and let B be the Baer radical of G. According to Theorem 9.31, the group R is radicable and nilpotent, so $R \leq B$. Now by Theorem 9.23 the subgroup $\zeta(R)$ is radicable, and clearly $\zeta(R) \lhd G$. But $\zeta(B)$ is reduced and by Lemma 9.36 every normal abelian subgroup of G is reduced. Hence $\zeta(R) = 1$ and consequently $R = 1$. ▯

Corollary (cf. Wehrfritz [10], Theorem R). Let G be a hyperabelian group with finite abelian section rank and assume that the centre of the Baer radical of G is reduced. Then the holomorph of G is residually finite.

Proof. By the theorem, G is residually finite: also each G/G^m is finite. The result now follows from the Corollary to Lemma 9.13. ▯

Nilpotent Groups with Finite Abelian Section Rank

By Theorem 9.37 a nilpotent $\mathfrak{S}_0$-group is residually finite if and only if its centre is reduced. However the following stronger result is valid.

Theorem 9.38 (Robinson [7], Theorem F). Let G be a nilpotent group with finite abelian section rank and let π be a set of primes. Then G is residually a finite π-group if and only if the centre of G is π-reduced.

Proof. If G is residually a finite π-group, $\zeta(G)$ is certainly π-reduced. Conversely let $\zeta(G)$ be π-reduced; by Lemmas 9.32 and 9.35 each factor $\zeta_{i+1}(G)/\zeta_i(G)$ is residually a finite π-group. We can assume that G has nilpotent class $c > 1$ and that $M = \zeta_{c-1}(G)$ is residually a finite π-group. Let $1 \neq x \in M$; it is sufficient to find a normal subgroup of index a π-number not containing x. Now $x \notin M^m$ for some positive π-number m. The method of Lemma 9.33 provides a normal subgroup of G with finite index which does not contain x. To show that this index is a π-number, it suffices to establish the following: if $N \lhd G$ and n is a positive π-number, then $G/C_G(N/N^n)$ is a finite π-group. This may be proved in a routine manner using the nilpotence of G and the fact that N/N^n is a finite π-group. ▯

From this we deduce an important theorem of Gruenberg ([2], Theorem 2.1).

Corollary. A finitely generated torsion-free nilpotent group is residually a finite p-group for every prime p.

Chief Factors and Maximal Subgroups

Let us now examine the nature of chief factors and maximal subgroups of hyperabelian groups with finite abelian section rank. We will prove

Theorem 9.39 (cf. Robinson [7], Theorems C and D). Let G be a hyperabelian group with finite abelian section rank. Then each chief factor of G is finite and each maximal subgroup of G has finite index.

Two elementary results of a general nature aid the proof.

Lemma 9.39.1 (P. Hall [10], p. 596). A finitely generated group A cannot be an irreducible group of automorphisms of a direct product G of finitely many copies of the additive of rational numbers.

Proof. For suppose that A is such a group of automorphisms and let G be written additively. Choose a basis $\{g_1, \ldots, g_n\}$ for G as a vector space over the field of rational numbers and let $A = \langle \alpha_1, \ldots, \alpha_m \rangle$. Then

$$g_i\alpha_j = \sum_{k=1}^{n} r_{i,j,k}g_k \quad \text{and} \quad g_i\alpha_j^{-1} = \sum_{k=1}^{n} r'_{i,j,k}g_k.$$

Let π be the set of prime divisors of the denominators of the $2mn^2$ rational numbers $r_{i,j,k}$ and $r'_{i,j,k}$. Then each element of the additive subgroup

$$G_1 = g_1A = \{g_1\alpha : \alpha \in A\}$$

is expressible as a linear combination of $g_1, \ldots, g_n$ in which the coefficients are rational numbers whose denominators are π-numbers. Since π is finite, $G_1 < G$ and the irreducibility of A is contradicted. ▯

Lemma 9.39.2. Let G be a hyper-(abelian or finite) group whose chief factors are finite. Then each maximal subgroup of G has finite index.

Proof. Suppose that M is a maximal subgroup with infinite index in G. By hypothesis there is an ascending normal series $\{G_\alpha\}$ in G whose factors are abelian or finite. There is a first ordinal α such that $G_\alpha \nleq M$ and clearly α cannot be a limit ordinal. Hence $G_{\alpha-1} \leq M$; evidently we may assume that $G_{\alpha-1} = 1$. Writing $N = G_\alpha$, we have $G = NM$. Since $|G: M| = |N: N \cap M|$, the subgroup N cannot be finite, so it must be abelian. Hence $N \cap M \lhd G$. If $N \cap M < X < N$ and $X \lhd G$, Dedekind's modular law shows that $M < XM < G$, which is impossible. It follows that $N/N \cap M$ is a chief factor of G, so it is finite and hence $|G: M|$ is finite. □

Proof of Theorem 9.39. Let G be a hyperabelian group with finite abelian section rank: for the first part of the theorem it is enough to show that a minimal normal subgroup of G is finite. Suppose that N is an infinite minimal normal subgroup of G; then N must be a direct product of finitely many copies of the additive group of rational numbers. Let $C = C_G(N)$ and $H = G/C$. If Q denotes the field of rational numbers, we may regard N as an irreducible QH-module. By the corollary to Zassenhaus' theorem (3.23), H is soluble. Let S be a subnormal abelian subgroup of H. By repeated application of Clifford's Theorem, N is completely reducible as a QS-module. But S has finite 0-rank, so Lemma 5.29.1 implies that S is finitely generated. Now a soluble group whose subnormal abelian subgroups are finitely generated is polycyclic—this the final step in the proof of Theorem 3.27. In particular, H is finitely generated: however H cannot be an irreducible group of automorphisms of N in this case by Lemma 9.39.1, so we have a contradiction.

That maximal subgroups of G have finite index is now an immediate consequence in view of Lemma 9.39.2. □

Corollary. If G is a hyperabelian group with finite abelian section rank and F is the Frattini subgroup of G, then G/F is residually finite and F contains the finite residual of G.

$\mathfrak{S}_1$-Groups

Let G be a hyperabelian group with finite abelian section rank and suppose that G contains elements of only finitely many distinct prime orders. We denote by

$$\mathfrak{S}_1$$

the class of all such groups G. If T is the maximal normal periodic subgroup of G, then G/T is soluble by Lemma 9.34. Also the Fitting subgroup F of T is easily seen to be a Černikov group. Hence $T/C_T(F)$ is a Černikov group by Theorem 3.29. Now $C_T(F) \leqq F$; hence *T is a Černikov group and G is soluble.**

By Lemma 9.34, *a group is an $\mathfrak{S}_1$-group if and only if it possesses an abelian series of finite length whose factors have finite 0-rank and Černikov*

* Mal'cev [7] has called $\mathfrak{S}_1$-groups "soluble A_3-groups".

torsion-subgroups. Also from Theorem 3.25 we have

$$\mathfrak{S}_1 \leqq \mathfrak{N}\mathfrak{A}\mathfrak{F},$$

a result of Mal'cev ([7], Theorem 4). It is clear from Lemma 1.44 that $\mathfrak{S}_1$ is a proper subclass of the class of soluble groups of finite rank, which is in turn a proper subclass of $\mathfrak{S}_0$.

We shall mention without proof some residual properties of $\mathfrak{S}_1$-groups. *If G is an $\mathfrak{S}_1$-group and the centre of the Baer radical of G is reduced, there is a finite set of primes π such that G is residually a finite π-group* (Robinson [7], Theorem E). A related result can be found in a paper of Merzljakov [6].

Thus in particular *a polycyclic group is residually a finite π-group for some finite set of primes π* (Learner [2]). Conversely, a hyperabelian group with finite abelian section rank which is residually a finite π-group for some finite set of primes π is an $\mathfrak{S}_1$-group: for it can contain no element of order a π'-number. It can also be shown that *if G is an $\mathfrak{S}_1$-group whose Baer radical has reduced centre, then* Aut G *is residually a finite π-group for some finite set of primes π* (Wehrfritz [10], Theorem R).

Completely Infinite Groups

Following Bowers ([1], p. 434) we shall call a soluble group *completely infinite* if it has a series of finite length whose factors are torsion-free abelian groups. Suppose that $N \lhd G$ and G/N is a torsion-free abelian group; if we define

$$M = \bigcap_{\alpha \in \operatorname{Aut} G} N^\alpha,$$

then G/M is also a torsion-free abelian group, and of course M is characteristic in G. This argument shows that *a completely infinite, soluble group has a characteristic series of finite length with torsion-free abelian factors.*

Naturally, every completely infinite, soluble group is torsion-free, but a torsion-free soluble group need not be completely infinite. For example, consider the group G with generators

$$x, y, z$$

and relations

$$z^{-1}xz = x^{-1}, \quad z^{-1}yz = y^{-1} \quad \text{and} \quad [x, y] = z^{4^\alpha}$$

where $\alpha \geqq 1$. It is not difficult so see that G is a torsion-free polycyclic group. But G has no non-trivial torsion-free abelian factor groups. For let G/N be such a factor group. Then N contains the commutator $[x, y] = z^{4^\alpha}$, and hence $z \in N$. Thus N contains $[z, x] = x^2$, which means that $x \in N$. For a similar reason $y \in N$ and $N = G$. This example is due to Hirsch

([5], p. 83).* On the other hand, Theorem 2.25 shows that for nilpotent groups "torsion-free" and "completely infinite" are identical properties.

We shall prove the following theorem.

Theorem 9.39.3 (Robinson [7], Theorem G). Let G be a soluble group with finite abelian section rank having elements of only finitely many distinct prime orders. If the centre of the Baer radical has reduced torsion-subgroup, then G has a characteristic, completely infinite subgroup of finite index.

Proof. The methods of Lemmas 9.35 and 9.36 are available to show that every normal abelian subgroup of G has reduced torsion-subgroup. The maximal normal periodic subgroup T of G is a Černikov group since $G \in \mathfrak{S}_1$: hence T is finite. It follows from Lemma 9.34 that there is a characteristic abelian series in G

$$1 = G_0 < G_1 < \cdots < G_n = G$$

such that each factor G_{i+1}/G_i is either finite or torsion-free.

It will suffice to establish the following fact: *if $H \lhd K$ with H finite and K/H torsion-free, abelian and of finite rank, then K contains a characteristic torsion-free abelian subgroup of finite index.* In the first place, H is the set of all elements of finite order in K, so H is characteristic in K. Let $C = C_K(H)$ and observe that K/C is finite. Let $m = |H|$. If $x, y \in C$, then, since K/H is abelian, $[x, y] \in H \cap C \leq \zeta(C)$; hence

$$1 = [x, y]^m = [x^m, y].$$

Consequently, $M = C^m \leq \zeta(C)$ and M is abelian. Evidently the torsion-subgroup of M is finite with order dividing m. Hence $N = M^m$ is torsion-free and abelian. Also N is characteristic in K because H is. Finally C/M and M/N are clearly finite, so K/N is finite. ▯

Corollary (Hirsch [3], [5]). A polycyclic group has a characteristic, completely infinite subgroup of finite index.

We remark that it is possible to prove that *if $G \in \mathfrak{S}_1$ and the Baer radical of G has reduced centre, then* Aut G *has a normal torsion-free subgroup of finite index.* A special case of this is in Wehrfritz ([10], Theorem T2).

Selberg [1] and Kargapolov [14] have shown that *a finitely generated linear group over a field of characteristic 0 has a characteristic torsion-free subgroup of finite index*: another proof has been given by Wehrfritz

* For further examples see Bowers [1] (Theorem 1).

([10], pp. 116–117). This is not true for linear groups of non-zero characteristic: see Platonov [4] in this connection.

Some other conditions sufficient to make a group torsion-free-by-periodic are in a paper of Simon [6].

Groups with a Finite Rational Series

A *rational series* in a group is a series whose factors are isomorphic with subgroups of the additive group of rational numbers. Clearly a group has a rational series of finite length if and only if it is a completely infinite soluble group with finite abelian section rank. Zaičev [5] has shown that the proper length of such a rational series—which is an invariant of the group by the Schreier refinement theorem—is equal to the rank of the group. See also Gluškov [4] (Theorem 1).

In 1949 Čarin proved that *a semi-radicable group with a finite rational series is nilpotent* (Čarin [2] and [3], Theorem 10)—this of course follows immediately from Theorem 9.31. On the other hand, as Černikov has pointed out ([26], § 1), a radicable group having an ascending rational series does not even have to be locally nilpotent: this is shown by the semi-direct product of the additive group of real numbers by the multiplicative group of positive real numbers with the natural automorphism representation of the latter (A more complicated example was later given by Baumslag [21]). However it is easy to see that *an $\mathfrak{F}$-perfect group has an ascending normal rational series if and only if it is hypercentral and radicable.*

Theorem 9.39.4 (Čarin [6]). Let the group G have a series of finite length whose infinite factors are non-cyclic torsion-free abelian groups of rank 1. Then G is a finite extension of a torsion-free nilpotent group.

Proof. Let $1 = G_0 \lhd G_1 \lhd \cdots \lhd G_n = G$ be the given series and assume that $n > 1$. By induction on n we may suppose the theorem true of $M = G_{n-1}$; thus M^m is torsion-free nilpotent and M/M^m is finite for some positive integer m. Naturally we can assume that G/M is infinite. Let $C = C_G(M/M^m)$; then G/C is finite and C/M^m is soluble. By Theorem 9.39.3 there is a torsion-free G-admissible subgroup F/M^m which has finite index in C/M. Thus G/F is finite. Evidently F/M^m is isomorphic with a subgroup of G/M, and it cannot by cyclic: for if it were, G/M would be finitely generated and therefore cyclic. We form the upper central series of M^m, note that its factors are torsion-free abelian groups of finite rank and refine it to a series of finite length of the same type on whose factors the action of F is rationally irreducible. Now F/M^m does not have an infinite cyclic factor group. Therefore, by Lemma 5.29.1, the group F/M^m induces a finite automorphism group in each factor of the series in M^m. Hence there exists an $L \lhd F$ such that F/L is finite and

$$[M^m, \underbrace{L, \ldots, L}_{r}] = 1 \tag{7}$$

for some r. Now F/M^m is abelian, so $L' \leq M^m$ and $\gamma_{r+2}(L) = 1$ by (7). Thus L is nilpotent and $|G: L|$ is finite; also it is clear that L is torsion-free. This implies that G is (torsion-free nilpotent)-by-finite. □

In conclusion we shall mention two stronger properties than residual finiteness. A group G is said to be *residually finite with respect to conjugacy* if, given two non-conjugate elements a and b of G, there is a finite homomorphic image H of G such that the images of a and b are not conjugate in H. Blackburn [5] and later Seksenbaev [3] showed that *finitely generated nilpotent groups are residually finite with respect to conjugacy*; Kargapolov [15] and Remeslennikov [4] have extended this result to supersoluble and polycyclic groups respectively. See also the paper [1] by Timošenko, where it is proved that *a finitely generated free metabelian group is residually finite with respect to conjugacy*, and Remeslennikov [1].

A group G is called *finitely separable* if, given an element g and a subgroup S such that $g \notin S$, there exists a homomorphism ϕ of G into a finite group such that $g^\phi \notin S^\phi$. Mal'cev [9] has studied finitely separable soluble groups, and Smirnov [8] has proved that *a nilpotent group is finitely separable if and only if each of its factor groups is residually finite.*

9.4 The Residual Finiteness of Some Relatively Free Groups

Following P. Hall [3] we call a group *relatively free* if it is a free group of some variety. We shall show that certain types of relatively free groups are residually finite-p for all primes p, in particular free soluble groups. This is easily deduced from

Theorem 9.41 (Baumslag [18], Theorem 2 and 3: Dunwoody [1]). Let $\mathfrak{X}$ be a root class, let v be a verbal mapping and let R be a normal subgroup of a free group F. If both F/R and $R/v(R)$ are residually $\mathfrak{X}$-groups, then $F/v(R)$ is a residually $\mathfrak{X}$-group.

The proof of the theorem presented here is due to Dunwoody [1]: for another version see Šmel'kin [9]. We begin with a simple result about Schreier systems.

Let F be a free group with a set of free generators X; let S be a (non-empty) subset of F and let $f \in S$. We write f in the reduced form

$$f = a_1 \cdots a_n$$

where $a_i \in X \cup X^{-1}$. Then S is called a (right) *Schreier system* with respect to X if it contains every initial segment

$$f^{(i)} = a_1 \cdots a_i$$

for $i = 0, 1, \ldots, n$ and all $f \in S$; here $f^{(0)}$ is taken to be 1.

Lemma 9.42 (Dunwoody [1]). Let F be a free group on a set X, let S be a Schreier system with respect to X and let H be a subgroup of F. If distinct elements of S lie in distinct right cosets of G, then S is contained in a right transversal to H which is also a Schreier system.

Proof. Consider the set $\mathscr{S}$ of all Schreier systems T such that $S \subseteq T$ and distinct elements of T lie in distinct right cosets of H. Then $\mathscr{S}$ is partially ordered by set inclusion and moreover the union of any chain in $\mathscr{S}$ belongs to $\mathscr{S}$. By Zorn's Lemma there is a maximal element of $\mathscr{S}$, say M.

Let $f \in F$ and let $f = a_1 \cdots a_n$ be the reduced form of f with $a_i \in X \cup X^{-1}$. Suppose that k is the largest integer such that $Hf^{(k)} = Hy$ for some $y \in M$; such an integer exists because $f^{(0)} = 1 \in M$. Then if $k < n$, we have

$$Hf^{(k+1)} = Hf^{(k)}a_{k+1} = Hya_{k+1}.$$

Hence $ya_{k+1} \notin Hy'$ for any $y' \in M$. Therefore the set

$$M \cup \{ya_{k-1}\}$$

is a Schreier system properly containing M whose distinct elements lie in distinct right cosets of H. This contradicts the maximality of M, so $k = n$ and $Hf = Hy$. It follows that M is a right transversal to H in G. ▯

We come now to another result of Dunwoody [1] which is vital in the proof of Theorem 9.41.

Theorem 9.43. Let F be a free group, let $\mathscr{S}$ be a set of subgroups of F which is closed with respect to forming finite intersections and let v be a verbal mapping. Then

$$v\Big(\bigcap_{H \in \mathscr{S}} H\Big) = \bigcap_{H \in \mathscr{S}} v(H).$$

Proof. Let I denote the intersection of all the subgroups in the set $\mathscr{S}$. Then it is clear that

$$v(I) \leq \bigcap_{H \in \mathscr{S}} v(H). \tag{8}$$

Let $u \in I \backslash v(I)$; it will be shown that $u \notin v(J)$ for some $J \in \mathscr{S}$; this implies the reverse inclusion of (8).

Let X be a set of free generators for F and choose a Schreier system T (with respect to X) which is also a right transversal to I in F: this exists by Lemma 9.42 for example. For any $f \in F$ we can write

$$Hf = H\phi(f)$$

where $\phi(f) \in T$. By the theorem of Schreier on subgroups of free groups the set

$$Y = \{tx\phi(tx)^{-1}: x \in X, t \in T, \phi(tx) \neq tx\}$$

freely generates I (see for example Kuroš [9], Vol. 2, p. 34). Write u as a reduced word $y_1^{\varepsilon_1} \cdots y_r^{\varepsilon_r}$ where $y_i \in Y$ and $\varepsilon_i = \pm 1$, and let $y_i = t_i x_i \phi(t_i x_i)^{-1}$. Then the set

$$A = \{t_1, \ldots, t_r, \phi(t_1 x_1), \ldots, \phi(t_r x_r)\}$$

is a subset of T. Denote by B the set of all $a^{(i)}$ where $a \in A$ and i does not exceed the length of a as a reduced word in the x's. Certainly B is a finite subset of T since the latter is a Schreier system. Finally, define

$$C = \{bb_1^{-1}: b, b_1 \in B, b \neq b_1\}.$$

If $1 \neq bb_1^{-1} \in I$ and $b, b_1 \in B$, then $Ib = Ib_1$ and, since $B \subseteq T$, this implies that $b = b_1$. Hence $C \cap I$ is empty. Therefore, to each $c \in C$ there corresponds an $H_c \in \mathscr{S}$ such that $c \notin H_c$. Let

$$J = \bigcap_{c \in C} H_c.$$

Since C is a finite set, $J \in \mathscr{S}$. Suppose that $Jb = Jb_1$ where b and b_1 are unequal elements of B: then $c = bb_1^{-1} \in C \cap J$ and hence $c \in H_c$. This is impossible, so different elements of B lie in different right cosets of J. Also it is clear from its construction that B is a Schreier system. Therefore B is contained in a right transversal M to J in F, by Lemma 9.42. If $f \in F$, we can write $Jf = J\phi'(f)$ for some $\phi'(f) \in M$.

Now

$$I\phi(t_i x_i) = It_i x_i \subseteq Jt_i x_i = J\phi'(t_i x_i),$$

so that $J\phi(t_i x_i) = J\phi'(t_i x_i)$. But

$$\phi(t_i x_i) \in A \subseteq B \subseteq M,$$

so $\phi(t_i x_i) = \phi'(t_i x_i)$ for $i = 1, \ldots, r$. The set

$$Y_1 = \{mx\phi'(mx)^{-1}: m \in M, x \in X, \phi'(mx) \neq mx\}$$

freely generates J. Also $t_i \in M$; thus

$$y_i = t_i x_i \phi(t_i x_i)^{-1} = t_i x_i \phi'(t_i x_i)^{-1} \in Y_1$$

and $\{y_1, \ldots, y_r\}$ is contained in $Y \cap Y_1$. This implies that $K = \langle y_1, \ldots, y_r \rangle$ is a free factor of both I and J, from which it follows that

$$K \cap v(I) = v(K) = K \cap v(J).$$

Finally, $u \in K$ and $u \notin v(I)$, so $u \notin v(J)$ as required. ☐

Proof of Theorem 9.41. We have to show that $F/v(R)$ is a residually $\mathfrak{X}$-group. Let $f \in F \backslash v(R)$ and define $\mathscr{S}$ to be the set of all normal subgroups N such that $R \leq N$ and $F/N \in \mathfrak{X}$. Since $\mathfrak{X} = \boldsymbol{D}_0\mathfrak{X} = \boldsymbol{S}\mathfrak{X}$ by definition of a root class, $\mathscr{S}$ is closed with respect to forming finite intersections. By hypothesis $F/R \in \boldsymbol{R}\mathfrak{X}$, so the intersection of all the subgroups in $\mathscr{S}$ is R. Hence by Theorem 9.43

$$v(R) = \bigcap_{N \in \mathscr{S}} v(N).$$

It follows that there is an N in $\mathscr{S}$ such that $f \notin v(N)$. Since $R \leq N$ and $F/N \in \mathfrak{X}$, we can assume that $f \in N$.

Suppose that F is a free group of infinite rank. If M is a normal subgroup with fewer generators than F, then each element of M is expressible in terms of the elements of a proper subset of the set of free generators of F; hence M is contained in a proper free factor of F. Since $M \lhd F$, it follows that $M = 1$. Now if $R = 1$, the theorem is obviously valid. Hence we can suppose that R, N and F are free groups with the same rank in this case.

Now suppose that F is a free group of finite rank. If $|F : R|$ is finite, then, since $R \leq N \leq F$, the group R requires at least as many generators as N, by the Reidemeister-Schreier Theorem. If $|F : R|$ is infinite and $R \neq 1$, then R requires a countably infinite number of generators (Magnus, Karrass and Solitar [1], Theorem 2.10, p. 104). Thus again the group R requires at least as many generators as N.

We may therefore assume that R has rank as a free group at least as great as N. Consequently N is isomorphic with a free factor T of R. Since $T \cap v(R) = v(T)$,

$$Tv(R)/v(R) \simeq T/T \cap v(R) = T/v(T) \simeq N/v(N).$$

Now $T \leq R$ and $R/v(R)$ is a residually $\mathfrak{X}$-group, so $N/v(N)$ is a residually $\mathfrak{X}$-group. Since $f \in N \backslash v(N)$, there is a subgroup K not containing f such that $v(N) \leq K \lhd N$ and $N/K \in \mathfrak{X}$. But $F/N \in \mathfrak{X}$ and $\mathfrak{X}$ is a root class, so there is a normal subgroup M of F such that $v(N) \leq M \leq K$ and $F/M \in \mathfrak{X}$. Now $R \leq N$, so $v(R) \leq M$; also $f \notin M$. Hence $F/v(R)$ is a residually $\mathfrak{X}$-group. ☐

Andreev and Ol'šanskiĭ [1] have established a partial converse of Theorem 9.41 as follows: *let R be a normal subgroup of a non-cyclic free* *group F, let $\mathfrak{X}$ be an* **H**-*closed class and let v be a verbal mapping. If $F/v(R)$ is a residually $\mathfrak{X}$-group, then so is F/R.*

In Theorem 9.41 we cannot take $\mathfrak{X}$ to be the class of nilpotent groups: indeed Gruenberg [5] has shown that *if R is a normal subgroup with finite index in a non-cyclic free group F, then F/R' is residually nilpotent if and only if F/R is a p-group for some prime p.* More generally, Lihtman [1] has shown that here R' can be replaced by

$$\gamma_{i_1}(\gamma_{i_2}(\cdots \gamma_{i_k}(R) \cdots))$$

for arbitrary positive integers $i_1, i_2, \ldots, i_k$.

The Residual Finiteness of Free Polynilpotent Groups

Let $i_1, \ldots, i_k$ be a set of positive integers. If G is any group we define

$$G_{i_1} = \gamma_{i_1+1}(G) \quad \text{and} \quad G_{i_1,\ldots,i_l} = \gamma_{i_l+1}(G_{i_1,\ldots,i_{l-1}}) \quad \text{if } l > 1.$$

Clearly $G_{i_1,\ldots,i_k} = 1$ if and only if

$$G \in \mathfrak{N}_{i_k} \cdots \mathfrak{N}_{i_2}\mathfrak{N}_{i_1};$$

this class is the variety of *polynilpotent groups of type* $(i_1, \ldots, i_k)$.

Theorem 9.44. If p is any prime, a free polynilpotent group of arbitrary type is residually a finite p-group.

Proof. Let F be a free group and let i be a non-negative integer. We will show first that the free nilpotent group $F/\gamma_{i+1}(F)$ is residually a finite p-group where p is an arbitrary prime. Let $f \in F\backslash\gamma_{i+1}(F)$. If X is a set of free generators for F, then f is expressible in terms of finitely many of these, say $x_1, \ldots, x_n$. The mapping $x_i \to x_i$, $(i = 1, \ldots, n)$, and $x \to 1$ if $x \in X\backslash\{x_1, \ldots, x_n\}$ extends to a homomorphism of F onto $F_0 = \langle x_1, \ldots, x_n\rangle$, and this induces a homomorphism of $F/\gamma_{i+1}(F)$ onto $F_0/\gamma_{i+1}(F_0)$ in which $f\gamma_{i+1}(F)$ is mapped to the non-identity element $f\gamma_{i+1}(F_0)$. Now $F_0/\gamma_{i+1}(F_0)$ is finitely generated and nilpotent, and it is also torsion-free by Witt's theorem that the lower central factors of a free group are torsion-free (Witt [1]; M. Hall [2], p. 175). It follows from Gruenberg's theorem (Corollary to Theorem 9.38) that $F_0/\gamma_{i+1}(F_0)$ is residually a finite p-group and $f\gamma_{i+1}(F_0)$ does not belong to some normal subgroup with index a power of p. By taking the complete inverse image of this subgroup, we obtain a normal subgroup not containing $f\gamma_{i+1}(F)$ with finite index in $F/\gamma_{i+1}(F)$ equal to a power of p.

The general result that $F/F_{i_1,\ldots,i_k}$ is residually a finite p-group, now follows by induction on k and application of Theorem 9.41. ∎

Corollary. A free soluble group is residually a finite p-group for every prime p.

Theorem 9.44 was first established by Gruenberg with certain limitations on the prime p: it was shown by P. Hall that these restrictions could be removed (Gruenberg [2], p. 31; also Gorčakov [8]). Theorem 9.44 can be applied to the study of subgroups of relatively free groups; see Baumslag [16]. A result of a more special nature—*every free metabelian group is residually a metacyclic p-group for each prime p*—has recently been obtained by Wehrfritz [7].

9.5 Finitely Generated Abelian-by-Nilpotent Groups

In the theory of infinite soluble groups, properties of group rings and algebras frequently turn out to be decisive. This is most strikingly illustrated in three important papers of P. Hall [4], [10], [11]. The second of these has as its chief result

Theorem 9.51 (P. Hall [10], Theorem 1). A finitely generated abelian-by-nilpotent group is residually finite.

It is our main object here to prove this theorem: we follow Hall's methods throughout.

Modules over Group Rings and Group Algebras of Polycyclic Groups

Let J be a principal ideal domain and let $\{p_\lambda : \lambda \in \Lambda\}$ be a complete set of primes for J; thus each non-zero element of J can be expressed uniquely up to order as a product of certain p_λ and a unit of J. Now let M be a (right) J-module. If $0 \neq a \in M$, the *order ideal of a*

$$0(a)$$

is the ideal consisting of all x in J for which $ax = 0$. Since J is principal, $0(a) = x_a J$ for some x_a in J: if $x_a \neq 0$ and the p_λ which divide x_a belong to a subset π of $\{p_\lambda : \lambda \in \Lambda\}$ for every non-zero element a of M, then M is a *π-torsion module.*

We define

$$\mathfrak{A}(J, \pi)$$

to be the class of all J-modules M which contain a free submodule S such that M/S is a π-torsion module. (A class of J-modules is understood to contain a zero J-module and all J-isomorphic images of its members). Thus if Z is the ring of rational integers, $\mathfrak{A}(Z, \pi)$ is the class of abelian groups which are extensions of free abelian groups by π-groups.

We begin with three simple properties of modules in the class $\mathfrak{A}(J, \pi)$.

Lemma 9.52. Let J be an integral domain and let π be a subset of a complete set of primes of J.

(i) The field of fractions of J, regarded as a right J-module, belongs to $\mathfrak{A}(J, \pi)$ if and only if π is the complete set of primes of J.

(ii) The class $\mathfrak{A}(J, \pi)$ is closed with respect to forming submodules.

(iii) If $\{M_\beta\colon 0 \leqq \beta \leqq \alpha\}$ is an ascending series of submodules of a J-module M and if each factor $M_{\beta+1}/M_\beta$ belongs to $\mathfrak{A}(J, \pi)$, then M belongs to $\mathfrak{A}(J, \pi)$.

Proof. The truth of (i) is evident and (ii) is a consequence of the well-known fact that a J-submodule of a free J-module is a free module if J is a principal ideal domain.

We will now prove (iii). By hypothesis there exists a J-module $\overline{M}_\beta$ such that $M_\beta \leqq \overline{M}_\beta \leqq M_{\beta+1}$ and $\overline{M}_\beta/M_\beta$ is a free module while $M_{\beta+1}/\overline{M}_\beta$ is a π-torsion module. Let $\{a_{\beta\lambda} + M_\beta\colon \lambda \in \Lambda(\beta)\}$ be a set of free generators for $\overline{M}_\beta/M_\beta$ and let S be the J-submodule of M generated by all the $a_{\beta\lambda}$, $(\lambda \in \Lambda(\beta),\ \beta < \alpha)$. If there were a non-trivial J-linear relation between the $a_{\beta\lambda}$, this would lead to a non-trivial relation between the $a_{\beta\lambda} + M_\beta$ for some $\beta < \alpha$. Hence S is a free J-module. It is clear that M/S is a π-torsion module; for each $(M_{\beta+1} + S)/(M_\beta + S)$ is a π-torsion module. □

Lemma 9.53. Let G be a polycyclic-by-finite group and let J be a principal ideal domain. Suppose that R is a right ideal of JG, the group ring of G over J. Then the J-module JG/R belongs to $\mathfrak{A}(J, \pi)$ where $\pi = \pi(R)$ is a finite set of primes in J.

Proof. There is a series of finite length in G,

$$1 = G_0 \lhd G_1 \lhd \cdots \lhd G_n = G,$$

whose factors are finite or cyclic. If $n = 0$, then $G = 1$ and $JG \simeq J$, so the result is obvious. Let $n > 0$ and write $N = G_{n-1}$, so that $N \lhd G$. Suppose first that $|G\colon N|$ is finite and let $\{x_1, \ldots, x_m\}$ be a transversal to N in G. Then

$$JG = \operatorname*{Dr}_{i=1}^{m} x_i(JN).$$

We define a JN-submodule A_i of JG as follows:

$$A_0 = R \quad \text{and} \quad A_i = A_{i-1} + x_i(JN)$$

where $i = 1, \ldots, m$. Then

$$R = A_0 \leqq A_1 \leqq \cdots \leqq A_m = JG.$$

Now

$$A_i/A_{i-1} \simeq x_i(JN)/(x_i(JN)) \cap A_{i-1} \simeq JN/(JN) \cap x_i^{-1}A_{i-1},$$

both isomorphisms being of right JN-modules. Consequently an induction hypothesis on n enables us to find a finite set of primes of J, say π, such that each A_i/A_{i-1} belongs to $\mathfrak{A}(J, \pi)$. By Lemma 9.52 it follows that $JG/R \in \mathfrak{A}(J, \pi)$.

We can therefore assume that G/N is an infinite cyclic group: let $G = \langle x, N \rangle$. Then

$$JG = \underset{i=0,\pm 1,\pm 2,\dots}{Dr} (JN)\, x^i. \tag{9}$$

For each non-negative integer i a JN-submodule L_i of JG is defined as follows:

$$L_0 = 0, \quad L_{2m} = \underset{j=-m+1}{\overset{m}{Dr}} (JN)\, x^j \quad \text{and} \quad L_{2m+1} = \underset{j=-m}{\overset{m}{Dr}} (JN)\, x^j.$$

Thus

$$L_{2m} = L_{2m-1} \oplus (JN)\, x^m \quad \text{and} \quad L_{2m+1} = (JN)\, x^{-m} \oplus L_{2m}.$$

Let

$$R_i = R + L_i$$

and observe that

$$R = R_0 \leqq R_1 \leqq R_2 \leqq \cdots$$

and

$$JG = \bigcup_{i=0,1,2,\dots} R_i.$$

There is a JN-isomorphism

$$R_k/R_{k-1} \simeq L_k/((R \cap L_k) + L_{k-1}). \tag{10}$$

By (9) every element f of $R \cap L_{2m}$ is uniquely expressible in the form $f = \sum_{i=-m+1}^{m} c_i x^i$ where $c_i \in JN$. Let Q_{2m} denote the set of all elements of JN which occur as coefficients of x^m in the expression for some element of $R \cap L_{2m}$. It is clear that Q_{2m} is a J-submodule of JN, and in fact it is even a right ideal of JN. For let $c_m \in Q_{2m}$ and $y \in JN$; suppose that $f = \sum_{i=-m+1}^{m} c_i x^i$ belongs to $R \cap L_{2m}$ where $c_i \in JN$. Then

$$g = f y^{x^m} = \sum_{i=-m+1}^{m} c_i^* \, x^i$$

where $c_i^* = c_i y^{x^{m-i}}$. Now $N \lhd G$, so $y^{x^{m-i}} \in JN$ and $c_i^* \in JN$. Therefore $g \in R \cap L_{2m}$ and $c_m^* = c_i y \in Q_{2m}$.

By definition of Q_{2m} we find that

$$(R \cap L_{2m}) + L_{2m-1} = Q_{2m}x^m + L_{2m-1},$$

so that by (10) there are J-isomorphisms

$$\begin{aligned} R_{2m}/R_{2m-1} &\simeq L_{2m}/(Q_{2m}x^m + L_{2m-1}) \\ &= ((JN)x^m + L_{2m-1})/(Q_{2m}x^m + L_{2m-1}) \\ &\simeq (JN)x^m/Q_{2m}x^m \simeq JN/Q_{2m}, \end{aligned}$$

since $(JN)x^m \cap L_{2m-1} = 0$.

If we define Q_{2m+1} to be the set of all elements of JN that occur as coefficients of x^{-m} in the expression for an element of $R \cap L_{2m+1}$, the same argument shows that Q_{2m+1} is a right ideal of JN and

$$R_{2m+1}/R_{2m} \simeq JN/Q_{2m+1},$$

as J-modules. Thus we have a J-isomorphism

$$R_k/R_{k-1} \simeq JN/Q_k$$

for each positive integer k.

Since $L_{2m-2}\, x \leqq L_{2m}$ and $L_{2m-1}\, x^{-1} \leqq L_{2m+1}$,

we have

$$(R \cap L_{2m-2})\, x \leqq R \cap L_{2m} \quad \text{and} \quad (R \cap L_{2m-1})\, x^{-1} \leqq R \cap L_{2m+1}.$$

Hence

$$Q_0 \leqq Q_2 \leqq Q_4 \leqq \cdots \quad \text{and} \quad Q_1 \leqq Q_3 \leqq Q_5 < \cdots$$

Now J, being a principal ideal domain, satisfies the maximal condition on ideals and consequently JN satisfies the maximal condition on right ideals (Corollary to Theorem 5.35). Hence there exists an integer $k_0 \geqq 0$ such that $Q_k = Q_{k+2}$ for all $k \geqq k_0$. This means that each factor R_i/R_{i-1} is isomorphic as a J-module with one of the modules

$$JN/Q_1, \ldots, JN/Q_{k_0+1}.$$

By induction hypothesis there is a finite set of primes π of J such that each $R_i/R_{i-1} \in \mathfrak{A}(J, \pi)$. Lemma 9.52 now shows that $JG/R \in \mathfrak{A}(J, \pi)$. □

Corollary 1. Let J be a principal ideal domain, let G be a polycyclic-by-finite group and let M be a finitely generated JG-module. Then $M \in \mathfrak{A}(J, \pi)$ where π is a finite set of primes of J.

Proof. Let $a_1, \ldots, a_r$ generate the JG-module M, and let $M_0 = 0$ and

$$M_i = a_1(JG) + \cdots + a_i(JG).$$

Then $0 = M_0 \leqq M_1 \leqq \cdots \leqq M_r = M$ is a series of JG-submodules of M. Clearly there are JG-isomorphisms

$$M_i/M_{i-1} \simeq a_i(JG)/a_i(JG) \cap M_{i-1} \simeq JG/R_i$$

where R_i is some right ideal of JG. It follows from the lemma that there is a finite set of primes π of J such that each M_i/M_{i-1} belongs to $\mathfrak{A}(J, \pi)$. Hence $M \in \mathfrak{A}(J, \pi)$ by Lemma 9.52. ☐

In applications J will be either the ring of integers or the group algebra

$$J = F\langle t\rangle$$

of an infinite cyclic group $\langle t\rangle$ over a field F. Notice that in both cases a complete set of primes for J is infinite. When $J = F\langle t\rangle$, this is because there are infinitely many monic irreducible polynomials over F: for F infinite this shown by the $t - a$ with $a \in F$, and if F is finite it follows from well-known facts about finite fields. These observations, combined with Lemma 9.52 and Corollary 1 above, yield

Corollary 2. Let J be either the ring of integers or the group algebra of an infinite cyclic group over a field. If G is a polycyclic-by-finite group, a finitely generated JG-module cannot contain a J-submodule which is isomorphic with the field of fractions of J.

We shall now explain how the case $J = F\langle t\rangle$ arises.

Let G be a polycyclic-by-finite group and let z be an element of the centre of G. Let F be an arbitrary field and M any FG-module. Our object is to turn M into a JG-module via the mapping $t \rightarrow z$. More precisely define

$$af = af(z)$$

where $a \in M$ and $f \in J = F\langle t\rangle$. Thus M becomes a J-module, and since z commutes with every element of G, we can make M into a JG-module via the rule

$$a\Bigl(\sum_{g\in G} r_g g\Bigr) = \sum_{g\in G} (ar_g)\, g$$

where $r_g \in J$. If M is finitely generated as an FG-module, it is certainly finitely generated as a JG-module.

Hence we may apply Corollaries 1 and 2 to Lemma 9.53 and obtain

Lemma 9.54. Let F be a field and let J be the group algebra over F of an infinite cyclic group $\langle t\rangle$. Let G be a polycyclic-by-finite group and choose an element z from the centre of G. Suppose that M is a finitely generated FG-module and that M is made into a JG-module by means of the mapping $t \to z$; then $M \in \mathfrak{A}(J, \pi)$ for some finite set π of primes in J. No J-submodule of M can be isomorphic with the field of fractions of J.

Irreducible Polycyclic-by-Finite Groups of Automorphisms

Let A be an irreducible group of automorphisms of an abelian group G. We have seen that if A is locally finite, G must be an elementary abelian p-group (Lemma 5.26), and that if A is finitely generated, G cannot be the direct product of finitely many copies of the additive group of rational numbers (Lemma 9.39.1).

We shall now discuss the nature of G under the stringent requirement that A be polycyclic-by-finite.

Theorem 9.55. Let A be an irreducible group of automorphisms of an abelian group G.

(i) If A is polycyclic-by-finite, G is an elementary abelian p-group for some prime p.

(ii) If A is finitely generated and nilpotent-by-finite, G is a finite elementary abelian p-group for some prime p.

Proof. Let A be polycyclic-by-finite. G is a ZA-module (where Z is the ring of rational integers) and $G = g^A$ if $1 \neq g \in G$. Thus G is a cyclic ZA-module. By Corollary 2 to Lemma 9.53, the group G cannot contain a copy of the additive group of rational numbers, so it must be an elementary abelian p-group.

Now assume that A is a finitely generated nilpotent-by-finite group. Then A is polycyclic-by-finite, so G is an elementary abelian p-group. It will therefore be sufficient if we can establish the following result.

Theorem 9.56 (P. Hall [10], Theorem 3.1). If G is a finitely generated nilpotent-by-finite group and if F is an absolutely algebraic field of prime characteristic p, then every irreducible FG-module M has finite dimension over F.

Proof. Evidently there is nothing to be lost in assuming that M is a faithful FG-module. For the present let G be a finitely generated nilpotent group. If the centre of G is periodic, then G is finite by Theorem 2.24; since $M = a(FG)$ for any non-zero a in M, this implies that M has finite dimension over F. We may therefore suppose that there is an element z of infinite order in the centre of G.

Let J be the group algebra of an infinite cyclic group $\langle t \rangle$ over F. Clearly M is a cyclic FG-module and it can be regarded as a JG-module as explained above. Lemma 9.54 assures us that M does not contain a J-submodule which is J-isomorphic with K, the field of fractions of J.

Since M is a faithful FG-module, we can regard G as a group of linear transformations of M. Denote by C the ring of all linear transformations of M which commute with every element of G. Then, by Schur's Lemma, C is a division algebra and its centre D is a field. Clearly $z \in D$. For convenience we will identify y in F with $y1$ in D, so that F is a subfield of D. Let L be the subring of D generated by F and z. Then the mapping $t \to z$ determines a ring homomorphism of J onto L, and $L \simeq J/J_0$ for a suitable ideal J_0 of J. Since L is an integral domain, J_0 is either 0 or a maximal ideal of J. In the former event $J \simeq L$ and this isomorphism can be extended in a canonical manner to a monomorphism $\alpha: K \to D$. Let $0 \neq a \in M$: since $f\alpha = f(z)$ for any $f \in K$ and since $K\alpha$ is a field, the mapping $x \to a(x\alpha)$ is a J-isomorphism of K with $a(K\alpha)$, and the latter is a J-submodule of M; however this is impossible.

Hence J_0 is a maximal ideal of J, and L, being ring isomorphic with the field J/J_0, is absolutely algebraic of characteristic p. But the multiplicative group of such a field is periodic; hence z has finite order, which is not the case.

Finally, let G be a finitely generated nilpotent-by-finite group; then there is an $N \lhd G$ such G/N is finite and N is a finitely generated nilpotent group. Let $\{t_1, \ldots, t_r\}$ be a transversal to N in G. Now M is an irreducible FG-module, so $M = a(FG)$ for any $0 \neq a \in M$. Hence

$$M = at_1(FN) + \cdots + at_r(FN)$$

and M is finitely generated as an FN-module. M must therefore contain a maximal FN-submodule, say S. Now M/S is an irreducible FN-module, so it has finite dimension over F, by the first part of the proof. If $g \in G$, then Sg is an FN-submodule of G since $N \lhd G$; also M/Sg is F-isomorphic with M/S. Thus M/Sg has finite dimension over F. Let

$$I = \bigcap_{i=1}^{r} St_i;$$

then M/I has finite dimension over F. Also, if $g \in G$, then $St_i g = St_{i'}$ where $i \to i'$ is a permutation of $\{1, \ldots, r\}$. Hence

$$Ig = \bigcap_{i=1}^{r} St_{i'} = \bigcap_{i=1}^{r} St_i = I,$$

and I is a proper FG-submodule of M. Since M is irreducible, $I = 0$ and M has finite dimension over F. $\square$

It is an open question whether Theorem 9.56 is true if G is polycyclic-by-finite. Positive answers have recently been announced by Levič [5] and Zalesskiĭ [4], but J. E. Roseblade has pointed out flaws in both papers.

We remark that it is not difficult to show that *if G is a finitely generated abelian-by-finite group and M is an irreducible FG-module, F being an arbitrary field, then M has finite dimension over F* (P. Hall [10], Thorem 3.2). On the other hand, *if G is a polycyclic group which is not abelian-by-finite and if F is a field which is not absolutely algebraic of characteristic a prime, there exists an irreducible FG-module of infinite dimension over F* (P. Hall [10], Theorem 3.3).

We can now prove the principal theorem of this section.

Proof of Theorem 9.51. It is to be proved that a finitely generated abelian-by-nilpotent group is residually finite. This will be done by means of a series of reductions.

(a) It is enough to show that a finitely generated monolithic abelian-by-nilpotent group is finite. Suppose that G is a finitely generated abelian-by-nilpotent group and let $1 \neq x \in G$. By Zorn's Lemma there is normal subgroup N of G which is maximal subject to $x \notin N$. A normal subgroup of G which properly contains N must contain x; hence G/N is monolithic. If the assertion above has been proved, then G/N is finite and G is residually finite as required.

(b) From now on we shall assume that G is a finitely generated monolithic abelian-by-nilpotent group and that M is the monolith of G. Since G is soluble, M is abelian. Also M lies in each non-trivial normal abelian subgroup of G, so M centralizes every normal abelian subgroup of G. Since G is abelian-by-nilpotent, it follows that there is a normal abelian subgroup A such that $M \leqq A$ and G/A is nilpotent. What is more, we can choose A to be maximal with respect to these properties. Now let $C = C_G(M)$. Since $A \leqq C$, the group G/C is finitely generated and nilpotent. G/C is an irreducible group of automorphisms of M, so M is a finite elementary abelian p-group for some prime p, by Theorem 9.55.

(c) We can assume that M lies in the centre of G and so has order p. For suppose that in this case the result stated in (a) has been proved. Let $1 \neq a \in M$ and choose K maximal subject to $K \lhd C$ and $a \notin K$. As before C/K is monolithic, and indeed its monolith is $\langle aK \rangle$ since $\langle a \rangle \lhd C$. Now $\langle aK \rangle$ lies in the centre of C/K; also $|G:C|$ is finite because M is finite, and therefore C is finitely generated. Consequently C/K is finite, so $|G:K|$ is finite. Since $a \notin K$, we have $a \notin \mathrm{Core}_G K$. But a belongs to the monolith of G, so $\mathrm{Core}_G K$ must be trivial and G is finite.

(d) Under the hypothesis of (c) there is a positive integer m such that $A^{p^m} = 1$. By Theorem 5.34 the group G satisfies Max-n, so A satisfies the maximal condition on characteristic subgroups. Let A_i be the subgroup of all a in A for which $a^{i!} = 1$; then $A_1 \leqq A_2 \leqq \cdots$ and A_i is

characteristic in A. Hence $A_i = A_{i+1} =$ etc. for some i and clearly $T = A_i$ is the torsion-subgroup of A. Thus $T^n = 1$ where $n = i!$. By a well-known result in the theory of abelian groups (see Kaplansky [1], Theorem 8, p. 18) T is a direct factor of A and we can write $A = T \times B$ where B is torsion-free. Obviously $A^n = B^n$, so A^n is torsion-free. Thus $A^n \cap M = 1$, because M is finite. Therefore $A^n = 1$. Let q be a prime other than p and denote by Q the q-component of A. Then $Q \lhd G$ and $Q \cap M = 1$, so $Q = 1$. Hence A is a p-group of finite exponent.

(e) Conclusion. Let $H = G/A$. We shall write A additively and regard it as a ZH-module. Let h be a non-trivial element of $\zeta(H)$ and let K_i be the kernel of the endomorphism $a \to a(h-1)^i$ of A. By hypothesis $[M, G] = 1$, so $M \leqq K_1$. Now $K_i \lhd G$ because $a(h-1)^i x = ax(h-1)^i$ if $x \in H$; also it is clear that $K_1 \leqq K_2 \leqq \cdots$. But G has Max-n, so $K_r = K_{r+1} =$ etc. for some $r > 0$. Let $A_r = A(h-1)^r$ and suppose that $A_r \neq 0$. Now $A_r \lhd G$ since $a(h-1)^r x = ax(h-1)^r$ if $x \in H$; thus $M \leqq A_r$. Let $0 \neq a \in M$; then there is an element b in A such that $a = b(h-1)^r$. Now $b(h-1)^{r+1} = a(h-1) = 0$ because $a \in \zeta(G)$. Hence $b \in K_{r+1} = K_r$ and $a = b(h-1)^r = 0$. It follows that $A_r = A(h-1)^r = 0$. Also $p^m A = 0$, so we conclude that h^{p^t} centralizes A where $t = m(r-1)$. Let $h = gA$; since $[G, g] \leqq A$, we have $\langle g^{p^t}, A\rangle \lhd G$; also $\langle g^{t^t}, A\rangle$ is abelian. By maximality of A it follows that $g^{p^t} \in A$ and h has finite order. We have proved that the centre of H is periodic; it follows from Theorem 2.24 that H is finite. Now $H = G/A$ and G is finitely generated; therefore A is finitely generated and consequently finite. Finally G is finite. This completes the proof of Theorem 9.51. ☐

Whether Theorem 9.51 is valid for finitely generated abelian-by-polycyclic groups is unknown.

Recently Remeslennikov [2] has proved a result which has a bearing on the theorem of P. Hall: *a finitely generated torsion-free metabelian group has for each prime p a normal residually finite-p subgroup of finite index*.

As a further application of these methods we will prove.

Theorem 9.57 (P. Hall [10], Theorem 5.2 and 6.1). Let G be a group which has a normal subgroup N of finite index such that N is an extension of a hypercentral group by a finitely generated nilpotent group. Then every chief factor of G is finite and every maximal subgroup of G has finite index.

Proof. By hypothesis there exist subgroups L and N such that $L \lhd N \lhd G$ and such that G/N is finite, N/L is finitely generated and nilpotent and L is hypercentral. Replacing L by its core in G if necessary, we can assume that $L \lhd G$.

Suppose that M is an infinite minimal normal subgroup of G. If $M \cap N = 1$, then $M \simeq MN/N$, which is finite. Hence $M \leqq N$. If $M \cap L = 1$, then M is G-isomorphic with ML/L. By Lemma 2.16 the subgroup ML/L must intersect the centre of N/L non-trivially, so it is contained in the centre of N/L and is abelian. Thus we have a representation of the finite group G/N as an irreducible group of automorphisms of ML/L. Clearly ML/L must be finitely generated and hence finite; therefore M is finite. Consequently $M \leqq L$ and so $M \leqq \zeta(L)$. It follows that $C_G(M) \geqq L$ and G/L is represented by an irreducible group of automorphisms of the abelian group M. Theorem 9.55 shows that M is finite.

It is now clear that every chief factor of G is finite; the second part of the theorem follows from Lemma 9.39.2. □

A Counterexample

By Theorems 9.51 and 9.57 a finitely generated metabelian group is residually finite, has finite chief factors and has its maximal subgroups of finite index. None of these statements holds for finitely generated soluble groups of derived length 3.

Theorem 9.58 (P. Hall [10], Theorem 2). There exists a 3-generator torsion-free soluble group G satisfying Max-n, with derived length 3, which has the following properties: (i) G has a minimal normal subgroup isomorphic with the direct product of a countable infinity of copies of the additive group of rational numbers, (ii) G has a maximal subgroup of infinite index, (iii) G is not residually finite.

Proof. Let V be a vector space of dimension $\aleph_0$ over the field of rational numbers Q and let $\{v_i : i = 0, \pm 1, \pm 2, \ldots\}$ be a basis for V. For each integer i we select a prime number p_i in such a way that $p_i \neq p_j$ if $i \neq j$, and every prime occurs among the p_i.

Let ξ and η be the linear transformations of V defined by

$$v_i\xi = v_{i+1} \quad \text{and} \quad v_i\eta = p_i v_i, \quad (i = 0, \pm 1, \pm 2, \ldots).$$

Clearly ξ and η are non-singular. Let

$$H = \langle \xi, \eta \rangle.$$

Writing $\eta^{\xi^j} = \eta_j$, we see that

$$v_i\eta_j = p_{i-j}v_i. \tag{11}$$

Hence the η_j commute with each other and η^H is a normal abelian subgroup of H. Therefore J is metabelian: indeed it is easy to show that H is isomorphic with the standard wreath product of two infinite cyclic groups.

Suppose that W is a non-zero H-admissible additive subgroup of V and let $0 \neq w \in W$. We can write

$$w = r_1 v_{i_1} + \cdots + r_k v_{i_k}$$

where the r_j are non-zero rational numbers and the integers $i_1, \ldots, i_k$ are distinct. Furthermore we shall assume that w has been chosen so that k is minimal. Suppose that $k > 1$. Then W contains the element

$$p_{i_k} w - w\eta = (p_{i_k} - p_{i_1}) r_1 v_{i_1} + \cdots + (p_{i_k} - p_{i_{k-1}}) r_{k-1} v_{i_{k-1}}.$$

Since the p_j are all distinct, this contradicts the minimality of k. Hence $k = 1$ and $r_1 v_{i_1} \in W$. Therefore, on applying ξ^{-i_1}, we find that W contains the element $r_1 v_0$.

Let p be any prime and let n be any integer. By construction $p = p_i$ for some integer i. By (11) the subgroup W contains

$$(r_1 v_0) \eta^n_{-i} = r_1 p_i^n v_0 = r_1 p^n v_0.$$

It follows that W contains every rational multiple of v_0. Since $v_0 \xi^i = v_i$, we deduce that W contains every rational multiple of v_i; hence $W = V$.

Now let A be the additive group of V—so that H becomes a finitely generated irreducible group of automorphisms of A (cf. Theorem 9.55)—and let G be the holomorph of A by H. Then

$$G = \langle v_0, \xi, \eta \rangle$$

and $G^{(3)} = 1$. Clearly A is a minimal normal subgroup of G; since $C_G(A) = A$, the subgroup A is actually the monolith of G. Now $G/A \simeq H$, a finitely generated metabelian group, so G/A is residually finite, by Theorem 9.51, and A is the finite residual of G. Suppose that $H < K \leq G$; then K must contain a non-trivial element of A, say a. Therefore K contains $a^H = A$ and $K = G$; hence H is maximal in G, and of course $|G : H|$ is infinite. Finally A satisfies Max-G, as does G/A by Theorem 5.34; hence G satisfies Max-n. □

Chapter 10

Some Topics in the Theory of Infinite Soluble Groups

10.1 The Unimodularity of Polycyclic Groups

Throughout this section Z will denote the ring of rational integers. By Theorem 3.27 a soluble subgroup of the unimodular group $GL(n, Z)$ is polycyclic, a result first found by Mal'cev. This fact led P. Hall to ask whether an arbitrary polycyclic group has a faithful representation as a subgroup of $GL(n, Z)$ for a suitable n. Partial solutions to this problem were obtained by Čarin ([3], Theorem 6), P. Hall ([6], p. 58), Jennings ([2], Theorem 8.1), Learner ([1], Theorem 3) and Wang [1]. Finally in 1967, a positive solution was obtained by L. Auslander [1], whose proof involved methods from the theory of Lie groups. We present here a purely algebraic proof of Hall's conjecture due to Swan [1].

Theorem 10.11. An arbitrary polycyclic-by-finite group has a faithful representation as a group of $n \times n$ unimodular matrices for some positive integer n.

We shall establish three preliminary results

Lemma 10.12 (Cartier [1], Exposé 8, § 1). Let G be a finitely generated group and let I be a (two-sided) ideal of the integral group ring ZG. Suppose that ZG/I is finitely generated as an additive abelian group. Then I is finitely generated as an ideal of ZG and for each positive integer i the additive abelian group ZG/I^i is finitely generated.

Proof (A. Learner). Let M be the additive subgroup of ZG generated by the generators of G and their inverses together with coset representatives of the generators of ZG/I. Then M is finitely generated as an additive subgroup and

$$ZG = M + I. \tag{1}$$

The subgroup M^2, generated by all the products xy where $x, y \in M$, is finitely generated, and so, therefore, is the subgroup

$$V = (M + M^2) \cap I.$$

Let J denote the ideal generated by V. Then $J \leqq I$ and J is finitely generated as an ideal of ZG. By (1)

$$M + M^2 = (M + M^2) \cap (M + I) = M + V,$$

so that $M^2 \leqq M + J$. It follows that $M + J$ is a subring containing G, by definition of M. Therefore $ZG = M + J$; hence $ZG/J \simeq M/M \cap J$, which shows that ZG/J is finitely generated as an additive group. Therefore I/J is a finitely generated additive group and I is finitely generated as an ideal of ZG. Hence I/I^2 is finitely generated as a (ZG/I)-bimodule; since ZG/I is finitely generated as an additive group, I/I^2 is too. Hence ZG/I^2 is a finitely generated additive group. This argument shows that ZG/I^{2^n} is finitely generated as an additive group for each $n \geqq 0$. Since $I^i \geqq I^{2^n}$ for a suitable n, the additive group ZG/I^i is finitely generated. □

Lemma 10.13. Let $N \leqq H < G$ where $N \lhd G$. Assume that G/N is abelian and G/H is infinite cyclic, and that H is finitely generated. If H has a faithful representation as a subgroup of $GL(n, Z)$ in which the elements of N are represented by unitriangular matrices, then G has a faithful representation as a subgroup of $GL(m, Z)$, for some integer m, with the same property.

Proof. Let ϱ be the given representation of H. The group monomorphism $\varrho\colon H \to GL(n, Z)$ can be extended in a natural way to a ring homomorphism (also denoted by ϱ) of ZH into $M(n, Z)$, the ring of all $n \times n$ matrices over Z. Let K be the kernel of this ring homomorphism; then K is an ideal of ZH and ZH/K is a finitely generated torsion-free additive abelian group; for it is isomorphic with a subgroup of $M(n, Z)$, a free abelian group of rank n^2.

Let I be the right ideal of ZH generated by all $a - 1$ where $a \in N$. If $a \in N$ and $h \in H$, we have

$$h(a - 1) = (a^{h^{-1}} - 1)\, h,$$

and since $N \lhd G$, it follows that I is actually a two-sided ideal of ZH. Now I^n is generated as a right ideal of ZH by all products of the form

$$(a_1 - 1) \cdots (a_n - 1), \quad (a_i \in N),$$

and, since N^ϱ is a group of unitriangular $n \times n$ matrices, the image under ϱ of such a product is 0. Therefore

$$I^n \leqq K.$$

Let

$$J_1 = (I + K)^n. \tag{2}$$

Then $J_1 \leqq I^n + K = K$, since K is an ideal of ZH. Let J/J_1 be the torsion-subgroup of the additive abelian group ZH/J_1. Then J is an ideal of ZH. Now $ZH/I + K$ is a finitely generated additive group since it is a homomorphic image of ZH/K; also H is finitely generated by hypothesis. Therefore ZH/J_1 is a finitely generated additive group, by Lemma 10.12. Hence ZH/J is a free abelian group of finite rank, say l. We allow the elements of H to act on ZH/J by right multiplication and in this way we obtain a representation σ of H as a subgroup of $GL(l, Z)$. Since ϱ is faithful, H acts faithfully on ZH/K by right multiplication. But $J \leqq K$ because ZH/K is torsion-free, so σ is faithful. Moreover $I^n \leqq J_1 \leqq J$, so σ, when restricted to N, represents that group by a group of unitriangular $m \times m$ matrices.

Since G/H is infinite cyclic, there is an element x such that $G = \langle x, H \rangle$ and $\langle x \rangle \cap H = 1$; thus G is a semi-direct product of H with the infinite cyclic group $\langle x \rangle$. We shall make ZH/J into a $Z\langle x \rangle$-module. Let ξ be the ring automorphism of ZH defined by

$$\left(\sum_{i=1}^{k} n_i h_i\right)\xi = \sum_{i=1}^{k} n_i h_i^x .$$

Then

$$h\xi = h^x = h([h, x] - 1) + h \equiv h \bmod I ,$$

since $[h, x] \in G' \leqq N$. Hence $(ZH)(\xi - 1) \leqq I$. Since $J_1 = (I + K)^n$, it follows that $J_1\xi = J_1$ and $J\xi = J$. Hence the mapping $\bar{\xi}$ in which $c + J \to c\xi + J$ is an automorphism of ZH/J. Now

$$(((a)\,\xi^{-1})\,h)\,\xi = ah^x .$$

Thus if we allow x to act on ZH/J in the same way as $\bar{\xi}$, then ZH/J will become a ZG-module. To obtain a faithful ZG-module we form the direct sum of ZH/J with a ZG-module which is trivial as a ZH-module and which affords the representation of $\langle x \rangle$

$$x \to \begin{pmatrix} 1 & 1 \\ \cdot & 1 \end{pmatrix} \qquad (3)$$ □

Theorem 10.14 (P. Hall [6], p. 56, Theorem 7.5). A finitely generated torsion-free nilpotent group G has a faithful representation as a group of unitriangular matrices with integral coefficients.

Proof. Let I_K denote the augmentation ideal of ZK for an arbitrary group K. Thus I_K is the additive subgroup of ZK generated by all the $k - 1$ where $k \in K$; since

$$k_1(k-1) = (k_1 - 1)(k-1) + (k-1) \quad \text{and}$$
$$(k-1)\,k_1 = (k-1)\,(k_1 - 1) + (k-1),$$

I_K is an ideal of ZK.

G has a central series of finite length with infinite cyclic factors. By induction on the length of this series we may suppose that there is a normal subgroup H such that G/H is infinite cyclic and H has a faithful representation as a group of $n \times n$ unitriangular matrices over Z. Let $G = \langle x, H\rangle$, so that $\langle x\rangle \cap H = 1$. We shall apply Lemma 10.13 and its proof with $N = H$. Then ZH becomes a ZG-module providing a representation σ such that $(a)\, h^\sigma = ah$ and $(a)\, x^\sigma = a^x$ where $a \in ZH$ and $h \in H$: moreover, there is a ZG-submodule J such that ZH/J is a free abelian group of finite rank furnishing a faithful representation of H by a group of unitriangular matrices over Z. Thus $I_H^n \leqq J$ for some integer $n > 0$. We need only prove that

$$(ZH)\,(I_G^r)^\sigma \leqq J$$

for some integer $r \geqq 0$. Thus it will be sufficient to obtain for each positive integer m a positive integer k such that

$$(ZH)\,(I_G^k)^\sigma \leqq I_H^m.$$

We shall prove this result by induction on the nilpotent class c of G. If $c \leqq 1$, then G is abelian and x acts trivially on ZH. Thus $(ZH)\,(I_G^m)^\sigma = I_H^m$. Let $c > 1$ and write $A = \gamma_c(G)$. Then $A \leqq H$. By induction hypothesis on c we have

$$(Z(H/A))\,(I_{G/A}^l)^\sigma \leqq I_{H/A}^m$$

for some $l > 0$: here, of course, σ indicates that, for example, xA acts on $Z(H/A)$ by transformation. The natural homomorphism of H onto H/A determines a homomorphism of ZH onto $Z(H/A)$ with kernel $(ZH)I_A$. Therefore

$$(ZH)\,(I_G^l)^\sigma \leqq I_H^m + (ZH)\,I_A. \tag{4}$$

Suppose that

$$(ZH)\,(I_G^{lr})^\sigma \leqq I_H^m + (ZH)\,I_A^r. \tag{5}$$

Then

$$\begin{aligned}(ZH)\,(I_G^{l(r+1)})^\sigma &\leqq (I_H^m)\,(I_G^l)^\sigma + ((ZH)\,I_A^r)\,(I_G^l)^\sigma\\ &\leqq I_H^m + (ZH)\,(I_A^r I_G^l)^\sigma\\ &= I_H^m + (ZH)(I_G^l I_A^r)^\sigma \end{aligned}\tag{6}$$

since I_H is a ZG-submodule of ZH and $[A, G] = 1$. From (4) and (6) we obtain

$$\begin{aligned}(ZH)\,(I_G^{l(r+1)})^\sigma &\leqq I_H^m + (I_H^m + (ZH)\,I_A)\,I_A^r\\ &= I_H^m + (ZH)\,I_A^{r+1}.\end{aligned}$$

Thus (5) is established for all integers $r \geqq 1$. Setting $r = m$ in (5) we obtain $(ZH)\,(I_G^{lm})^\sigma \leqq I_H^m + (ZH)\,I_A^m = I_H^m$ since $A \leqq H$. □

Proof of Theorem 10.11. Let G be a polycyclic-by-finite group. Every polycyclic group is nilpotent-by-abelian-by-finite (Theorem 3.25) and has a torsion-free normal subgroup with finite index (Corollary to Theorem 9.39.3). Hence there is a normal subgroup H with finite index such that H is an extension of a torsion-free nilpotent group by a free abelian group. Suppose that M is a right ZH-module; we form the so-called *induced ZG-module*

$$T = M \otimes_{ZH} (ZG):$$

here G acts on T according to the rule

$$(m \otimes a)\,g = m \otimes (ag),$$

where $m \in M$, $a \in ZG$ and $g \in G$. If M is free abelian of finite rank r as an additive group, then T is free abelian of rank ri where $i = |G: H|$. Moreover, if M yields a faithful representation of H, then T yields a faithful representation of G (see Curtis and Reiner [1], § 12 D).

Hence we may assume that G contains a normal torsion-free nilpotent subgroup N such that G/N is free abelian and finitely generated. By Lemma 10.14 the subgroup N has a faithful representation as a group of unitriangular matrices over Z. Now use Lemma 10.13 and induction on the rank of G/N to obtain the result. □

We remark that *the holomorph of a finitely generated nilpotent group has a faithful representation as a subgroup of* $GL(n, Z)$ *for a suitable positive integer* n (P. Hall [6], p. 58; Learner [1], Lemma 3.4; see also Auslander and Baumslag [1]). This may be established by the methods of Lemma 10.13.

Recently Kopytov and Merzljakov have obtained necessary and sufficient conditions for the linearity of soluble groups. For example, Kopytov [1] has proved that *a soluble group with finite rank which has a torsion-free subgroup of finite index may be faithfully represented as a subgroup of some* $GL(n, Q)$; here Q is the field of rational numbers. Thus by Theorem 9.39.3 any residually finite $\mathfrak{S}_1$-group has such a representation. Merzljakov has proved that *a soluble group G with finite rank has a faithful representation over a field of characteristic* 0 *if and only if there is a normal torsion-free nilpotent subgroup N such that G/N is abelian-by-finite* (Merzljakov [5], Theorem 10). For further results we refer the reader to the papers of Kopytov and Merzljakov cited and also to Smirnov [9], Levič [4] and Remeslennikov [3].

10.2 Theorems of Černikov, Muhammedžan and Blackburn on Černikov p-Groups

A Černikov p-group is an extension of a direct product of finitely many p^∞-groups by a finite p-group. Among the non-nilpotent soluble p-groups, the Černikov p-groups may be considered those closest to finite p-groups. We recall that a Černikov p-group is hypercentral (Corollary 2 to Theorem 5.27) and note that the hypercentral length is less than ω^2. Černikov p-groups can be characterized as the locally finite p-groups satisfying Min-ab or Min-n, and as the locally finite p-groups with finite abelian subgroup rank (see Sections 3.3 and 5.2).

Much of the theory of Černikov p-groups was developed by Černikov himself; further contributions to the subject were made by Muhammedžan, and also by Blackburn whose paper [4] collects many earlier results. We present some criteria for a group to be a Černikov p-group drawn from Blackburn's paper.

Lemma 10.21 (Černikov [9], Theorem 3). Let G be a p-group and suppose that G is abelian-by-finite. Then G is a Černikov group if and only if its centre satisfies Min.

Proof. We assume that $\zeta(G)$ satisfies Min. By hypothesis there is a normal abelian subgroup A with finite index in G. Let B denote the subgroup of all b in A such that $b^p = 1$; then $B \lhd G$. To prove that G is a Černikov group it suffices to show that A satisfies Min, and this will certainly be the case if B is finite.

Let $\{t_1, \dots, t_n\}$ be a transversal to A in G and let

$$H = \langle t_1, \dots, t_n, B\rangle .$$

Then H/B is finite since it is finitely generated and G is locally finite. Now B is elementary abelian, so by Lemma 6.34 the group H is nilpotent. It will therefore be sufficient to show that $B_i = B \cap \zeta_i(H)$ is finite for each integer $i \geqq 0$. Clearly $G = AH$. Now $B_1 = B \cap \zeta(H)$ centralizes H and also A since $B \leqq A$ and A is abelian. Therefore $B_1 \leqq \zeta(G)$ and B_1 satisfies Min; but B is elementary abelian, so B_1 is finite. Let us assume that B_i is finite where $i > 0$. The mapping $a \to [a, t_j]\, B_{i-1}$ is a homomorphism of B_{i+1} into B_i/B_{i-1} with kernel $K_j = C_{B_{i+1}}(t_j B_{i-1})$. Hence B_{i+1}/K_j is finite. But

$$K_1 \cap \cdots \cap K_n \leqq C_B(H/B_{i-1}) \leqq B_i, \tag{7}$$

since A is abelian; (7) shows that B_{i+1}/B_i is finite, and consequently B_{i+1} is finite. It follows by induction on i that each B_i is finite as required. The converse is obvious. ☐

Theorem 10.22 (Blackburn [4], Theorem 3.1). Let G be a p-group and let N be a normal subgroup of G contained in the hypercentre of G. Then N is a Černikov group if and only if both $\zeta(G) \cap N$ and $G/C_G(N)$ are Černikov groups.

Proof. Since a periodic group of automorphisms of a Černikov group is a Černikov group (Theorem 3.29), the conditions on N are necessary if it is to be a Černikov group. Conversely, let us suppose that these conditions are satisfied. Let

$$C = C_G(N).$$

Since CN/C is a Černikov group and $C \cap N = \zeta(N)$, it is sufficient to prove that $\zeta(N)$ is a Černikov group. We observe that the conditions on N are inherited by any normal subgroup of G that is contained in N, for example by the subgroup of all a in $\zeta(N)$ such that $a^p = 1$. Consequently we can assume that N is an elementary abelian p-group.

Let R/C be the finite residual of G/C. Since G/C is a Černikov group, G/R is finite and R/C is a radicable abelian group. Let $N_\alpha = N \cap \zeta_\alpha(G)$, so that $N = N_\beta$ for some β. Suppose that $[N_\alpha, R] = 1$. Now $[N_{\alpha+1}, R] \leq N_\alpha$, so $[N_{\alpha+1}, R, R] = 1$ and the mapping $n \otimes (rC) \to [n, r]$ determines a homomorphism of $T = N_{\alpha+1} \otimes (R/C)$ onto $[N_{\alpha+1}, R]$. However T is trivial because R/C is radicable, so $[N_{\alpha+1}, R] = 1$; thus $[N, R] = 1$. Hence $R = C$ and G/C is finite. Let K be the semi-direct product of N by G/C. Then K is abelian-by-finite. Now $\zeta(K) \cap N = \zeta(G) \cap N$, which satisfies Min and hence is finite. Also $\zeta(K)\, N/N \leqq K/N \simeq G/C$, so $\zeta(K)\, N/N$ is finite. Therefore $\zeta(K)$ is finite. Lemma 10.21 implies that K, and therefore N, is a Černikov group. ∎

Corollary (Černikov [18]). A periodic hypercentral group G is a Černikov group if and only if one of its maximal abelian subgroups satisfies Min.

Proof. Only the sufficiency of the condition is in doubt. Let A be a maximal abelian subgroup of G and let A satisfy Min. Now G is the direct product of its p-components G_p and hence

$$A = \underset{p}{Dr}\, (A \cap G_p).$$

Thus $A \cap G_p$ is a maximal abelian subgroup of G_p. Since A satisfies Min, so does each $A \cap G_p$ and, moreover, only finitely many of these are non-trivial; hence only finitely many of the G_p are non-trivial. We may therefore assume that G is a p-group. Let N be a maximal *normal* abelian subgroup of G and write $H = AN$. Since A is a maximal abelian subgroup of H, we have $\zeta(H) \leqq A$. Hence $\zeta(H) \cap N$ is a Černikov group. Also $H/N \simeq A/A \cap N$, so H/N is a Černikov group. By Lemma 2.19.1

we have $N = C_G(N)$; since $N \leqq H$, this implies that $N = C_H(N)$ and $H/C_H(N)$ is a Černikov group. Theorem 10.22 now shows that N is a Černikov group. Finally G/N is essentially a periodic group of automorphisms of N, so G/N is a Černikov group by Theorem 3.29, and G is a Černikov group by the Corollary to Theorem 3.12. ▯

We can now prove the main result of this section.

Theorem 10.23 (Blackburn [4], Theorem 3.2). Let N be a normal subgroup of a periodic hypercentral group G. Then the following assertions about N are equivalent.

(i) N is a Černikov group.

(ii) $N \cap \zeta_i(G)$ satisfies Min for each non-negative integer i.

(iii) If F is a normal finite elementary abelian p-subgroup of G contained in N, then $(N/F) \cap \zeta(G/F)$ satisfies Min.

Proof. It is obvious that (i) implies (ii). Suppose that N satisfies (ii) and let F be a subgroup of the type described in (iii). If F has order p^m, then $F \leqq \zeta_m(G)$ by Lemma 2.16. Hence

$$(N/F) \cap \zeta(G/F) \leqq (N \cap \zeta_{m+1}(G))/F,$$

which satisfies Min. Thus (iii) is valid.

Finally, suppose that N satisfies (iii). Taking $F = 1$ in (iii), we find that $N \cap \zeta(G)$ is a Černikov group. Hence by Lemma 2.16 the subgroup N has elements of only finitely many distinct prime orders. G is the direct product of its primary components and it is easy to see that we can assume that G is a p-group. By Zorn's Lemma there is subgroup M which is maximal subject to being abelian, normal in G and contained in N. Let L be the subgroup consisting of all a in M such that $a^p = 1$; then $L \lhd G$. It will be sufficient to prove that L is finite. For then M will be a Černikov group and, since $C_N(M) = M$ (by Lemma 2.19.1), N/M, and hence N, will be a Černikov group by Theorem 3.29.

We shall therefore assume that L is infinite. Let us write

$$Z_i = \zeta_i(G).$$

If $L \cap Z_j$ is finite, then $L \cap Z_{j+1}$ is finite; for $L \cap Z_j$ is elementary abelian and

$$(L \cap Z_{j+1})/(L \cap Z_j) \leqq (N/L \cap Z_j) \cap \zeta(G/L \cap Z_j),$$

which has Min by hypothesis. Since $L \cap Z_0 = 1$, we conclude that every $L \cap Z_i$ is finite.

Consider a chain of normal infinite subgroups of G contained in L, say $\{L_\lambda : \lambda \in \Lambda\}$, and denote by I the intersection of this chain. Let i be a fixed integer and let $\lambda \in \Lambda$. Then $L_\lambda \cap Z_i \leqq L \cap Z_i$, which is finite;

hence $L_\lambda \not\leq Z_i$ and consequently $L_\lambda Z_i/Z_i$ intersects Z_{i+1}/Z_i non-trivially. Therefore $L_\lambda \cap Z_i < L_\lambda \cap Z_{i+1}$. Since $L \cap Z_{i+1}$ is finite, we can choose $\lambda_0 \in \Lambda$ such that $L_{\lambda_0} \cap Z_{i+1}$ is minimal (i being fixed). Then

$$I \cap Z_{i+1} = L_{\lambda_0} \cap Z_{i+1} > L_{\lambda_0} \cap Z_i \geqq I \cap Z_i.$$

This is valid for each integer $i \geqq 0$, so I must be infinite. Zorn's Lemma can now be invoked to produce a normal subgroup A of G which is minimal with respect to being infinite and contained in L.

Let $C = C_G(A)$. If $C = G$, then $A \leqq L \cap Z_1$, which yields the contradiction that A is finite. Hence $C \neq G$ and, since G is hypercentral, there is a normal subgroup of G/C with order p, say B/C. Then $A \cap \zeta(B) < A$ since $B \not\leq C$. Also $A \cap \zeta(B) \lhd G$; thus $A \cap \zeta(B)$ is finite by minimality of A. Now $C \leqq B$, so $C_B(A) = C$; also B/C is finite, so Theorem 10.22 may be applied to show that A is finite. This contradiction completes the proof. □

Corollary 1 (Muhammedžan [3], Theorem 8). A periodic hypercentral group is a Černikov group if and only if each upper central factor with finite ordinal type satisfies Min.

This follows from Theorem 10.23 on setting $N = G$. Notice that the hypothesis of periodicity is essential in this corollary: for let G be the holomorph of a group of type p^∞ by the automorphism group generated by $a \to a^{p+1}$; then G is hypercentral and $\zeta_{i+1}(G)/\zeta_i(G)$ has order p for each integer i, but G is not a Černikov group. On the other hand we note

Corollary 2. A hypercentral group is a Černikov group if and only if each upper central factor satisfies Min.

We recall from Theorem 5.22 that in a hypercentral Černikov group every upper central factor with the possible exception of the first is finite.

Baumslag and Blackburn [1] have determined all the p-groups G such that $G/\zeta_\omega(G)$ is abelian and $\zeta_{i+1}(G)/\zeta_i(G)$ has order p for each integer $i \geqq 0$: such groups are Černikov p-groups by Corollary 1 above. Finally, we mention that Muhammedžan has proved the following analogue of Corollary 1. *A group G is a soluble Černikov group if and only if there is an ascending normal series $\{G_\alpha\}$ in G such that G_{i+1}/G_i satisfies Min and is a maximal normal abelian subgroup of G/G_i for each integer $i \geqq 0$* (Muhammedžan [3], Theorem 14).

10.3 Soluble Minimax Groups I

We shall employ the notation

$$\hat{\mathfrak{M}} \text{ and } \check{\mathfrak{M}}$$

for the classes of groups which satisfy

$$\text{Max} \quad \text{and} \quad \text{Min}$$

respectively: here the "$\wedge$" and "$\vee$" are intended to suggest "maximal" and "minimal" respectively.

A *minimax group* is a group G which has a *minimax series*, i.e. a series of finite length each of whose factors satisfies either Max or Min. So the class of minimax groups is

$$\boldsymbol{P}(\hat{\mathfrak{M}} \cup \check{\mathfrak{M}}).$$

The length of a shortest minimax series in a minimax group G is called the *minimax length* of G and is written

$$m(G).$$

It is evident that the class of minimax groups is $\boldsymbol{S}$, $\boldsymbol{H}$ and $\boldsymbol{P}$-closed. We shall be almost exclusively concerned with soluble minimax groups: it is worth noting that if it is true that a group satisfying Max is polycyclic-by-finite and a group satisfying Min is Černikov, then every minimax group is a finite extension of a soluble minimax group.

Soluble minimax groups were first studied by Baer [51], Robinson [6], [7] and Zaičev [4], [8]; the first author has called these groups *polyminimax groups*.

Since every polycyclic group and every Černikov group has finite rank, Lemma 1.44 shows that *a soluble minimax group has finite rank.* Since a periodic polycyclic group is finite, *a periodic soluble minimax group is a Černikov group.* Hence a soluble minimax group is an $\mathfrak{S}_1$-group in the sense of Section 9.3; by Theorem 3.25, therefore, every soluble minimax group is nilpotent-by-abelian-by-finite.

We note a characterization of soluble minimax groups due to Baer and Zaičev. A group is said to satisfy *the weak maximal condition* if, given an ascending chain of subgroups $H_1 < H_2 < \cdots$, only a finite number of the indices $|H_{\alpha+1} : H_\alpha|$ are infinite. *The weak minimal condition* is defined analogously. It is not hard to show that *for soluble groups the weak maximal condition, the weak minimal condition and "minimax" are equivalent properties* (Baer [51], Lemma 1.2; Zaičev [4], [8]). Zaičev has also proved that *a locally soluble group satisfying the weak minimal condition is a soluble minimax group* (Zaičev [6], Theorem 5).

Abelian Minimax Groups

Lemma 10.31. Let A be an abelian group.

(i) If A is a minimax group, its torsion-subgroup is a direct factor.

(ii) A is a minimax group if and only if there exists a finitely generated subgroup X such that A/X is a direct product of finitely many quasicyclic groups. Moreover, in this event the factor group A/X is uniquely determined up to isomorphism.

Proof. If A is a minimax group, its torsion-subgroup satisfies Min, and this implies that it is a direct factor of A—see Kaplansky [1], Theorem 8, p. 18. To prove the first part of (ii) it is enough, in view of the structure of abelian groups satisfying Min, to show that if A is a minimax group, then $A \in \hat{\mathfrak{M}}\check{\mathfrak{M}}$. By induction on $m(A)$ we may suppose that $A \in \check{\mathfrak{M}}\hat{\mathfrak{M}}\check{\mathfrak{M}}$. From (i) we obtain $A \in \hat{\mathfrak{M}}\check{\mathfrak{M}}\check{\mathfrak{M}}$, so $A \in \hat{\mathfrak{M}}\check{\mathfrak{M}}$ as required.

Finally, suppose that A is a minimax group and X and X_1 are finitely generated subgroups of A such that A/X and A/X_1 are direct products of quasicyclic groups. XX_1/X is finite, so the numbers of p^∞-factors in the direct decomposition of A/X and A/XX_1 are the same. Hence

$$A/X \simeq A/XX_1 \simeq A/X_1. \quad \square$$

Corollary. If A is an abelian minimax group, $A \in \hat{\mathfrak{M}}\check{\mathfrak{M}}$ and $m(A) \leqq 2$.

Notice, however, that an abelian minimax group need not belong to the class $\check{\mathfrak{M}}\hat{\mathfrak{M}}$, as one may see from the example of the additive group of all rational numbers whose denominators are powers of a fixed prime.

If A is an abelian minimax group and X is a finitely generated subgroup such that A/X is a direct product of quasicyclic groups, we define the *spectrum* of A,

$$Sp(A),$$

to be the finite set of primes which divide orders of elements of A/X. Lemma 10.31 shows that $Sp(A)$ is an invariant of A. Clearly $p \in Sp(A)$ if and only if A has an infinite p-group as a homomorphic image, and A is finitely generated if and only if $Sp(A)$ is empty.

If G is a soluble minimax group and π is a set of primes, G is called a *π-minimax group* if there is an abelian series of finite length such that the spectrum of each factor is a subset of π. Evidently the class of soluble π-minimax groups is $\boldsymbol{S}$, $\boldsymbol{H}$ and $\boldsymbol{P}$-closed. Also an abelian group A is π-minimax if and only if $Sp(A) \subseteq \pi$.

If π is a finite set of primes, let

$$Q_\pi$$

denote the additive group of all rational numbers whose denominators are π-numbers. Evidently Q_π is a torsion-free minimax group of rank 1 and $Sp(Q_\pi) = \pi$. Moreover, if A is any torsion-free abelian minimax group with rank r and spectrum contained in π, the mapping

$$a \to a \otimes 1$$

is a monomorphism of A into $\bar{A} = A \otimes Q_\pi$ and $\bar{A}$ is a direct product of r copies of Q_π. Hence the torsion-free abelian minimax groups with rank $\leqq r$ and spectrum contained in the set π are just those groups that can be embedded in a direct product of r copies of Q_π. Also it is clear how Aut A may be identified with a subgroup of $GL(n, Q_\pi)$.

We remark that torsion-free abelian minimax groups form a complicated class: one indication of this is the existence of torsion-free abelian minimax groups of rank 2 which are indecomposable, the well-known example of Pontryagin being of this type (Fuchs [3], p. 151, C).

Nilpotent Minimax Groups

We shall find it convenient (especially in Section 10.4) to employ the following notation. If $\mathfrak{X}$ and $\mathfrak{Y}_j$ are classes of groups and i_j is a non-negative integer, define recursively

$$\mathfrak{X}\mathfrak{Y}_1^{(i_1)} \cdots \mathfrak{Y}_r^{(i_r)} = (\mathfrak{X}\mathfrak{Y}_1^{(i_1)} \cdots \mathfrak{Y}_r^{(i_r-1)})\, \mathfrak{Y}_r$$

if $i_r > 0$, and

$$\mathfrak{X}\mathfrak{Y}_1^{(i_1)} \cdots \mathfrak{Y}_{r-1}^{(i_{r-1})}\mathfrak{Y}_r^{(0)} = \mathfrak{X}\mathfrak{Y}_1^{(i_1)} \cdots \mathfrak{Y}_{r-1}^{(i_{r-1})}.$$

So, for example,

$$\mathfrak{X}\mathfrak{Y}^{(i)} = \mathfrak{X}\underbrace{\mathfrak{Y} \cdots \mathfrak{Y}}_{\leftarrow i \rightarrow}.$$

Lemma 10.32. Let G be a nilpotent π-minimax group with class c. Then there exists a series $X = G_0 \lhd G_1 \lhd G_2 \cdots \lhd G_c = G$ in which X is finitely generated and G_{i+1}/G_i is a radicable abelian π-group satisfying Min. In particular, $G \in \hat{\mathfrak{M}}\check{\mathfrak{M}}^{(c)}$ and $m(G) \leqq c + 1$.

Proof. The class of soluble π-minimax groups is $\boldsymbol{S}$ and $\boldsymbol{H}$-closed, so each $\zeta_{i+1}(G)/\zeta_i(G)$ is an abelian π-minimax group. Hence there is a finitely generated subgroup X_{i+1} of $\zeta_{i+1}(G)$ such that $\zeta_{i+1}(G)/X_{i+1}\zeta_i(G)$ is a radicable abelian π-group satisfying Min. Let X be the subgroup generated by $X_1, \ldots, X_c$. Then X is finitely generated; if $G_i = X\zeta_i(G)$, then $X = G_0 \lhd G_1 \lhd \cdots \lhd G_c = G$ is a series of the required type. ∎

Corollary. Let G be a soluble minimax group and suppose that G is poly-nilpotent of type $(i_1, \ldots, i_k)$. Then

$$G \in (\hat{\mathfrak{M}}\check{\mathfrak{M}}^{(i_k)}) \cdots (\hat{\mathfrak{M}}\check{\mathfrak{M}}^{(i_1)})$$

and

$$m(G) \leqq \sum_{j=1}^{k} (i_j + 1).$$

In particular, if G has derived length d, we may take $k = d$ and each $i_j = 1$. Hence

$$G \in (\hat{\mathfrak{M}}\check{\mathfrak{M}})^d \quad \text{and} \quad m(G) \leqq 2d.$$

The Structure of Soluble Minimax Groups

Theorem 10.33. Let G be an $\mathfrak{S}_1$-group, let R be the subgroup generated by all the quasicyclic subgroups of G and let F/R be the Fitting subgroup of G/R. Then

(i) R is the direct product of finitely many quasicyclic subgroups of G and, if G is a minimax group, R is the finite residual of G;
(ii) F/R is nilpotent;
(iii) G/F is polycyclic and abelian-by-finite.

Proof. Let S denote the finite residual of G. Then $R \leqq S$ and by Theorem 9.31 the group S is radicable and nilpotent. Now periodic subgroups of S lie in its centre, by Theorem 9.23. Hence R is abelian and consequently it is the direct product of finitely many quasicyclic groups. Now let G be a minimax group. An upper central factor of S is radicable (by Theorem 9.23 again), and the additive group of rational numbers is not a minimax group. Hence S is periodic and $S = R$. Thus (i) is proved.

Let T be the maximal normal periodic subgroup of G; then $R \leqq T$ and T is a Černikov group, so in fact T/R is finite. Moreover, Lemma 9.34 may be applied to show that G/R has a normal series of finite length whose factors are either finite or torsion-free and abelian; we may suppose that the action of G is rationally irreducible on an infinite factor. Let F be a factor of this series and set $\bar{G} = G/C_G(F)$. If F is finite, so is $\bar{G}$. If F is torsion-free, then $\bar{G}$ can be regarded as an irreducible linear group over the field of rational numbers. Subnormal abelian subgroups of $\bar{G}$ are completely reducible by Clifford's Theorem, and hence are finitely generated by Lemma 5.29.1. The final step in the proof of Theorem 3.27 shows that $\bar{G}$ is polycyclic. It follows that there is a normal subgroup $N \geqq R$ such that G/N is polycyclic and N/R is nilpotent. Now $N \leqq F$ since F/R is the Fitting subgroup of G/R, and F/R is nilpotent since G/N satisfies Max. Finally, G/F is clearly polycyclic and it is also abelian-by-finite because $G/R \in \mathfrak{N}\mathfrak{A}\mathfrak{F}$ by Theorem 3.25. ▯

Corollary (Robinson [6], Corollary to Theorem 2.11). Let G be a soluble minimax group, let R be the finite residual of G and let c be the nilpotent class of the Fitting subgroup of G/R. Then G belongs to the class

$$\check{\mathfrak{M}}\,(\hat{\mathfrak{M}}\,\check{\mathfrak{M}}^{(c)})\,\hat{\mathfrak{M}}$$

and $m(G) \leqq c + 3$.

This follows from Lemma 10.32 and Theorem 10.33

Residual Properties of Soluble Minimax Groups

The criteria for residual finiteness derived in Section 9.3 (Theorems 9.31 and 9.37) may be applied to soluble minimax groups. For example, a soluble minimax group is residually finite if and only if it has no quasicyclic groups. By similar methods the following sharper results may be established: *let G be a soluble minimax group and assume that the centre of the Baer radical of G is reduced; then for all but a finite number of primes p both G and Aut G are finite extensions of residually finite p-groups* (Wehrfritz [10], Theorem R). When G is polycyclic, this is true for all primes p (cf. Šmel'kin [8]).

Theorem 10.34. Let G be a torsion-free nilpotent minimax group. If the prime p does not belong to the spectrum of the centre of G, then G is residually a finite p-group. This conclusion is therefore valid for all but a finite number of primes p.

Proof. Let Z be the centre of G. In view of Theorem 9.38, the group G will be residually finite-p if Z is. Let $\pi = Sp(Z)$ and let p be a prime not in π. By hypothesis there is a finitely generated subgroup X such that Z/X is a π-group. Since $p \notin \pi$, we have $Z^{p^i} \cap X = X^{p^i}$ and

$$\left(\bigcap_{i=0}^{\infty} Z^{p^i}\right) \cap X = \bigcap_{i=0}^{\infty} X^{p^i} = 1, \tag{8}$$

because X is free abelian. But Z is torsion-free while Z/X is periodic; consequently (8) implies that

$$\bigcap_{i=0}^{\infty} Z^{p^i} = 1$$

and hence that Z is residually finite-p. □

We shall mention some comparable results. *If G is a torsion-free nilpotent minimax group and π is an infinite set of primes, then* $\bigcap_{p\in\pi} G^p = 1$ (Robinson [7], Theorem 4.31): for finitely generated groups this is due to Higman [5]; see also Baer [18], Hauptsatz 1. In addition, *a soluble group with finite abelian section rank which is residually finite-p for infinitely many primes p is nilpotent* (Seksenbaev [1] for polycyclic groups; for the general case and related results see Robinson [11] (p. 170) and [16].

Radical Groups whose Abelian Subgroups are Minimax Groups

Theorem 10.35 (Baer [51], Hauptsatz 4.2). Let G be a radical group and let π be a set of primes. G is a soluble π-minimax group if and only if each abelian subgroup of G is a π-minimax group.

When π is empty, this reduces to the Mal'cev-Baer-Plotkin theorem (see Theorem 3.31). Taking π to be the set of all primes, we deduce that *a radical group all of whose abelian subgroups are minimax groups is itself a soluble minimax group*: this theorem has also been proved by Zaičev [4], [8]).

In order to prove Theorem 10.35 we require a result on groups of units of certain integral domains which plays a role in the proof similar to the Dirichlet Unit Theorem in the proof of Theorem 3.31.

Lemma 10.36 (Baer [51], Lemma 3.1). Let R be an integral domain and assume that the additive group of R is a torsion-free minimax group. Then R^*, the multiplicative group of units of R, is finitely generated. *

Proof. Let F be the field of fractions of R and let P be the prime subfield of F. Clearly F has characteristic 0, so P is isomorphic with Q, the field of rational numbers. We shall identify P with Q, so that $Q \leqq F$. By hypothesis R contains an additive subgroup X such that X is free abelian of finite rank and R/X is a π-group for some finite set of primes π. The multiplicative group

$$M = \langle p : p \in \pi \rangle$$

is therefore free abelian of finite rank and $M \leqq F^*$. Let $0 \neq a \in R$ and let S be the subring generated by a. Since $S \cap X$ is finitely generated as an additive group, it is contained in the additive subgroup generated by $a, \ldots, a^n$ for some positive integer n. Also

$$S/S \cap X \simeq (S + X)/X \leqq R/X.$$

Hence there exists a positive π-number m such that $ma^{n+1} \in S \cap X$ and

$$ma^{n+1} = \sum_{i=1}^{n} l_i a^i$$

where the l_i are integers. Hence

$$(ma)^{n+1} = \sum_{i=1}^{n} m^{n-i} l_i (ma)^i. \tag{9}$$

It follows that each a in R is algebraic over Q. The mapping $t \to a$ determines a homomorphism of the polynomial ring $Q[t]$ onto the subring T of R generated by Q and a; the kernel of this homomorphism has

* For a converse see Ayoub [1].

the form $fQ[t]$, where the polynomial f has to be irreducible since R is an integral domain. Hence $T \simeq Q[t]/fQ[t]$, a field, and it follows that $a^{-1} \in T$. Hence every element of F is a sum of elements of the form qa where $q \in Q$ and $a \in R$. Therefore

$$F = QR.$$

If $a \in R$, then $la \in X$ for some positive integer l. Hence

$$F = QX. \tag{10}$$

Since X is a finitely generated additive abelian group, we deduce from (10) that F has finite degree over Q. Therefore F is an algebraic number field.

Let I be the subring of all algebraic integers in F. If $a \in R$, there exists a positive π-number m such that $ma \in I$, in view of (9). By definition of M we see that R is contained in the ring MI and

$$R^* \leqq (MI)^*. \tag{11}$$

Let $a \in (MI)^*$; then aI is a non-zero finitely generated I-submodule of F, that is, aI is a *fractional ideal* of F. If A and B are fractional ideals of F, then AB is a fractional ideal of F and so is

$$A^{-1} = \{x : x \in F, xA \leqq I\}.$$

It is easy to verify that the set of all fractional ideals in F is a multiplicative abelian group G with respect to these operations of multiplication and inversion: the identity element of G is, of course, I (see Curtis and Reiner [1], p. 112, Theorem 18.13).

The mapping $a \to aI$ is a homomorphism of $(MI)^*$ into G. The kernel is I^*, the group of algebraic units of F, and by the Dirichlet Unit Theorem this is finitely generated. By (11) it is sufficient to show that the multiplicative group G_1 of all fractional ideals aI with $a \in (MI)^*$ is finitely generated. Let $a \in (MI)^*$; by definition of M there is a positive π-number m such that $ma \in I$. Clearly $m \in (MI)^*$, so $ma \in (MI)^*$ and there exists an element $a' \in (MI)^*$ such that $maa' = 1$. Also there is a positive π-number m' such that $m'a' \in I$. Now $(ma)(m'a') = m'$ and consequently aI belongs to the multiplicative group G_2 generated by all integral ideals dividing ideals of the form lI where l is a positive π-number. Hence $G_1 \leqq G_2$. Now π is finite, so G_2 is finitely generated—here we use the primary decomposition theorem for integral ideals (Curtis and Reiner [1], p. 110, Theorem 18.10). Hence G_1 is finitely generated. ▯

Lemma 10.37 (Baer [51], Folgerung 3.2). Let A be an abelian group of automorphisms of a torsion-free abelian minimax group G and suppose that A is rationally irreducible. Then A is finitely generated.

Proof. Let R be the ring of endomorphisms of G generated by A. Then R is a commutative ring with unity. Let $0 \neq \varrho \in R$ and let $K(\varrho)$ be the kernel of ϱ. Then $K(\varrho)$ is an A-admissible subgroup of G and $G/K(\varrho) \simeq G^{\varrho}$, so the rational irreducibility of A implies that $K(\varrho) = 1$ or G. The latter is impossible since $\varrho \neq 0$; thus $K(\varrho) = 1$ and ϱ is a monomorphism. It follows that R is an integral domain. Let $1 \neq g \in G$ and observe that the mapping $\varrho \to g^{\varrho}$ is a group monomorphism of R into G. Hence the additive group of R is a torsion-free minimax group. We can now apply Lemma 10.36, concluding that R^* is finitely generated. But $A \leqq R^*$, so the lemma follows. ▯

Corollary. An abelian group of automorphisms A of a torsion-free abelian π-minimax group G is itself a π-minimax group.

Proof. When A is rationally irreducible, this result is clear from the lemma. When G is rationally reducible, we obtain the result by induction on the rank of G, utilizing the fact that Hom (X, Y) is a torsion-free abelian π-minimax group if both X and Y are. ▯

For another version of Lemma 10.37 see Hulse [1] (Theorems B and C).

Proof of Theorem 10.35

(i) The locally nilpotent case

Let G be a locally nilpotent group and assume that all abelian subgroups of G are π-minimax groups. Denote by T the torsion-subgroup of G. Then T is a Černikov group, by Theorem 3.32, and from this it is evident that T is a soluble π-minimax group. Thus it will be sufficient to prove that G/T is a nilpotent π-minimax group.

Let R denote the finite residual of T; then T/R is finite and R is the direct product of a finite set of quasicyclic subgroups; hence R is a π_0-group where π_0 is a finite subset of π. We will show first that any abelian subgroup A/R of G/R is a π-minimax group. Now A/R has finite torsion-subgroup, so by replacing A/R by a suitable power, we may assume that A/R is torsion-free and R is the torsion-subgroup of A. Let $A_1 = \{x_1, x_2, \ldots\}$ be a countable subgroup of A containing R. For any pair of positive integers i and j the subgroup $\langle x_i, x_j\rangle$ is finitely generated and nilpotent, so its torsion-subgroup S is a finite abelian π_0-group. Since

$$[y, \underbrace{x_j, \ldots, x_j}_{t}] = 1$$

for all $y \in S$ and some integer t, there is a positive π_0-number r such that x_j^r centralizes S. Now $[x_i, x_j^r] \in \langle x_i, x_j\rangle \cap R = S$, so

$$[x_i, x_j^r, x_j^r] = 1. \qquad (12)$$

The element $[x_i, x_j^r]$ has order equal to a positive π_0-number s; hence

$$[x_i, x_j^{rs}] = [x_i, x_j^r]^s = 1,$$

by (12). Thus we have found a positive π_0-number $m_{ij} = rs$ such that x_i and $x_j^{m_{ij}}$ commute. Defining m_k to be $m_{1k} \cdots m_{k-1k}$ if $k > 1$ and $m_1 = 1$, we observe that the subgroup H generated by the $x_k^{m_k}$ is abelian, and hence a π-minimax group. Now $HR \lhd A_1$ since A_1/R is abelian; also every m_k is a positive π_0-number, so A_1/HR is a π_0-group. $HR/R \simeq H/H \cap R$, so A_1/R has finite 0-rank. Also A_1/R has finite torsion-subgroup because T/R is finite. Since π_0 is finite, it follows that A_1/HR has Min. Hence A_1/R is a π-minimax group. This clearly implies that A/R is a π-minimax group.

It is therefore permissible to assume that $R = 1$ and T is finite. Once again let A/T be an abelian subgroup of G/T. The argument of the last paragraph shows that A has finite rank. Let $C = C_A(T)$; then A/C is finite. Let $x, y \in C$ and let $n = |T|$. Then $[x, y] \in T$ and hence $[x, y] \in \zeta(C)$. Consequently

$$[x, y^n] = [x, y]^n = 1.$$

Hence $C^n \leq \zeta(C)$. Since A has finite rank, C/C^n is finite and therefore $A/\zeta(C)$ is finite. It follows that A is a π-minimax group, so the same is true of A/T.

We can now assume that G is torsion-free—and hence nilpotent by Theorem 6.36. Let M be a maximal normal abelian subgroup of G. Then M is a torsion-free, abelian π-minimax group, so it can be regarded as a subgroup of the direct sum of a finite number (say r) of copies of Q_{π_1} where π_1 is a finite subset of π; also $\bar{G} = G/C_G(M)$ can be regarded as a subgroup of $GL(r, Q_{\pi_1})$. Since G is nilpotent, $M = C_G(M)$ and $\bar{G}$ may even be considered a subgroup of $U = U(r, Q_{\pi_1})$, the group of all $r \times r$ unitriangular matrices over Q_{π_1}. But U is a π_1-minimax group because each lower central factor is a direct product of finitely many copies of Q_{π_1}. Hence G is a π_1-minimax group, and therefore a π-minimax group.

(ii) Let $N \lhd G$ where N is a Černikov group. If all the abelian subgroups of G are π-minimax, the same is true of G/N

Let A/N be an abelian subgroup of G/N and denote by R the finite residual of N. Let R_i be the subgroup generated by all $a \in R$ such that $a^{p^i} = 1$ for some prime p. Then $R_1 \leq R_2 \leq \cdots$ and R is the union of the R_i. Let

$$C = C_A(N/R) \cap C_A(R_1).$$

If $x \in C$, then x centralizes R_1 and this is easily seen to imply that x

centralizes every R_{i+1}/R_i. Thus $\langle x, R_i \rangle \lhd \langle x, R_{i+1} \rangle$. Also x centralizes N/R, so we have $\langle x \rangle$ *asc* $\langle x, R \rangle \lhd \langle x, N \rangle$. Since A/N is abelian, $\langle x, N \rangle \lhd A$ and we obtain finally $\langle x \rangle$ *asc* C. Hence C is a Gruenberg group and therefore it is locally nilpotent. By *(i)* the subgroup C is π-minimax and since A/C is finite, A is π-minimax. Thus A/N is a π-minimax group, as required.

(iii) Let $N \lhd G$ where N is a torsion-free, abelian group. If all the abelian subgroups of G are π-minimax, the same is true of G/N

Let A/N be an abelian subgroup of G/N and let $C = C_A(N)$. Then $[C', C] = 1$, and C is nilpotent. By *(i)* C is a π-minimax group and by the Corollary to Lemma 10.37 the group A/C is π-minimax. It follows that A, and therefore A/N, is a π-minimax group, as required.

(iv) If G is soluble and all its abelian subgroups are π-minimax groups, then G is a π-minimax group.

There is normal abelian series in G of finite length whose factors are torsion-free or periodic. Let N be the least non-trivial term of this series. Then N is either a Černikov group or a torsion-free, abelian π-minimax group. By *(ii)* and *(iii)* each abelian subgroup of G/N is a π-minimax group, and by induction on the length of the series G/N is a π-minimax group. Hence G is also a π-minimax group.

(v) The general case

Let G be a radical group whose abelian subgroups are π-minimax groups. Let H be the Hirsch-Plotkin radical of G. By Corollary 1, Theorem 6.36 we see that H has finite rank. Thus Theorem 8.16 may be applied to show that G is hypercentral-by-abelian-by-finite. Now periodic subgroups of G are Černikov and therefore soluble, by Theorem 3.32. Thus Theorem 6.36 implies that G is soluble. The theorem now follows from *(iv)*. □

By methods similar to those used above, the following result may be proved: *an SN*-group all of whose ascendant abelian subgroups are π-minimax groups is itself a π-minimax group* (Baer [51], Hauptsatz 7.1); there is a slight complication in the locally nilpotent case. On the other hand, there is no corresponding theorem for subsoluble or hyperabelian groups; for example, let H be the holomorph of a group A of type p^∞; then it is easy to see that H is metabelian and that every subnormal abelian subgroup lies in A: however H is not a minimax group since Aut A is isomorphic with the multiplicative group of units of the ring of p-adic integers and thus is uncountable.

We mention also a theorem of Amberg [1]: *if G is a group possessing an ascending series with locally nilpotent or finite factors and if all abelian subgroups of G are π-minimax groups, then G is a finite extension of a soluble π-minimax group.*

Here is a sketch of the proof: by Baer's theorem the maximal normal radical subgroup R of G is a soluble π-minimax group. G/R inherits the hypothesis on G by steps *(ii)* and *(iii)* in the proof of Theorem 10.35; thus we can assume $R = 1$. If G is infinite, then one shows with the aid of Lemma 5.44 and Corollary, Theorem 5.45 that G contains a normal infinite hyperfinite subgroup H. But clearly H satisfies Min-*ab*, so it is a Černikov group, by Theorem 4.39.2 for example. Since G contains no non-trivial normal locally nilpotent subgroups, H is finite, which is not the case.

Radical groups whose abelian subgroups belong to the class $\check{\mathfrak{M}}\hat{\mathfrak{M}}$ are the subject of two papers of Newell [1], [3].

Finitely Generated Soluble Groups of Finite Rank

The main point which we wish to make here is that *a finitely generated soluble group of finite rank is a minimax group* (Robinson [12]; for the torsion-free case see Zaičev [5]). In fact we shall prove

Theorem 10.38 (Robinson [12]). A finitely generated group with finite rank which belongs to the class

$$\langle \acute{P}, L\rangle \mathfrak{A}$$

is a soluble minimax group.

Let us first note

Lemma 10.39. Let G be a locally soluble group with finite rank r. Then there is a non-negative integer n depending only on r such that $G^{(n)}$ is a periodic hypercentral group with Černikov primary components.

Proof. Let H be a finitely generated subgroup of G. Since H is soluble and has finite rank $\leqq r$, there exists an ascending normal abelian series in H whose factors are either torsion-free of rank $\leqq r$ or elementary abelian p-groups of order $\leqq p^r$ for certain primes p. Each factor affords a representation of H as a linear group of degree r. By the well-known theorem of Zassenhaus (Theorem 3.23) there is an integer $n = n(r)$ for which $H^{(n)}$ centralizes each factor of this series. Hence $H^{(n)}$ is hypercentral. Since n does not depend on H, the subgroup $G^{(n)}$ is locally hypercentral, i.e., locally nilpotent. The required result follows from the structure of locally nilpotent groups of finite rank (Corollary 2, Theorem 6.36). ▯

Corollary (Čarin [8]). A torsion-free locally soluble group of finite rank is soluble.

Proof of Theorem 10.38. (a) Let us first of all deal with the case where G is a finitely generated soluble group with finite rank. Here *it is enough to show that G is nilpotent-by-abelian-by-finite.* For suppose that this has been established. Since subgroups with finite index in G are finitely generated, we can assume without prejudice that G is nilpotent-by-abelian, and $N = G'$ is nilpotent. G/N is finitely generated, so we have only to show that N is a minimax group. Now the tensor product of two abelian minimax groups is evidently a minimax group. Thus N will be a minimax group if N/N' is, by the Corollary to Theorem 2.26. Accordingly let N be abelian. G is now a finitely generated metabelian group, so by Theorem 5.34 it satisfies Max-n. Hence the torsion-subgroup of N must be finite and there is no loss in assuming that N is torsion-free. The property Max-n shows also that

$$N = a_1^G \cdots a_k^G$$

for some finite subset $\{a_1, \ldots, a_k\}$. Thus N will be a minimax group if a^G is, for an arbitrary $a \in N$: for the class of minimax groups is $\boldsymbol{H}$-closed and $\boldsymbol{P}$-closed, so it is $\boldsymbol{N}_0$-closed.

Let us identify a^G with an additive subgroup of a rational vector space V of dimension r. We can then represent G by a group of linear transformations of V. Choose a basis for V. Then each element g of G is represented by an $r \times r$ matrix $M(g)$ with rational entries in the usual manner. Let the components of a in terms of the basis be $a_1, \ldots, a_r$ and let G be generated by elements $g_1, \ldots, g_n$. The primes which divide the denominator of an a_i or of one of the entries in an $M(g_j)$ or $M(g_j^{-1})$ (when all rationals are reduced to their lowest terms) form a finite set π. Moreover, since $A = a^G$, each element b in A has components $b_1, \ldots, b_r$ whose denominators are π-numbers. Hence A is isomorphic with a subgroup of the direct sum of r copies of Q_π, so A is a minimax group.

We assume now that G is an arbitrary finitely generated soluble group with finite rank r. Then G possesses a normal abelian series

$$1 = G_0 \leqq G_1 \leqq \cdots \leqq G_n = G$$

in which each G_{i+1}/G_i is either torsion-free or a direct product of p-groups for different primes p. Naturally we shall assume that $n > 1$ and use induction on n; thus if $A = G_1$, then G/A is a minimax group. If A is torsion-free, then Theorem 3.25 shows that $G \in \mathfrak{N}\mathfrak{A}\mathfrak{F}$: by the first part of this proof G is a minimax group.

We assume that A is periodic and that G is not a minimax group. Then A cannot be a minimax group, and since A has finite rank, it follows that A must have an infinite number of non-trivial primary components. Hence there exists a normal subgroup B of G contained in

A such that for infinitely many primes p the p-component of A/B is non-trivial and is either elementary of order $\leqq p^r$ or is a direct product of at most r groups of type p^∞. Clearly we can suppose that $B = 1$. The action of G on the p-component of A yields a representation of G as a linear group over either $GF(p)$ or the field of p-adic numbers. By Mal'cev's theorem on soluble linear groups, there exists an integer $m = m(r)$ such that $R = (G^m)'$ is represented on each component by a group of $r \times r$ unitriangular matrices. Hence

$$[A, \underbrace{R, \ldots, R}_{r}] = 1. \tag{13}$$

Since G/A is a minimax group, it is nilpotent-by-abelian-by-finite, by Theorem 3.25. Hence there is a positive integer l divisible by m such that $(G^l)'A/A$ is nilpotent. Now $(G^l)' \leqq R$, so the group $(G^l)'A$ is nilpotent, by(13). Also G/G^l is finite; thus $G \in \mathfrak{N}\mathfrak{A}\mathfrak{F}$ and G is a minimax group by the first part of the proof.

(b) It remains to show that a group with finite rank which belongs to the class $\mathfrak{X} = \langle \acute{P}, L \rangle \mathfrak{A}$ is locally soluble. Suppose that this is false. Now $\mathfrak{X}$ is the union of the classes

$$\mathfrak{X}_\alpha = (\acute{P}L)^\alpha \mathfrak{A}$$

where α is an ordinal. Hence there is a first ordinal α for which the class $\mathfrak{X}_\alpha$ contains a non-locally soluble group with finite rank. Clearly α is not a limit ordinal. Let $G \in \mathfrak{X}_\alpha$ and suppose that G has finite rank. Since $\mathfrak{X}_\alpha = \acute{P}L\mathfrak{X}_{\alpha-1}$, the group G has an ascending series whose factors are $L\mathfrak{X}_{\alpha-1}$-groups and hence are locally soluble. Denote this series by $\{G_\beta\}$ and let β be the first ordinal for which G_β is not locally soluble. Again β cannot be a limit ordinal, so both $G_{\beta-1}$ and $G_\beta/G_{\beta-1}$ are locally soluble.

Let H be a finitely generated subgroup of G_β. Then $H/H \cap G_{\beta-1}$ is soluble and $H \cap G_{\beta-1}$ is locally soluble. By Lemma 10.39 there is an integer $n \geqq 0$ such that $S = H^{(n)}$ is a periodic hypercentral group. Now H/S' is soluble, so by part (a) of this proof it is a minimax group and this implies that S/S' is a Černikov group. Since S is the direct product of its primary components, it follows that all but a finite number of these are perfect. But a perfect hypercentral group is trivial by Grün's Lemma. Hence all but a finite number of the primary components of S are trivial and S is a Černikov group. It follows that H is soluble and G_β is locally soluble, a contradiction which completes the proof. ☐

With the aid of Lemma 10.39 we can draw the following conclusion.

Corollary 1. A group with finite rank which belongs to the class $\langle \acute{P}, L \rangle \mathfrak{A}$ is locally soluble and hyperabelian. Its structure is described by Theorem 8.16 (cf. Kargapolov [4]).

Thus for groups of finite rank the three properties radical, locally soluble and hyperabelian coincide.

Corollary 2. Let θ be an outer commutator word and let G be a group which is locally a soluble group of finite rank. If θ is finite-valued on G, then $\theta(G)$ is finite.

Proof. Since θ is finite-valued, we may assume that G is finitely generated: hence G is a soluble minimax group. Let R be the finite residual of G; now θ is finite-valued on G/R, so $\theta(G/R) = \theta(G)R/R$ is finite, by the Corollary to Theorem 4.29.1. Clearly $\theta(G)$ is finitely generated, so $\theta(G) \cap R$ is finitely generated. By Theorem 10.33 the group R is the direct product of finitely many quasicyclic groups, so $\theta(G) \cap R$ is finite and $\theta(G)$ is finite. □

An Example of a Locally Soluble Group with Finite Rank which is Insoluble

In contrast to the Corollary to Lemma 10.39, *a periodic locally soluble or even hypercentral group with finite rank need not be soluble.* The following example is due to Kegel (see Baer [51], pp. 27—28).

Let p be a fixed prime and let r be a positive integer; denote by H_r the group of all 2×2 matrices x over Z/p^rZ, the ring of integers modulo p^r, such that $x \equiv 1 \bmod p$. Evidently H_r is a finite group of order $p^{4(r-1)}$. Suppose that the derived length of H_r does not exceed some number d for all r. Let F be the group generated by the *integral* matrices

$$\begin{pmatrix} 1 & p \\ \cdot & 1 \end{pmatrix} \text{ and } \begin{pmatrix} 1 & \cdot \\ p & 1 \end{pmatrix}.$$

The mapping $x \to x \bmod p^r$ is a homomorphism of F into H_r with kernel K_r consisting of all x in F such that $x \equiv 1 \bmod p^r$. Hence

$$F^{(d)} \leqq \bigcap_{r=1}^{\infty} K_r = 1;$$

but this is not the case, for Brenner [1] has shown that F is a free group.

Hence for each prime p we can select a positive integer $r(p)$ such that if $H(p)$ is the group of all 2×2 matrices over $Z/p^{r(p)}Z$ which are congruent to 1 modulo p, the derived length of $H(p)$ increases to ∞ with p. Let G be the direct product of these $H(p)$ for all primes p. Then G is insoluble, although it is periodic, hypercentral and a Baer group.

By definition $H(p)$ can be regarded as a group of automorphisms of an abelian p-group of rank 2. Hence $H(p)$ has rank at most 9, by Lemma 7.44. Therefore G has finite rank not exceeding 9.

10.4 Soluble Minimax Groups II. Computing Minimax Lengths

In this section we shall consider two questions. If G is a soluble minimax group, what can be said of the factors of a shortest minimax series and how can $m(G)$ be computed? The present discussion is based on a paper of Robinson [6].

The following remark is nearly obvious and we omit its proof.

Lemma 10.41. Let N be a normal subgroup of a group G and assume that N and G/N are minimax groups.

(i) G is a minimax group and $m(G) \leqq m(N) + m(G/N)$.

(ii) If N is a proper finite subgroup of G, then $m(G) = m(G/N)$.

(iii) If N is a non-trivial subgroup with finite index in G, then $m(G) = m(N)$.

Upper-Finite Groups

It turns out that the class of groups all of whose finitely generated homomorphic images are finite plays a crucial part in the resolution of both of the above questions. We shall call such groups *upper-finite*; these are, in a sense, dual to locally finite groups (so "co-locally finite" might be a more appropriate term). By Lemma 6.17 a group is upper-finite if and only if it has no just-infinite homomorphic images.

The class of upper-finite groups has reasonable closure properties.

Lemma 10.42. If $\mathfrak{X}$ denotes the class of upper-finite groups, then

$$\mathfrak{X} = \boldsymbol{H}\mathfrak{X} = \boldsymbol{P}\mathfrak{X} = \boldsymbol{L}\mathfrak{X}.$$

Proof. $\boldsymbol{H}$-closure is obvious and $\boldsymbol{P}$-closure is easily verified. Let $G \in \boldsymbol{L}\mathfrak{X}$ and suppose that G/N is a finitely generated factor group of G. Then $G = HN$ for some upper-finite subgroup H. Since $G/N \simeq H/H \cap N$, we conclude that G/N is finite and G is upper-finite. □

Corollary. The class of upper-finite groups is a radical class.

Proof. By Theorem 1.32 we only have to establish $\boldsymbol{N}$-closure; this follows from the relation $\boldsymbol{N} \leqq \langle \boldsymbol{P}, \boldsymbol{H}, \boldsymbol{L} \rangle$. □

We shall write

$$\sigma(G)$$

for the upper-finite radical of G. Thus $\sigma(G)$ is upper-finite and of course $\sigma(G/\sigma(G))$ is trivial by $\boldsymbol{P}$-closure. By $\boldsymbol{L}$-closure and Lemma 1.31, the subgroup $\sigma(G)$ contains all ascendant upper-finite subgroups of G.

A subgroup of an upper-finite group need not be upper-finite—consider for example the additive group of rational numbers. However there are two weak forms of $\boldsymbol{S}$-closure which will prove to be important.

Lemma 10.43. If G is an upper-finite group and H is a subgroup with finite index in G, then H is upper-finite.

Proof. Without loss of generality we can assume that $H \lhd G$. If H is not upper-finite, it has a finitely generated just-infinite factor group H/K. Let $K_1, \ldots, K_n$ be the finitely many conjugates of K in G; the intersection of the K_i is $\text{Core}_G K$, which we may assume to be trivial. Suppose that every intersection of $i+1$ distinct conjugates of K is finitely generated and let I be the intersection of a set of i distinct conjugates of K. If I is contained in every K_j, then $I = 1$. Otherwise there is a j such that $I \nleq K_j$, and our hypothesis shows that $I \cap K_j$ is finitely generated. Also IK_j/K_j is a non-trivial normal subgroup of the just-infinite group H/K_j, so it has finite index and must be finitely generated. Thus I is finitely generated. By induction on $n - i$ we conclude that K is finitely generated. However this means that G is finitely generated and therefore finite, an evident contradiction. □

Lemma 10.44. Let G be a group and let $N \lhd G$. If N is an upper-finite soluble minimax group, then $[N, G]$ is upper-finite.

Proof. If $g \in G$, then $[N, g]\, N'/N'$ is a homomorphic image of N, so it is upper-finite. Thus $[N, G]/N'$, being a product of normal upper-finite subgroups, is upper-finite. Therefore we need only prove that N' is upper-finite. By Theorem 10.33 there is a normal series $1 \leqq L \leqq M \leqq N$ in which N/M is finite, M/L is nilpotent and L is a Černikov group. The group $N/[M, N]$ is central-by-finite, so Schur's theorem (4.12) shows that $N'/[M, N]$ is finite. Also M is upper-finite because $|N : M|$ is finite (by Lemma 10.43); therefore $[M, N]/M'$ is upper-finite. Thus we need only show that M' is upper-finite. Now $M/M'L$ is upper-finite and M/L is nilpotent. Moreover, it is easy to verify that the tensor product of two abelian upper-finite groups is upper-finite. Thus Theorem 2.26 implies that each lower central factor of M/L is upper-finite. Hence $M'L/L$ is upper-finite. Finally, L is a Černikov group, so $M' \cap L$ is certainly upper-finite and M' is upper-finite. □

Corollary. If G is a soluble minimax group and G is upper-finite, then $\gamma_i(G)$ is upper-finite for every positive integer i.

The next result is fundamental for our investigation.

Lemma 10.45. Let G be a soluble group. If $G/\sigma(G)$ contains no subnormal free abelian subgroups of infinite rank, it is polycyclic.

Proof. Let $S = \sigma(G)$ and suppose that G/S is not polycyclic; then it contains a subnormal abelian subgroup which cannot be finitely generated (see Part 1, p. 82). Now G/S has no subnormal periodic subgroups except 1 because every locally finite group is upper-finite. Hence G/S contains a subnormal free abelian subgroup of infinite rank. ▯

Corollary. If G is a soluble minimax group, then $G/\sigma(G)$ is polycyclic.

Since a subsoluble group whose subnormal abelian subgroups are finitely generated is polycyclic (Baer [50], Satz N. B.), Lemma 10.45 is still true if G is merely subsoluble.

The Structure of a Shortest Minimax Series

Let $1 = G_0 \lhd G_1 \lhd \cdots \lhd G_n = G$ be a series of finite length in a group G and suppose that

$$G_i/G_{i-1} \in \mathfrak{X}_i$$

for some class of groups $\mathfrak{X}_i$. Then we will say that the series is of *type*

$$(\mathfrak{X}_1, \ldots, \mathfrak{X}_n).$$

Theorem 10.46 (Robinson [6], Theorem 3.12). A shortest minimax series in a soluble minimax group has type

$$(\check{\mathfrak{M}}, \hat{\mathfrak{M}}, \underbrace{\check{\mathfrak{M}}, \ldots, \check{\mathfrak{M}}}_{i_k}, \ldots, \hat{\mathfrak{M}}, \underbrace{\check{\mathfrak{M}}, \ldots \check{\mathfrak{M}}}_{i_1}, \hat{\mathfrak{M}},)$$

where $k \geqq 0$, the integers i_j satisfy $1 \leqq i_j \leqq 4$ and the initial $\check{\mathfrak{M}}$ and the final $\hat{\mathfrak{M}}$ may not occur.

The point of the theorem is, of course, that there cannot be two consecutive factors satisfying Max and there cannot be a sequence of five consecutive factors satisfying Min.

The proof requires two further lemmas.

Lemma 10.47. Let $\mathfrak{X}$ be a class of groups such that $\mathfrak{X} = \boldsymbol{S}_n\mathfrak{X}$ and let G be a soluble minimax group. If $G \in \mathfrak{X}\hat{\mathfrak{M}}^{(r)}$ for some integer r, then $G \in \mathfrak{X}\hat{\mathfrak{M}}$.

Proof. Let $S = \sigma(G)$. Since S is upper-finite, a factor group of S that satisfies Max is finite: this implies that $S \in \mathfrak{X}\mathfrak{F}$ since $\mathfrak{X} = \boldsymbol{S}_n\mathfrak{X}$. Hence there is a positive integer m such that $S^m \in \mathfrak{X}$; also S/S^m is finite. Now G/S has Max (Corollary, Lemma 10.45). Thus G/S^m has Max and $G \in \mathfrak{X}\hat{\mathfrak{M}}$. ▯

This rules out the possibility of two consecutive factors satisfying Max in a shortest minimax series.

Lemma 10.48. Let $\mathfrak{X}$ be a class of groups such that $\mathfrak{X} = \boldsymbol{S}_n\mathfrak{X}$ and let G be a soluble minimax group. If $G \in \mathfrak{X}\check{\mathfrak{M}}^{(r)}$, then

$$G \in \mathfrak{X}\mathfrak{N}_r\mathfrak{F}.$$

Proof. Since a soluble group satisfying Min is abelian-by-finite, the result is true for $r \leqq 1$. Let $r > 1$, so that there is a normal subgroup N belonging to the class $\mathfrak{X}\check{\mathfrak{M}}^{(r-1)}$ such that G/N satisfies Min. By induction on r

$$N \in \mathfrak{X}\mathfrak{N}_{r-1}\mathfrak{F}.$$

Since $\mathfrak{X} = \boldsymbol{S}_n\mathfrak{X}$, it follows that there is a positive integer m for which $\gamma_r(N^m) \in \mathfrak{X}$. Also N/N^m is finite and $N^m \lhd G$, so G/N^m satisfies Min. Thus we can assume that $M = N^m$ is nilpotent of class $\leqq r - 1$ and G/M is a radicable abelian group satisfying Min, provided that we prove G to be nilpotent of class at most r. Form the upper central series of M, refine it by inserting the torsion-subgroup in each factor and then adjoin G. Let the resulting normal abelian series in G be

$$1 = M_0 \leqq M_1 \leqq \cdots \leqq M_{2r-2} = M \leqq M_{2r-1} = G.$$

Here M_{i+1}/M_i satisfies Min if i is even and is torsion-free of finite rank if i is odd. Since M_{i+1}/M_i is central with respect to M, we obtain a representation of G/M by automorphisms of M_{i+1}/M_i. Now G/M is radicable and periodic, so G must centralize each M_{i+1}/M_i, by Lemmas 4.45 and 5.29.1.

Let $g \in G$; the mapping τ_g in which $aM_{2i} \to [a, g]\, M_{2i-1}$ is a well-defined homomorphism of M_{2i+1}/M_{2i} into M_{2i}/M_{2i-1}. Now the first of these groups satisfies Min while the second is torsion-free. Therefore τ_g is trivial and consequently M_{2i+1}/M_{2i-1} is central in G for $i = 1, \ldots, r-1$. Thus, as M_1 is already central in G, it follows that $\gamma_{r+1}(G) = 1$ and G is nilpotent of class at most r. ▯

Proof of Theorem 10.46. Let G be a soluble minimax group. The most general type possible for a given shortest minimax series of G is

$$(\check{\mathfrak{M}}, \hat{\mathfrak{M}}, \underbrace{\check{\mathfrak{M}}, \ldots, \check{\mathfrak{M}}}_{i_k}, \ldots, \hat{\mathfrak{M}}, \underbrace{\check{\mathfrak{M}}, \ldots, \check{\mathfrak{M}}}_{i_1}, \hat{\mathfrak{M}}) \tag{14}$$

in view of Lemma 10.47; here $k \geqq 0$ and $i_j \geqq 1$, while the initial $\check{\mathfrak{M}}$ and the final $\hat{\mathfrak{M}}$ may not be present. If $m(G) \leqq 1$, then $G \in \hat{\mathfrak{M}}$ or $G \in \check{\mathfrak{M}}$, and the theorem is obviously true. Let $m(G) > 1$. We have to show that $i_j \leqq 4$.

We assume first of all that the shortest minimax series in question "ends with" Max, so that the final $\hat{\mathfrak{M}}$ *does* occur. Let $S = \sigma(G)$, the upper-finite radical of G. Then G/S satisfies Max (Corollary, Lemma 10.45). Intersecting the given series in G with S, term by term, we obtain a minimax series in S which is of type (14) but with the final $\hat{\mathfrak{M}}$ replaced by $\mathfrak{F}$; this can clearly be omitted. This must be a *shortest* minimax series of S since G/S satisfies Max. Hence

$$m(G) = m(S) + 1. \tag{15}$$

By induction on $m(G)$ we conclude that $1 \leqq i_j \leqq 4$.

Now suppose that the shortest minimax series of G ends with a $\check{\mathfrak{M}}$, so that $k > 0$ and the series is of type

$$(\check{\mathfrak{M}}, \hat{\mathfrak{M}}, \underbrace{\check{\mathfrak{M}}, \ldots, \check{\mathfrak{M}}}_{i_k}, \ldots, \hat{\mathfrak{M}}, \underbrace{\check{\mathfrak{M}}, \ldots, \check{\mathfrak{M}}}_{i_1}),$$

where now the initial $\check{\mathfrak{M}}$ may not occur. Let us write

$$\mathfrak{X} = (\check{\mathfrak{M}})\hat{\mathfrak{M}}\check{\mathfrak{M}}^{(i_k)} \cdots \hat{\mathfrak{M}}\check{\mathfrak{M}}^{(i_2)} \tag{16}$$

if $k > 1$ and

$$\mathfrak{X} = (\check{\mathfrak{M}})$$

if $k = 1$. Here the bracketed $\check{\mathfrak{M}}$ may not appear. Thus $G \in \mathfrak{X}\hat{\mathfrak{M}}\check{\mathfrak{M}}^{(i_1)}$ and Lemma 10.48 shows that

$$G \in (\mathfrak{X}\hat{\mathfrak{M}})\, \mathfrak{N}_{i_1}\mathfrak{F}.$$

It follows that there is a positive integer m for which $\gamma_{i_1+1}(G^m)$ belongs to the class $\mathfrak{X}\hat{\mathfrak{M}}$. Hence

$$T = \sigma(\gamma_{i_1+1}(G^m)) \in \mathfrak{X}\mathfrak{F}. \tag{17}$$

By the Corollary to Lemma 10.45 we see that

$$G/T \in \hat{\mathfrak{M}}\mathfrak{N}_{i_1}\mathfrak{F} \tag{18}$$

and consequently

$$m(G) \leqq m(T) + m(G/T) \leqq m(T) + i_1 + 1: \tag{19}$$

n this last step we have used Lemma 10.32.

If T is finite, then $m(G) = m(G/T)$, by Lemma 10.41, and so $m(G) \leqq i_1 + 1$. Consequently $k = 1$ and

$$m(G) = i_1 + 1. \tag{20}$$

If on the other hand T is infinite, then $\mathfrak{X}$ cannot be the class of trivial groups, by (17); hence $T \in \mathfrak{X}\mathfrak{F} = \mathfrak{X}$. Now (16) shows that $m(T) \leqq m(G) - (i_1 + 1)$; a comparison of this with (19) yields

$$m(G) = m(T) + (i_1 + 1). \tag{21}$$

We can now proceed to show that $i_1 \leqq 4$. The derived length of a nilpotent group of class $\leqq i_1$ is at most $d = [\log_2 i_1] + 1$ (Corollary, Lemma 2.21), so every minimax group in the class $\mathfrak{N}_{i_1}$ belongs to the class $(\hat{\mathfrak{M}}\check{\mathfrak{M}})^d$ (Corollary to Lemma 10.32). By equation (18) we have

$$G/T \in \hat{\mathfrak{M}}(\hat{\mathfrak{M}}\check{\mathfrak{M}})^d \,\mathfrak{F} = (\hat{\mathfrak{M}}\check{\mathfrak{M}})^d,$$

since $d > 0$. Therefore $m(G) \leqq m(T) + 2d$. If T is finite, then $G \in (\hat{\mathfrak{M}}\check{\mathfrak{M}})^d$ and $m(G) \leqq 2d$ in the same way. Comparing these inequalities for $m(G)$ with (21) and (20) respectively, we obtain

$$i_1 + 1 \leqq 2d = 2[\log_2 i_1] + 2,$$

which implies that $i_1 \leqq 5$.

Now suppose that $i_1 = 5$. If $X \in \mathfrak{N}_5$, then $\gamma_3(X)$ is abelian. Therefore a minimax group which is nilpotent of class $\leqq 5$ belongs to

$$(\hat{\mathfrak{M}}\check{\mathfrak{M}})\,(\hat{\mathfrak{M}}\check{\mathfrak{M}}^{(2)}).$$

Hence by (18)

$$G/T \in \hat{\mathfrak{M}}(\hat{\mathfrak{M}}\check{\mathfrak{M}})\,(\hat{\mathfrak{M}}\check{\mathfrak{M}}^{(2)})\,\mathfrak{F} = (\hat{\mathfrak{M}}\check{\mathfrak{M}})\,(\hat{\mathfrak{M}}\check{\mathfrak{M}}^{(2)})$$

and $m(G) \leqq m(T) + 5$: if T is finite, $m(G) \leqq 5$. However, these inequalities contradict (21) and (20) respectively. Hence $i_1 \leqq 4$. By induction on $m(G)$ all the i_j are at most equal to 4. $\square$

It is not difficult to show that each of the integers 1, 2, 3 and 4 is a possible multiplicity in a shortest minimax series: also, every soluble minimax group has a shortest minimax series in which each $i_j = 1$ or 2 (Robinson [6], Theorem 3.12).

Computing the Minimax Length of a Soluble Minimax Group

The argument of the proof of Theorem 10.46 provides a method of computing minimax lengths. Let G be a soluble minimax group and let G have a shortest minimax series of type

$$(\check{\mathfrak{M}}, \hat{\mathfrak{M}}, \underbrace{\check{\mathfrak{M}}, \ldots, \check{\mathfrak{M}}}_{i_k}, \ldots, \hat{\mathfrak{M}}, \underbrace{\check{\mathfrak{M}}, \ldots, \check{\mathfrak{M}}}_{i_1}, \hat{\mathfrak{M}}),$$

where the first and last factors may not be present.

We shall assume that $k > 0$. By Theorem 10.46 we have $1 \leqq i_j \leqq 4$. If the series ends with Max, it was shown that

$$m(G) = m(\sigma(G)) + 1$$

(equation (15)). Suppose that the series ends with a Min. Then there is a positive integer m such that if $T = \sigma(\gamma_{i_1+1}(G^m))$,

$$m(G) = i_1 + 1 \quad \text{or} \quad m(G) = (i_1 + 1) + m(T) \qquad (22)$$

according as T is finite or infinite. If T is infinite but satisfies Min, $m(T) = 1$ and $m(G) = (i_1 + 1) + 1$. Suppose that T does not satisfy Min. Now T belongs to the class $\mathfrak{X}\mathfrak{F} = \mathfrak{X}$ defined by equation (16). Moreover T is upper-finite and therefore has a shortest minimax series with type equal to that of G with the "top" segment $(\hat{\mathfrak{M}}, \check{\mathfrak{M}}, \ldots, \check{\mathfrak{M}})$ removed.
$\longleftarrow i_1 \longrightarrow$

We can therefore apply the same method to the group T. By repeated application of the argument we arrive at a series of characteristic subgroups

$$G_{k+1} < G_k < \cdots < G_1 = G$$

where G_{k+1}, but not G_k, satisfies Min and where

$$G_{j-1} = \sigma(\gamma_{i_j+1}(G_j^{m_j})), \quad (0 < j \leqq k),$$

for certain positive integers m_j. Moreover, by (22),

$$m(G) = \sum_{j=1}^{k} (i_j + 1) + \varepsilon \tag{23}$$

where $\varepsilon = m(G_{k-1}) = 1$ if G_{k+1} is infinite and $\varepsilon = 0$ if G_{k+1} is finite. Of course (23) is still valid when $k = 0$ and G is infinite, for trivial reasons.

The Set $\Delta(G)$ and the Function δ

Let G be any group. We denote by

$$\Delta(G)$$

the set of all vectors $v = (v_1, \ldots, v_k)$ where $v_i = 1$, 2, 3 or 4 and $k > 0$, such that there exist positive integers $m_1, \ldots, m_k$ with the following property: define

$$G_1 = G \quad \text{and} \quad G_{j-1} = \sigma(\gamma_{v_j+1}(G_j^{m_j}));$$

then G_{k+1} is finite. Notice that if G is a soluble minimax group, Lemma 10.43 and the Corollary to Lemma 10.44 show that in fact

$$G_{j-1} = \gamma_{v_j+1}(G_j^{m_j})$$

provided $j > 1$: should G be upper-finite, this is also true when $j = 1$.

If $v = (v_1, \ldots, v_k)$ is a vector with positive integral components v_i, write

$$\delta(v) = \sum_{i=1}^{k} (v_i + 1).$$

If Σ is a set of vectors with positive integral components, let

$$\delta_\Sigma = \min \{\delta(v) : v \in \Sigma\},$$

with the convention that should Σ be empty,

$$\delta_\Sigma = 0.$$

Finally, we define

$$\delta_G = \delta_{\Delta(G)} \text{ if } G \text{ is infinite}$$

and

$$\delta_G = 0 \text{ if } G \text{ is finite}.$$

The next theorem reduces the computation of minimax length to that of the invariant δ_G. We shall write

$$\tau(G)$$

for the maximal normal periodic subgroup a group G.

Theorem 10.49 (Robinson [6], Theorem 4.11). Let G be an infinite soluble minimax group.

(i) If G has a shortest minimax series ending with Max, then

$$m(G) = 1 + m(\sigma(G)),$$

unless G is polycyclic, when $m(G) = 1$.

(ii) If G has a shortest minimax series ending with Min, then

$$m(G) = \min \{\delta_G, 1 + \delta_{G/\tau(G)}\}.$$

Proof. (i) has already been proved. Let G satisfy the hypothesis of (ii). If $m(G) = 1$, then G satisfies Min and $G = \tau(G)$; thus $\delta_{G/\tau(G)} = 0$ and the formula is valid. Assume therefore that G does not satisfy Min. Let $v = (v_1, \ldots, v_k) \in \Delta(G)$. By definition of $\Delta(G)$

$$G \in \mathfrak{F}(\widehat{\mathfrak{M}\mathfrak{N}_{v_k}}\mathfrak{F}) \cdots (\widehat{\mathfrak{M}\mathfrak{N}_{v_1}}\mathfrak{F}),$$

and hence

$$m(G) \leqq \sum_{i=1}^{k} (v_i + 1) = \delta(v).$$

Thus $m(G) \leqq \delta_G$. Since $\tau(G)$ satisfies Min,

$$m(G) \leqq m(\tau(G)) + m(G/\tau(G)) \leqq 1 + \delta_{G/\tau(G)},$$

showing that $m(G) \leqq \min \{\delta_G, 1 + \delta_{G/\tau(G)}\}$. Now from (23) we know that

$$m(G) = \sum_{i=1}^{k} (v_i + 1) + \varepsilon$$

for some $v = (v_1, \ldots, v_k)$ in $\Delta(G/\tau(G))$: here $\varepsilon = 0$ if $v \in \Delta(G)$ and $\varepsilon = 1$ if $v \in \Delta^* = \Delta(G/\tau(G)) \backslash \Delta(G)$. Now it is easy to show that

$$m(G) \leqq \min\{\delta_G, 1 + \delta_{\Delta^*}\},$$

so that

$$m(G) = \min\{\delta_G, 1 + \delta_{\Delta^*}\}. \tag{24}$$

If the restriction of the function δ to $\Delta(G/\tau(G))$ assumes its minimum value in Δ^*, then $\delta_{G/\tau(G)} = \delta_{\Delta^*}$ and the result follows at once from (24). Otherwise $\delta_{G/\tau(G)} = \delta_G \leqq \delta_{\Delta^*}$, since $\Delta(G) \subseteq \Delta(G/\tau(G))$. By (24) we obtain $m(G) = \delta_G = \min\{\delta_G, 1 + \delta_{G/\tau(G)}\}$. ☐

Corollary. If G is an infinite soluble minimax group,

$$m(G) = \min\{\delta_G, 1 + \delta_{\sigma(G)}, 1 + \delta_{G/\tau(G)}, 2 + \delta_{\sigma(G)/\tau(G)}\}.$$

Proof. Since $\tau(G)$ is locally finite, $\tau(G) \leqq \sigma(G)$. The ***P***-closure of the class of upper-finite groups implies that

$$\sigma(G/\tau(G)) = \sigma(G)/\tau(G).$$

If G is polycyclic, $m(G) = 1$ and $\sigma(G)$ is finite, so $\delta_{\sigma(G)} = 0$; hence the formula is valid in this case. If G is not polycyclic, the result follows from (i) and (ii) of the theorem. ☐

Notice that since a soluble group satisfying Min is abelian-by-finite, the inequalities

$$\delta_{G/\tau(G)} \leqq \delta_G \leqq 2 + \delta_{G/\tau(G)}$$

are valid in any soluble minimax group.

The Minimax Length of a Nilpotent Minimax Group

Although the computation of $m(G)$ is reduced by Theorem 10.49 to that of δ_G, it is clear from the definition that the calculation of δ_G is likely to be a difficult business. However, there is a substantial simplication when G is nilpotent—and by Theorem 10.33 not too much is lost in restricting attention to nilpotent minimax groups.

The key result here is a very simple one indeed.

Lemma 10.49.1. Let G be a nilpotent group with class c and let N be a normal subgroup of G such that G/N has finite exponent e. If i is a positive integer, then $\gamma_i(G)/\gamma_i(N)$ has finite exponent dividing $e^{i(c-i+1)}$.

Proof. Let $x \in \gamma_i(G)$: we will prove that

$$x^{e^{ij}} \in \gamma_{i+j}(G)\, \gamma_i(N) \tag{25}$$

for each integer $j \geqq 0$. When $j = 0$ this is clear, so let $j > 0$. Then by induction on j the element $y = x^{e^{i(j-1)}}$ belongs to $\gamma_{i+j-1}(G)\, \gamma_i(N)$. Now, modulo $\gamma_{i+j}(G)\,\gamma_i(N)$, the element y is a product of commutators like $[x_1, \ldots, x_{i+j-1}]$; also

$$[x_1, \ldots, x_{i+j-1}]^{e^i} \equiv [x_1^e, \ldots, x_i^e, x_{i+1}, \ldots, x_{i+j-1}] \bmod \gamma_{i+j}(G).$$

Since $x_r^e \in N$ for all r, it follows that

$$x^{e^{ij}} = y^{e^i} \in \gamma_{i+j}(G)\,\gamma_i(N).$$

To prove the lemma set $j = c - i + 1$ in (25). ☐

Corollary. Let G be a nilpotent minimax group; if G/N is finite, then $\gamma_i(G)/\gamma_i(N)$ is finite for all positive integers i.

If, in addition, G is upper-finite, a vector $v = (v_1, \ldots, v_k)$ with integral components satisfying $1 \leqq v_i \leqq 4$ belongs to $\Delta(G)$ if and only if $\gamma_{v_k+1}(\cdots \gamma_{v_1+1}(G) \cdots)$ is finite.

Proof. The first part is evident. If G is upper-finite, we see from Lemma 10.43 and the Corollary to Lemma 10.44 that $v = (v_1, \ldots, v_k) \in \Delta(G)$ if and only if each $v_i = 1, 2, 3$ or 4 and

$$G \in \mathfrak{F}\mathfrak{N}_{v_k}\mathfrak{F} \cdots \mathfrak{F}\mathfrak{N}_{v_1}\mathfrak{F}.$$

The result follows by induction and k applications of the lemma. ☐

We will now use the corollary to obtain upper and lower bounds for δ_G when G is a nilpotent minimax group. It will turn out that δ_G coincides with the upper bound for a significant class of groups.

Theorem 10.49.2. (Robinson [6], Theorem 4.22). Let G be a nilpotent minimax group and let c and d be the least integers such that $\gamma_{c+1}(G)$ and $G^{(d)}$ are finite respectively.

(i) $\delta_G \leqq 3\,[\log_3(c+1)] + \varepsilon(c) = F(c)$

where $\varepsilon(c) = 0, 1, 2$ or 3 according as the fractional part of $\log_3(c+1)$ is 0 or lies in the intervals $(0, \log_3 \frac{4}{3}]$, $(\log_3 \frac{4}{3}, \log_3 2]$, or $(\log_3 2, 1)$ respectively.

(ii) If G is upper-finite,

$$\delta_G \geqq \left[\frac{3d+1}{2}\right] = f(d).$$

Proof. If $c = 0$ or $d = 0$, then G is finite: in this case $\delta_G = 0$ and both inequalities hold. We shall therefore assume that $c > 0$ and $d > 0$.

Suppose that G is upper-finite. Let $v = (v_1, \ldots, v_k) \in \Delta(G)$; then $\gamma_{v_k+1}(\cdots(\gamma_{v_1+1}(G))\cdots)$ is finite (Corollary, Lemma 10.49.1.) Since the derived length of a nilpotent group of positive class a is at most $[\log_2 a] + 1$, it follows by definition of d that

$$d \leqq \sum_{i=1}^{k} ([\log_2 v_i] + 1). \tag{26}$$

Let $\Delta^{(d)}$ denote the set of all vectors $v = (v_1, \ldots, v_k)$ with $v_i = 1, 2, 3$ or 4 for which (26) holds. Then $\Delta(G) \subseteq \Delta^{(d)}$ and

$$\delta_G \geqq \delta_{\Delta^{(d)}} = f(d) \tag{27}$$

say. Thus $f(d)$ is the minimum value of $\sum_{i=1}^{k} (v_i + 1)$ subject to the requirement (26) and $v_i = 1, 2, 3$ or 4. Hence $f(d)$ is the minimum value of

$$2r + 3s + 4t + 5u$$

where the non-negative integers r, s, t and u satisfy

$$r + 2s + 2t + 3u \geqq d.$$

Now

$$2r + 3s + 4t + 5u = 2(r + 2t + u) + 3(s + u)$$

and

$$r + 2s + 2t + 3u = (r + 2t + u) + 2(s + u).$$

Therefore $f(d)$ is also the minimum value of

$$2r + 3s$$

where the non-negative integers r and s satisfy $r + 2s \geqq d$. The minimum value of $2r + 3s$ for non-negative *real* r and s is $\frac{3}{2} d$, occurring at $r = 0$ and $s = \frac{1}{2} d$. Thus if d is even, $f(d) = \frac{3}{2} d = \left[\frac{3d+1}{2}\right]$. Let d be odd. If $r = 1$ and $s = \frac{1}{2}(d - 1)$, then $r + 2s = d$; hence

$$f(d) \leqq 2 + \frac{3}{2}(d - 1) = \frac{3d + 1}{2} = \left[\frac{3d + 1}{2}\right].$$

Hence $f(d) = \left[\frac{3d + 1}{2}\right]$ in both cases and the second inequality follows from (27).

We now drop the assumption that G is upper-finite. Let $v = (v_1, \ldots, v_k)$ be a vector with $v_i = 1, 2, 3$ or 4 and suppose that

$$(v_1 + 1) \cdots (v_k + 1) \geqq c + 1. \tag{28}$$

By Lemma 2.21 (iv)

$$\gamma_{v_k+1}(\cdots \gamma_{v_1+1}(G) \cdots) \leqq \gamma_{c+1}(G),$$

which is finite. Hence $v \in \Delta(G)$. Let Δ_c denote the set of all vectors $v = (v_1, \ldots, v_k)$ with $v_i = 1, 2, 3$ or 4 which satisfy (28); then $\Delta_c \subseteq \Delta(G)$ and

$$\delta_G \leqq \delta_{\Delta_c} = F(c)$$

say. We have to compute $F(c)$, the minimum value of $\sum_{i=1}^{k} (v_i + 1)$ subject to (28) and $v_i = 1, 2, 3$ or 4. Clearly $F(c)$ is also the minimum value of

$$2r + 3s + 4t + 5u$$

where the non-negative integers r, s, t and u satisfy

$$2^r 3^s 4^t 5^u \geqq c + 1.$$

Once again

$$2r + 3s + 4t + 5u = 2(r + 2t + u) + 3(s + u)$$

and also

$$2^{r+2t+u} 3^{s+u} = 2^r 3^s 4^t 6^u \geqq 2^r 3^s 4^t 5^u.$$

Hence $F(c)$ is the minimum value of

$$2r + 3s$$

where the non-negative integers r and s satisfy

$$2^r 3^s \geqq c + 1.$$

The minimum value for non-negative real r and s is $3 \log_3 (c + 1)$, occurring when $r = 0$ and $s = \log_3 (c + 1)$. Since

$$3^{[\log_3(c+1)]+1} \geqq 3^{\log_3(c+1)} = c + 1,$$

$r = 0$ and $s = [\log_3 (c + 1)] + 1$ are admissible values. Therefore

$$3 \log_3 (c + 1) \leqq F(c) \leqq 3 [\log_3 (c + 1)] + 3. \tag{29}$$

Let x and y denote the integral and fractional parts of $\log_3 (c + 1)$ respectively. It follows from (29) that $F(c) = 3x, 3x + 1, 3x + 2$ or $3x + 3$.

Case (i). $y = 0$; then certainly $F(c) = 3x$, this minimum occurring at $r = 0$, $s = x$.

Case (ii). $0 < y \leqq \log_3 \frac{4}{3}$. In the first place, $F(c) \geqq 3 \log_3 (c + 1) > 3x$, since $y > 0$. Now $x > 0$, for $x = 0$ would imply that $y = \log_3 (c + 1) < 1$ and hence that $c = 1$ and $y = \log_3 2$. Moreover

$$2^2 3^{x-1} = \frac{4}{3} 3^{-y} (c + 1) \geqq \frac{4}{3} 3^{-\log_3\left(\frac{4}{3}\right)} (c + 1) = c + 1,$$

so $F(c) \leqq 2 \cdot 2 + 3(x - 1) = 3x + 1$. Hence $F(c) = 3x + 1$.

Case (iii). $\log_3 \frac{4}{3} < y \leqq \log_3 2$. Since

$$2 \cdot 3^x = 2 \cdot 3^{-y}(c+1) \geqq 2 \cdot 3^{-\log_3 2}(c+1) = c+1,$$

$F(c) \leqq 2 \cdot 1 + 3 \cdot x = 3x + 2$. Also $F(c) > 3x$. Suppose that $F(c) = 3x + 1$ and that this minimum occurs at $r = r_0$ and $s = s_0$. Then $2r_0 + 3s_0 = 3x + 1$, so $2r_0 - 1$ is divisible by 3 and $r_0 = 3m - 1$ for some positive integer m; also $s_0 = \frac{1}{3}(3x + 1 - 2r_0) = x - 2m + 1$. Now

$$2^{3m-1}3^{x-2m+1} = \left(\frac{2^{3m-1}}{3^{2m-1}}\right) 3^{-y}(c+1) \geqq c+1$$

and $3^{-y} < 3^{-\log_3\left(\frac{4}{3}\right)} = \frac{3}{4}$. Hence $\left(\frac{8}{9}\right)^{m-1} > 1$, which is impossible since m is a positive integer. Therefore $F(c) = 3x + 2$.

Case (iv). $\log_3 2 < y < 1$. Here $F(c) \geqq 3\log_3(c+1) = 3x + 3y > 3x + 1$ since $3\log_3 2 > 1$. Suppose that $F(c) = 3x + 2$ and that this minimum occurs at $r = r_0$ and $s = s_0$. Then $2r_0 + 3s_0 = 3x + 2$, so $r_0 - 1$ is divisible by 3 and $r_0 = 3m + 1$ for some non-negative integer m; also $s_0 = x - 2m$. Now

$$2^{3m+1}3^{x-2m} = \frac{2^{3m+1}}{3^{2m}} 3^{-y}(c+1) \geqq c+1$$

and $3^{-y} < 3^{-\log_3 2} = \frac{1}{2}$. Therefore $\left(\frac{8}{9}\right)^m > 1$, again impossible. Therefore $F(c) = 3x + 3$. □

We shall call a group G *exact* if

$$\gamma_i(\gamma_j(G)) = \gamma_{ij}(G)$$

for all positive integers i and j. Clearly an exact soluble group is nilpotent. Suppose that the nilpotent minimax group G is exact and upper-finite; then, with the notation of the preceding proof,

$$\Delta(G) = \Delta_c$$

and

$$\delta_G = F(c);$$

here c is the least non-negative integer such that $\gamma_{c+1}(G)$ is finite. Thus we can state

Theorem 10.49.3. If the nilpotent minimax group G is exact and upper-finite, then $\delta_G = F(c)$ where c is the least positive integer such that $\gamma_{c+1}(G)$ is finite and F is the function of Theorem 10.49.2.

Examples. As an illustration let us consider the group

$$U = U(n, Q_\pi)$$

of all $n \times n$ upper unitriangular matrices with coefficients in Q_π, the ring of rational numbers whose denominators are π-numbers: here π is a finite set of primes. It is easy to verify that if $0 \leqq i < n$, then

$$U_i = \gamma_{i+1}(U) = \zeta_{n-i-1}(U)$$

coincides with the subgroup of all matrices in U which have their first i superdiagonals consisting entirely of zeros. This is easily seen to imply that $\gamma_i(\gamma_j(U)) = \gamma_{ij}(U)$, so that U is exact. Also, of course, U is nilpotent of class $n - 1$. Now U_i/U_{i+1} is isomorphic with a direct sum of $n - i - 1$ copies of the additive group Q_π. Since Q_π is an upper-finite π-minimax group, so is U. Clearly U is torsion-free. Thus

$$\tau(U) = 1 \quad \text{and} \quad \sigma(U) = U.$$

Theorem 10.49.3 and the Corollary to Theorem 10.49 combine to show that

$$m(U) = F(n - 1).$$

Clearly $F(n) \to \infty$ as $n \to \infty$. Hence *there exists torsion-free nilpotent minimax groups of arbitrary minimax length.* It is evident that $F(n)$ diverges very slowly; for example

$$m(U(10^6, Q_\pi)) = F(10^6 - 1) = 38.$$

We denote by

$$T = T(n, Q_\pi)$$

the group of invertible triangular $n \times n$ matrices over Q_π. Since $U \lhd T$ and T/U is finitely generated (π being finite), T is a soluble minimax group. Although T is neither exact nor upper-finite, it is not hard to prove that

$$m(T) = 1 + F(n-1).$$

It is noteworthy that T satisfies Max-n and is finitely generated, yet still $m(T)$ becomes unbounded with n.

For these and further results we refer the reader to Robinson [6].

10.5 Two Theorems on Finitely Generated Soluble Groups

In this final section we present two theorems which depend on and illustrate the power of P. Hall's theory of finitely generated abelian-by-nilpotent groups (Section 9.5). The first of these tells us that a finitely generated soluble group G which is not nilpotent must have a finite non-nilpotent homomorphic image, a result which enables one to deduce properties of G from those of its finite homomorphic images.

Theorem 10.51 (Robinson [15]). Let G be a finitely generated hyper-(abelian or finite) group and assume that G is not nilpotent. Then G has a finite non-nilpotent homomorphic image.

When G is polycyclic, the theorem—which does not now depend on Hall's theory—is due to Hirsch [3] (Theorem 3.26). We mention in passing a comparable result: *a non-nilpotent linear group over a finitely generated integral domain has a finite non-nilpotent homomorphic image* (Platonov [2], Wehrfritz [4]).

Proof of Theorem 10.51. Let G be a finitely generated hyper-(abelian-or-finite) group and assume that every finite homomorphic image of G is nilpotent, yet G itself is not nilpotent. By Lemma 6.17 we can even assume that every proper homomorphic image of G is nilpotent. By hypothesis there is a non-trivial normal subgroup N which is either finite or abelian. Suppose that N is finite and let $D = C_G(N)$. Then G/D is finite and hence nilpotent, while G/N is nilpotent since $N \neq 1$. Therefore $G/(N \cap D)$ is nilpotent and G is soluble because $N \cap D$ is abelian. Moreover G satisfies Max because N and G/N do, so G is polycyclic. Let F be the Fitting subgroup of G. Then $C_G(F) \leqq F$ by Lemma 2.17, so F is infinite. Theorem 2.24 shows that $\zeta(F)$ is also infinite. A suitable power of $\zeta(F)$ is a non-trivial free abelian normal subgroup of G.

We conclude that there exists an infinite abelian subgroup B such $B \lhd G$ and B is either torsion-free or an elementary abelian p-group for some prime p. Choose an element a other than 1 from B and put

$$A = a^G.$$

Then, of course, $A \leqq B$. Let us write $C = C_G(A)$ and

$$H = G/C.$$

H is finitely generated and nilpotent since $1 \neq A \leqq C$. Also A can be regarded in a natural way as a cyclic ZH-module where Z is the ring of rational integers.

Suppose that B is torsion-free. Corollary 1 to Lemma 9.53 shows that there is a free abelian subgroup X of A such that A/X is a π-group where π is a finite set of primes. Let p be a prime which is not in π. Then $A^p \cap X = X^p$. Let I be the intersection of all the A^p with $p \notin \pi$. Then

$$I \cap X = \bigcap_{p \notin \pi} X^p = 1, \tag{30}$$

since X is a free abelian group and π is finite. But A is torsion-free and A/X is periodic; hence (30) shows that $I = 1$. Now $1 \neq A^p \lhd G$; thus

G/A^p is a finitely generated nilpotent group and consequently A/A^p is finite. Since $X/X^p \simeq XA^p/A^p$, the rank of X is finite, and equal to r say. Then A has rank r and A/A^p has order dividing p^r. Therefore A/A^p lies in the rth term of the upper central series of G/A^p (Lemma 2.16). Hence

$$[A, \underbrace{G, \ldots, G}_{r}] \leqq \bigcap_{p \notin \pi} A^p = I = 1;$$

but this implies that G is nilpotent since we know that G/A is nilpotent.

By this contradiction B is an elementary abelian p-group. Let F_p be a field of p elements. Then A is clearly a cyclic F_pH-module. If the centre of H is periodic, then H is finite by Theorem 2.24; in this case A, being a cyclic F_pH-module, is finite. Hence G is polycyclic and B is finite. By this contradiction there is an element $\bar{z} = zC$ in the centre of H with infinite order. Let $J = F_p\langle t\rangle$, the group algebra over F_p of an infinite cyclic group $\langle t\rangle$. The mapping $t \to z$ allows us to turn A into a (cyclic) JH-module (see Section 9.5).

Regarding A as a J-module, we observe that the torsion-elements of A form a J-submodule T and that $T \lhd G$: here we use the fact that $\bar{z} \in \zeta(H)$. Suppose that $T \neq 1$, so that G/T is nilpotent. To each b in T there corresponds a non-zero polynomial f over F_p such that $b^{f(z)} = 1$. If f has degree n, the subgroup

$$L = \langle b^{z^i} : i = 0, \pm 1, \pm 2, \ldots\rangle$$

may be generated by $b, b^z, \ldots, b^{z^{n-1}}$ and hence is finite. Since $L = L^z$, there is a positive integer $m = m(b)$ such that z^m centralizes L and, in particular, b. Indeed, z^m even centralizes b^G since $\bar{z} \in \zeta(H)$. Now G/T is finitely generated and nilpotent, so it is finitely presented and there exists a finite set $\{b_1, \ldots, b_k\}$ such that

$$T = b_1^G \cdots b_k^G,$$

(see Lemma 1.43). If l is the least common multiple of $m(b_1), \ldots, m(b_k)$, then z^l centralizes T. Therefore, since G/T is nilpotent,

$$[A, \underbrace{z^l, \ldots, z^l}_{t}] = 1 \tag{31}$$

for some positive integer t. Since $A^p = 1$, equation (31) implies that $z^{lp^{t-1}}$ centralizes A and $\bar{z} = zC$ has finite order.

By this contradiction A is torsion-free as a J-module. By Lemma 9.54 there is a free J-submodule X of A such that A/X is a π-torsion module where π is a finite subset of a complete set of primes of J. Let f be a prime in the complete set which is not in π. Now $A^f \neq 1$ since A is torsion-free as a J-module, and $A^f \lhd G$ since $\bar{z} \in \zeta(H)$. Hence G/A^f is a finitely generated nilpotent group and A/A^f is finite. Also $A^f \cap X = X^f$ by choice of f, and

$$\bigcap_{f \notin \pi} X^f = 1;$$

for π is finite, a complete set of primes for J is infinite and X is a free module over the principal ideal domain J. It follows as before that

$$\bigcap_{f \notin \pi} A^f = 1.$$

Since $X/X^f \simeq XA^f/A^f$ if $f \notin \pi$, the J-module X/X^f is finite and X is therefore a free J-module of finite rank r. Now A/A^f is a (J/fJ)-module and J/fJ is a field because f is a prime. Hence A/A^f is a vector space over J/fJ with dimension at most r. Let us regard A/A^f as a $(J/fJ)G$-module, observing that the submodules are just the G-admissible subgroups. Since G/A^f is nilpotent, it follows that

$$[A, \underbrace{G, \ldots, G}_{r}] \leqq \bigcap_{f \notin \pi} A^f = 1$$

and G is nilpotent, our final contradiction. ☐

Corollary 1. A finitely generated hyper-(abelian or finite) group is nilpotent if and only if each of its maximal subgroups is normal.

This follows at once from the theorem and the corresponding result (due to Wielandt) for finite groups.

Corollary 2 (cf. Baer [7], Theorem 2.2, and [18], § 2, Satz 2). A hyper-abelian Baer-nilpotent group is a Gruenberg group.

Proof. Theorem 10.51 implies that the group is locally nilpotent: the corollary now follows directly from Lemma 7.33. ☐

We note in passing two comparable results. *A Baer-nilpotent group G has a normal series (ascending normal series) whose factors are abelian groups of finite rank if and only if G is a Z-group (a hypercentral group).* For special cases of these see Baer [7] (Theorem 2.2) and Kuroš and Černikov [1] (§ 10).

The crucial point for both theorems is the following: if A is a normal abelian subgroup with finite rank r in a Baer-nilpotent group G and if A is either torsion-free or an elementary abelian p-group, then $A \leqq \zeta_r(G)$. Let $a \in A$ and $g \in G$; then it can be shown by elementary arguments that $[a, {}_r g] = 1$. Wedderburn's theorem, or further elementary arguments, yields $A \leqq \zeta_r(G)$.

Corollary 3. If the Frattini subgroup of a group is finitely generated and hyper-(abelian or finite), then it is nilpotent.

Proof. Let F be the Frattini subgroup of a group G and suppose that F is finitely generated and hyper-(abelian or finite), but is not nilpotent. According to Theorem 10.51 there is a finite non-nilpotent factor group F/N. Since F is finitely generated, it has only finitely many factor groups of the same order as F/N. Hence there exists a characteristic subgroup M of F such that $M \leq N$ and F/M is finite. Hence $M \lhd G$ and F/M is not nilpotent. Now F/M is the Frattini subgroup of G/M and by the well-known "Sylow argument" a finite Frattini subgroup is always nilpotent (see M. Hall [2], p. 157, Theorem 10.4.2). Thus we have a contradiction. □

We mention a deep theorem of P. Hall which contrasts strongly with Corollary 3. *If G is a finitely generated soluble group of nilpotent length $l+1$ where $l > 0$, the Frattini subgroup of G has nilpotent length $\leq l$ and this bound is attained for all l* (P. Hall [11], Theorem 1). For a discussion of Frattini problems in infinite groups the reader may consult Hall's paper [11]: see also Dlab and Kořinek [1], P. Hall [10], Higman and Neumann [1], Hirsch [6], Itô [1], Robinson [7], Sokolov [1], Wehrfritz [4], [10], Whittemore [1], Cossey and Whittemore [1].

For comparison with Theorem 10.51 we note a similar but easier result of Baur which can be proved by a direct application of P. Hall's residual finiteness theorem (9.51).

Theorem 10.52 (Baur [1]). Let G be a group in which every finite section is abelian. If G belongs to the class $(\hat{P}_n)^\alpha \mathfrak{A}$ for some ordinal α, then G is abelian.

Proof. Let α be the first ordinal for which the theorem is false. Then α is certainly not a limit ordinal. Also there exists a finitely generated non-abelian group G whose finite sections are abelian and which belong to $(\hat{P}_n)^\alpha \mathfrak{A}$. Now

$$(\hat{P}_n)^\alpha \mathfrak{A} = \hat{P}_n((\hat{P}_n)^{\alpha-1} \mathfrak{A}),$$

so by minimality of α the group G belongs to $\hat{P}_n\mathfrak{A}$, i.e., G is an SI-group.

G/G'' is residually finite, by Theorem 9.51, and hence is abelian. Therefore $G' = G''$. Since G is an SI-group, there is a normal G-admissible abelian series in G', say $\{\Lambda_\sigma, V_\sigma : \sigma \in \Sigma\}$. Now G is finitely generated, so G' may be generated by the conjugates of finitely many non-trivial commutators $c_1, \ldots, c_n$. Let $c_i \in \Lambda_{\sigma_i} \backslash V_{\sigma_i}$ and let $\sigma = \max\{\sigma_1, \ldots, \sigma_n\}$. Then $N = V_\sigma < \Lambda_\sigma = G'$, since $\Lambda_\sigma \lhd G$. Hence G'/N is abelian and $G'' \leq N < G'$, contradicting our previous conclusion $G' = G''$. □

It is, of course, unknown whether a group whose finite sections are abelian is itself abelian.

Uniform Automorphisms

Next we note a somewhat different application of Theorem 10.51. An automorphism α of a group G is called *uniform*—the term is due to Zappa [3]—if each element of G can be written in the form $[g, \alpha] = g^{-1}g^{\alpha}$ for some $g \in G$. When G is a finite group, it is easy to see that α is uniform if and only if it is fixed-point-free. The application we have in mind is

Theorem 10.53. (Robinson [15]; for polycyclic groups see Zappa [3]). Let G be a finitely generated hyper-(abelian or finite) group and suppose that G possesses a uniform automorphism of prime order p. Then G is a finite nilpotent group with order prime to p.

Proof. Suppose that G is not nilpotent. Then, for reasons given during the proof of Corollary 3 to Theorem 10.51, there is a *characteristic* subgroup N such that G/N is finite and non-nilpotent. But α induces in G/N a uniform and hence fixed-point-free automorphism of order p. By the celebrated theorem of Thompson [1], the group G/N is nilpotent. This shows that G must be nilpotent. Next let $A = G/G'$ and let α^* be the automorphism induced in A by α. Then α^* is uniform, so $\alpha^* = 1 + \beta$ where $\beta: A \to A$ is an epimorphism. Now $1 = (\alpha^*)^p = 1 + p\gamma + \beta^p$ where γ is an endomorphism of A. Therefore $\beta^p = -p\gamma$; since β is an epimorphism, this means that $A = A^p$. However A is also finitely generated, so it must be a finite p'-group. The Corollary to Theorem 2.26 implies that G is a finite p'-group. □

On the other hand, Zappa [3] has pointed out that an infinite dihedral group $G = \langle x, a: x^2 = 1, x^{-1}ax = a^{-1}\rangle$ has an automorphism of order 2 which is fixed-point-free, namely the automorphism in which $a \to a^{-1}$ and $x \to xa$.

Soluble Groups with the Subnormal Intersection Property

We recall from Section 5.4 that a group has *the subnormal intersection property* (s.i.p.) if the intersection of any collection of subnormal subgroups is itself a subnormal subgroup. Not every soluble group has the s.i.p.—a simple example is the infinite dihedral group—and it seems to be difficult to determine the effect of imposing the s.i.p. (which is a finiteness condition) on soluble groups.

For finitely generated soluble groups, however, this problem has been solved.

Theorem 10.54 (Robinson [3]). A finitely generated soluble group has the subnormal intersection property if and only if it is finite-by-nilpotent.

The sufficiency of the condition in the theorem is easy to prove; it follows from the simple fact that a group with bounded subnormal indices has the s.i.p. (see Part 1, p. 175) and from

Lemma 10.55. Let N be a normal subgroup of a group G and assume that N has a subnormal composition series of finite length. If G/N has bounded subnormal indices, then so has G.

Proof. Let l be the subnormal composition length of N and let $G/N \in \mathfrak{B}_r$. If $H \ sn \ G$, then certainly $HN \lhd^r G$. Let s be the subnormal index of H in HN and write

$$H = H_s < \cdots < H_0 = HN$$

for the series of successive normal closures of H in HN. If $N_i = H_i \cap N$, then $H_i = H_i \cap (HN) = HN_i$. Since $H_{i+1} < H_i$, it follows that $N_{i+1} < N_i$; also of course, $N_{i+1} \lhd N_i$. The series

$$H \cap N = N_s < N_{s-1} < \cdots < N_0 = N$$

may be refined to a subnormal composition series of N and the Jordan-Hölder Theorem implies that $s \leqq l$; thus $H \lhd^l HN$. Hence $H \lhd^{l+r} G$ and $G \in \mathfrak{B}_{l+r}$. ☐

The crucial result for the proof of Theorem 10.54 is the following technical lemma.

Lemma 10.56. Let A be a normal abelian subgroup of a group G and assume that A satisfies Max-G. Assume also that all homomorphic images of G are residually finite. If x is an element of G such that $H = \langle x, A\rangle$ has the subnormal intersection property and xA belongs to the centre of G/A, then $\langle x^l, A\rangle$ is nilpotent for some positive integer l.

Proof. We shall suppose the lemma to be false; certainly this means that $\langle x\rangle \cap A = 1$. If t is a positive integer, let ξ_t be the endomorphism of A in which $a \to [a, x^t]$. If i is a positive integer, the kernel of the endomorphism ξ_t^i consists of all a in A such that $[a, {}_i x^t] = 1$ and is denoted by $K_t(i)$. By hypothesis A is abelian and xA lies in the centre of G/A, so $[a, x^t]^g = [a^g, x^t]$ for all $a \in A$ and $g \in G$. Therefore

$$K_t(i) \lhd G.$$

Now clearly $K_t(1) \leqq K_t(2) \leqq \cdots \leqq A$; thus by the property Max-G there is a positive integer $s = s(t)$ such that $K_t(s) = K_t(s+1) =$ etc.; let this subgroup be denoted by K_t.

Next $X_t = \langle x^t, A\rangle \lhd \langle x, A\rangle$ and $\langle x, A\rangle$ has the s.i.p.; therefore so does X_t. It follows that there is a positive integer $r = r(t)$ for which

$$(x^t)^{X_{t,r}} = (x^t)^{X_{t,r+1}} = \text{etc.};$$

now $L^{M,i} = \langle L, [M, \underbrace{L, \ldots, L}_{i}] \rangle$ —see Part 1, p. 173. Hence

$$(x^t)^{X_{t,i}} = \langle x^t, [A, \underbrace{x^t, \ldots, x^t}_{i}] \rangle = \langle x^t \rangle A^{\xi_t^i}$$

Since $\langle x \rangle \cap A = 1$, this implies that

$$A^{\xi_t^r} = A^{\xi_t^{r+1}} = \text{etc.}$$

a subgroup which we denote by I_t. By Fitting's Lemma (see, for example, Zassenhaus [3], p. 113, Theorem 5)

$$A = I_t \times K_t.$$

We choose the positive integer t so that K_t is maximal; it is possible to do so because A satisfies Max-G. If $K_t = A$, then $\langle x^t, A \rangle$ is nilpotent and the lemma is true. Therefore $K_t < A$. By hypothesis G/K_t is residually finite, so A/K_t contains a proper subgroup of finite index which is normal in G/K_t, say A_1/K_t. Since A/A_1 is finite, there is a positive integer m such that x^{tm} centralizes A/A_1. Hence $I_{tm} \leqq [A, x^{tm}] \leqq A_1$. Now for obvious reasons $K_t \leqq K_{tm}$, and the maximality of K_t implies that $K_t = K_{tm}$. Hence $A = I_{tm} \times K_{tm} \leqq A_1 K_t = A_1 < A$, a contradiction which completes the proof of the lemma. ☐

Let us now deal with the simplest case of Theorem 10.54.

Lemma 10.57. Let A be a normal subgroup of a group G and assume that $G = \langle x, A \rangle$ and A is a free abelian group of finite rank. If G has the subnormal intersection property, then it is nilpotent.

Proof. Let $C = C_A(x)$; since A is abelian, $C \leqq \zeta(G)$, while $\langle x \rangle \cap A \leqq C$. Hence $\langle xC \rangle$ intersects A/C trivially. Also $A/C \simeq [A, x]$, so A/C is free abelian of finite rank. Now G will be nilpotent if G/C is. For this reason we may assume that

$$\langle x \rangle \cap A = 1.$$

The group G is obviously polycyclic, so all homomorphic images of G are residually finite, by a theorem of Hirsch (Corollary 2, Theorem 9.31). The hypotheses of Lemma 10.56 are fulfilled, so there exists a least positive integer l such that $\langle x^l, A \rangle$ is nilpotent. If $l = 1$, there is nothing further to prove. Let $l > 1$ and select a prime p which divides l. If $y = x^{l/p}$, then $\langle y^p, A \rangle$ is certainly nilpotent. We define

$$A_i = A^p [A, \underbrace{y^p, \ldots, y^p}_{i}].$$

Then $A_i \lhd G$ and by the nilpotence of $\langle y^p, A\rangle$ there is an integer n such that $A_n = A^p$. Thus $A^p = A_n \leqq A_{n-1} \leqq \cdots \leqq A_0 = A$. Since $[A_i, y^p] \leqq A_{i+1}$, we can assert that $\langle y^p, A_{i+1}\rangle \lhd \langle y, A_i\rangle$ and

$$\langle y, A_i\rangle / \langle y^p, A_{i+1}\rangle$$

is a finite p-group and therefore nilpotent. It follows that, for some positive integer m,

$$[A_i, \underbrace{y, \ldots, y}_{m}] \leqq A_{i+1}, \quad (i = 0, 1, \ldots, n-1),$$

or $A_i^{(y-1)^m} \leqq A_{i+1}$. Repeated applications of the endomorphism $a \to a(y-1)^m$ lead to us

$$A^{(y-1)^{mnj}} \leqq A^{pj} \tag{32}$$

for $j = 1, 2, \ldots$ Since G has the s.i.p., so has $\langle y, A\rangle$ and

$$A^{(y-1)^t} = A^{(y-1)^{t+1}} = \text{etc.}, \tag{33}$$

for some positive integer t. By (32) and (33) we obtain

$$A^{(y-1)^{mnt}} \leqq \bigcap_{j=t}^{\infty} A^{pj} = 1,$$

A being free abelian. Therefore $\langle y, A\rangle$ is nilpotent, which contradicts the minimality of l. □

Proof of Theorem 10.54. Let G be a finitely generated soluble group with the s.i.p. We have to show that G is finite-by-nilpotent.

First of all consider the case where G is polycyclic. Then G has a cyclic series of finite length

$$1 = G_n \lhd G_{n-1} \lhd \cdots \lhd G_1 \lhd G_0 = G.$$

Evidently we may assume that $n > 1$. Now $N = G_1$ inherits the s.i.p. from G, so N is finite-by-nilpotent by induction on n. This implies that the elements with finite order in N form a characteristic finite subgroup. Hence we can assume that N is torsion-free and nilpotent. Let

$$1 = Z_0 < Z_1 < \cdots < Z_c = N$$

be the upper central series of N; then each Z_{i+1}/Z_i is torsion-free, by Theorem 2.25, and $Z_i \lhd G$. Since G/N is cyclic, we can write $G = \langle x, N\rangle$. Suppose that G/Z_{i+1} is nilpotent; then $\langle x, Z_{i+1}\rangle$ *sn* G and consequently $\langle x, Z_{i+1}\rangle/Z_i$ has the s.i.p. Now Z_{i+1}/Z_i is a free abelian group with finite rank. Thus Lemma 10.57 is applicable and we conclude that $\langle x, Z_{i+1}\rangle/Z_i$ is nilpotent. Since Z_{i+1}/Z_i is a central factor of N and $G = \langle x, N\rangle$, it follows that $Z_{i+1}/Z_i \leqq \zeta_m(G/Z_i)$ for some finite m. Therefore G/Z_i is nilpotent. By induction on $c-i$, it follows that G is nilpotent. Thus the theorem has been proved.

We revert now to the original hypothesis—G is a finitely generated soluble group with the s.i.p. By the first part of the proof it is sufficient to show that G is polycyclic. Let G have derived length $d > 1$ and let $A = G^{(d-1)}$. By induction on d the group G/A is finite-by-nilpotent, as well as finitely generated. Hence G/A is polycyclic. By a theorem of Hirsch, G/A contains a torsion-free normal subgroup G_1/A with finite index (Corollary to Theorem 9.39.3). Since G/A is finite-by-nilpotent, G_1/A is nilpotent. Also $|G: G_1|$ is finite, so G_1 is finitely generated; moreover, G will be polycyclic if G_1 is. Hence we may assume that G/A is nilpotent.

Let F be the Fitting subgroup of G; then $A \leq F$ and G/F is a finitely generated nilpotent group. If the centre of G/F were periodic, G/F would be finite (Theorem 2.24) and G would be polycyclic. Hence we may assume that xF is an element of infinite order in the centre of G/F. Thus $\langle x, F\rangle \lhd G$ and in consequence $\langle x, F\rangle$ has the s.i.p. Now G is a finitely generated abelian-by-nilpotent group. By Theorem 5.34 we conclude that G satisfies Max-n and therefore F/F' satisfies Max-G. Also, by Theorem 9.51, every homomorphic image of G is residually finite. We can now apply Lemma 10.56 to the group G/F'. It follows that there is a positive integer l such that $\langle x^l, F\rangle/F'$ is nilpotent. Since N is itself nilpotent, a well-known theorem of P. Hall (Theorem 2.27) implies that $\langle x^l, F\rangle$ is nilpotent. But $\langle x^l, F\rangle \lhd G$, so $x^l \in F$, contradicting the fact that xF has infinite order. This completes the proof of Theorem 10.54. □

We remark, for what it is worth, that the s.i.p. has not been fully used in the proof, but only the following weaker property: if $x \in G$, then $x^{G,r} = x^{G,r+1}$ for some $r = r(x)$. This is evidently equivalent to requiring each element of G to lie in a unique minimal subnormal subgroup of G.

Corollary 1. The finitely generated soluble groups which have the subnormal intersection property are precisely the groups which can be embedded in the direct product of a finite soluble group and a finitely generated torsion-free nilpotent group.

Proof. Let G be a finitely generated finite-by-nilpotent group. It is sufficient to obtain an embedding of G of the specified type. Clearly the elements in G with finite order form a normal subgroup T and G/T is finitely generated, torsion-free and nilpotent. Since G is polycyclic, there is a normal torsion-free subgroup N with finite index, by Hirsch's theorem. $N \cap T = 1$, so the mapping $g \to (gN, gT)$ is a monomorphism of G into $(G/N) \times (G/T)$. □

Corollary 2. A finitely generated hyperabelian group has the subnormal intersection property if and only if it is finite-by-nilpotent.

Proof. This follows from Lemma 6.17 and Theorem 10.54. ▯

It would be interesting to determine the structure of soluble groups of finite rank, or at least of soluble minimax groups, which have the s.i.p. Some results on this problem are in the recent paper [1] of McDougall.

In conclusion we shall mention briefly some other types of problems which have been studied recently. Rhemtulla [1] has proved the following theorem: *let G be a finitely generated soluble group; then G is polycyclic if and only if for each subgroup H the set of all intersections of finite sets of conjugates of H in G satisfies the minimal condition.*

Milnor [1], [2] and Wolf [1] have characterized those finitely generated soluble groups that are nilpotent-by-finite by means of the asymptotic behaviour of the number of elements of given length in terms of a set of generators. Rhemtulla [4] considers finitely generated soluble groups in which each element of the derived group is the product of a bounded number of commutators. For details we must refer the reader to the works cited.

Bibliography

The following is a list of items which are referred to in the text or which are relevant to material in the text. (For an extensive bibliography of the theory of infinite groups up to 1966 the reader should consult Kuroš [9] and [13].) Russian language papers are marked with an asterisk and, as is customary, their titles have been translated into English. Where an English translation of such a work exists, this is given immediately after the original.

Abramovskiĭ, I. N.

*1. Locally generalized Hamiltonian groups. Sibirsk. Mat. Ž. **7**, 481–485 (1966) = Siberian Math. J. **7**, 391–393 (1966).

*2. The structure of locally generalized Hamiltonian groups. Leningrad Gos. Ped. Inst. Učen. Zap. **302**, 43–49 (1967).

Abramovskiĭ, I. N. Kargapolov, M. I.

*1. Finite groups with the property of transitivity for normal subgroups. Uspehi Mat. Nauk. **13**, 232–243 (1958).

Ado, I. D.

*1. On nilpotent algebras and p-groups. Dokl. Akad. Nauk. SSSR **40**, 299–301 (1943).

*2. On subgroups of the countable symmetric group. Dokl. Akad. Nauk. SSSR **50**, 15–18 (1945).

*3. On locally finite p-groups with the minimal condition for normal subgroups. Dokl. Akad. Nauk. **54**, 471–473 (1946).

*4. Proof of the countability of a locally finite p-group with the minimal condition for normal subgroups. Dokl. Akad. Nauk. **58**, 523–524 (1947).

Alperin, J. L.

1. Groups with finitely many automorphisms. Pacific J. Math. **12**, 1–5 (1962).

2. A classification of n-abelian groups. Canad. J. Math. **21**, 1238–1244 (1969).

Alperin, J. L., Brauer, R., Gorenstein, D.

1. Finite groups with quasi-dihedral and wreathed Sylow 2-subgroups. Trans. Amer. Math. Soc. **151**, 1–261 (1970).

Amberg, B.

1. Fast-Polyminimaxgruppen. Math. Ann. **175**, 44–49 (1968).

2. Noethersche Gruppen mit Normalisatorbedingungen. Arch. Math. (Basel) **19**, 265–278 (1968).

3. Gruppen mit Minimalbedingung für Subnormalteiler. Arch. Math. (Basel) **19**, 348—358 (1968).
4. Gruppentheoretische Eigenschaften und Normalisatorbedingungen. Rend. Sem. Mat. Univ. Padova **41**, 97—118 (1968).
5. Groups with maximum conditions, Pacific J. Math. **32**, 9—19 (1970).
6. On groups with chain conditions. Canad. J. Math. **23**, 151—159 (1971).
7. Abelian factorisations of infinite groups. Math. Z. **123**, 201—214 (1971).

Amberg B., Scott, W. R.
1. Products of abelian subgroups. Proc. Amer. Math. Soc. **26**, 541—547 (1970).

Amitsur, S.
1. A general theory of radicals. Amer. J. Math. **74**, 774—786 (1952); **76**, 100—125 (1954); **76**, 126—136 (1954).

Andreadakis, S.
1. On semicomplete groups. J. London Math. Soc. **44**, 361—364 (1969).

Andreev, K. K., Ol'sanskii, A. Yu.
*1. The approximation of groups. Vestnik Moskov. Univ. Ser. I Mat. Meh. **23**, 60—62 (1968).

Asiatiani, R. V.
*1. On a certain characteristic subgroup. Soobšč. Akad. Nauk. Gruzin SSR **41**, 3—10 (1960).

Auslander, L.
1. On a problem of Philip Hall. Ann. of Math. (2) **86**, 112—116 (1967).
2. The automorphism group of a polycyclic group. Ann. of Math. (2) **89**, 314—322 (1970).

Auslander, L., Baumslag, G.
1. Automorphism groups of finitely generated nilpotent groups. Bull. Amer. Math. Soc. **73**, 716—717 (1967).

Ayoub, C.
1. On the units in certain integral domains. Arch. Math. (Basel) **19**, 43—46 (1968).
2. On properties possessed by solvable and nilpotent groups. J. Austral. Math. Soc. **9**, 218—227 (1969).

Bachmuth, S., Heilbronn, H. A., Mochizuki, H. Y.
1. Burnside metabelian groups. Proc. Roy. Soc. Ser. A **307**, 235—250 (1968).

Bachmuth, S., Lewin, J.
1. The Jacobi identity in groups. Math. Z. **83**, 170—176 (1964).

Bachmuth, S., Mochizuki, H. Y.
1. The class of the free metabelian group with exponent p^2. Comm. Pure Appl. Math. **21**, 385—399 (1968).
2. Kostrikin's theorem on Engel groups of prime power exponent. Pacific J. Math. **26**, 197—213 (1968).

Bachmuth, S., Mochizuki, H. Y., Walkup, D.
1. A nonsolvable group of exponent 5. Bull. Amer. Math. Soc. **76**, 638—640 (1970).

Bačurin, G. F.
*1. Groups with an ascending central series. Mat. Sb. **45**, 105—112 (1958).
*2. On a class of nilpotent groups. Trudy Magnitogorsk Gorno. Metall. Inst. **16**, 99—112 (1958).
*3. Mixed ZA-groups with finite centre. Mat. Sb. **52**, 879—890 (1960).

*4. A criterion for finiteness of rank of a nilpotent torsion-free group. Izv. Vysš. Učebn. Zaved. Mat. **6**, 25–28 (1963).

*5. On one-sheeted p-groups with an ascending central series. Izv. Vysš. Učebn. Zaved. Mat. **5**, 27–30 (1965).

*6. On torsion-free almost nilpotent groups. Algebra i Logika **7**, 27–32 (1968) = Algebra and Logic **7**, 14–16 (1968).

*7. On multipliers of torsion-free nilpotent groups. Mat. Zametki **3**, 541–544 (1969) = Math. Notes **3**, 325–327 (1969).

Baer, R.

1. Situation der Untergruppen und Struktur der Gruppe. Sitz. Ber. Heidelberg Akad. **2**, 12–17 (1933).
2. Der Kern, eine charakteristische Untergruppe. Compositio Math. **1**, 254–283 (1934).
3. Die Kompositionsreihe der Gruppe aller eineindeutigen Abbildungen einer unendliche Menge auf sich. Studia Math. **5**, 15–17 (1934).
4. Groups with abelian central quotient group. Trans. Amer. Math. Soc. **44**, 357–386 (1938).
5. Groups with preassigned central and central quotient group. Trans. Amer. Math. Soc. **44**, 387–412 (1938).
6. Almost hamiltonian groups. Compositio Math. **6**, 382–406 (1939).
7. Nilpotent groups and their generalizations. Trans. Amer. Math. Soc. **47**, 393–434 (1940).
8. Sylow theorems for infinite groups. Duke Math. J. **6**, 598–614 (1940).
9. The higher commutator subgroups of a group. Bull. Amer. Math. Soc. **50**, 143–160 (1944).
10. Groups without proper isomorphic quotient groups. Bull. Amer. Math. Soc. **50**, 267–278 (1944).
11. Representations of groups as quotient groups. Trans. Amer. Math. Soc. **58**, 295–419 (1945).
12. Finiteness properties of groups. Duke Math. J. **15**, 1021–1032 (1948).
13. Groups with descending chain condition for normal subgroups. Duke Math. J. **16**, 1–22 (1949).
14. Endlichkeitskriterien für Kommutatorgruppen. Math. Ann. **124**, 161–177 (1952).
15. Factorization of n-soluble and n-nilpotent groups. Proc. Amer. Math. Soc. **4**, 15–26 (1953).
16. The hypercenter of a group. Acta Math. **89**, 165–208 (1953).
17. Das Hyperzentrum einer Gruppe II. Arch. Math. (Basel) **4**, 86–96 (1953).
18. Das Hyperzentrum einer Gruppe III. Math. Z. **59**, 299–338 (1953).
19. Das Hyperzentrum einer Gruppe IV. Arch. Math. (Basel) **5**, 56–59 (1954).
20. Nil-Gruppen. Math. Z. **62**, 402–437 (1955).
21. Supersoluble groups. Proc. Amer. Math. Soc. **6**, 16–32 (1955).
22. Burnsidesche Eigenschaften. Arch. Math. **6**, 165–169 (1955).
23. Finite extensions of abelian groups with minimum condition. Trans. Amer. Math. Soc. **79**, 521–540 (1955).
24. Auflösbare Gruppen mit Maximalbedingung. Math. Ann. **129**, 139–173 (1955).
25. Norm and hypernorm, Publ. Math. Debrecen **4**, 347–350. (1956)
26. Noethersche Gruppen. Math. Z. **66**, 269–288 (1956/1957).
27. Lokal Noethersche Gruppen. Math. Z. **66**, 341–363 (1956/1957).
28. Engelsche Elemente Noetherscher Gruppen. Math. Ann. **133**, 256–270 (1957).

29. Die Potenzen einer gruppentheoretischen Eigenschaft. Abh. Math. Sem. Univ. Hamburg **22**, 276–294 (1958).
30. Überauflösbare Gruppen. Abh. Math. Sem. Univ. Hamburg **23**, 11–28 (1959).
31. Kriterien für die Endlichkeit von Gruppen. Jahresber. Deutsch. Math. Verein. **63**, 53–77 (1960).
32. Abzählbar erkennbare gruppentheoretische Eigenschaften. Math. Z. **79**, 344–363 (1962).
33. Gruppentheoretische Eigenschaften. Math. Ann. **149**, 181–210 (1963).
34. Gruppen mit Minimalbedingung. Math. Ann. **150**, 1–44 (1963).
35. Irreducible groups of automorphisms of abelian groups. Pacific J. Math. **14**, 385–406 (1964).
36. Erreichbare und engelsche Gruppenelemente. Abh. Math. Sem. Univ. Hamburg **27**, 44–74 (1964).
37. The hypercenter of functorially defined subgroups. Illinois J. Math. **8**, 177–230 (1964).
38. Der reduzierte Rang einer Gruppe. J. reine angew. Math. **214**, 146–173 (1964).
39. Groups with minimum condition. Acta Arith. **9**, 117–132 (1964).
40. Die Sternbedingung: eine Erweiterung der Engelbedingung. Math. Ann. **162**, 54–73 (1965).
41. Noethersche Gruppen II. Math. Ann. **165**, 163–180 (1966).
42. Endlich definierbare gruppentheoretische Funktionen. Math. Z. **87**, 163–213 (1965).
43. Group theoretical properties and functions. Colloq. Math. **14**, 285–328 (1966).
44. Local and global hypercentrality and supersolubility. Nederl. Akad. Wetensch. Proc. Ser. A **69**, 93–126 (1966).
45. Nilpotency. Proc. Internat. Conf. Theory of Groups, Canberra 1965, pp. 11–15 (1967).
46. Noetherian soluble groups. Proc. Internat. Conf. Theory of Groups, Canberra 1965, pp. 17–32 (1967).
47. Noetherian groups. Proc. Internat. Conf. Theory of Groups, Canberra 1965, pp. 33–36 (1967).
48. Soluble artinian groups. Canad. J. Math. **19**, 904–923 (1967).
49. Normalisatorreiche Gruppen. Rend. Sem. Mat. Univ. Padova **38**, 358–450 (1967).
50. Auflösbare, artinsche, noethersche Gruppen. Math. Ann. **168**, 325–363 (1967).
51. Polyminimaxgruppen. Math. Ann. **175**, 1–43 (1968).
52. Gruppen mit abzählbaren Automorphismengruppen. Hamburg Math. Einzelschr. 2 (1969).
53. Lokal endliche-auflösbare Gruppen mit endlichen Sylowuntergruppen. J. reine angew. Math. **239/240**, 109–144 (1969).
54. Fast-zyklische Gruppen. Arch. Math. (Basel) **21**, 225–239 (1970).
55. The determination of groups by their groups of automorphisms. Studies in Pure Mathematics (1971).

Baer, R., Heineken, H.

1. Radical groups with finite abelian subgroup rank. Illinois J. Math.

Balcerzyk, S.

1. On classes of abelian groups. Bull. Acad. Polon. Sci. Ser. Sci. Math. Astronom. Phys. **9**, 327–329 (1961).

2. On classes of abelian groups. Fund. Math. **51**, 149–178 (1962).

Bass, H., Lazard, M., Serre, J.-P.
1. Sous-groupes d'indice fini dans $SL(n, Z)$. Bull. Amer. Math. Soc. **70**, 385–392 (1964).

Baumslag, G.
1. A theorem on infinite groups. Proc. Cambridge Philos. Soc. **53**, 545–548 (1957).
2. Finite factors in infinite ascending derived series. Math. Z. **68**, 465–478 (1958).
3. Wreath products and p-groups. Proc. Cambridge Philos. Soc. **55**, 224–231 (1959).
4. Some aspects of groups with unique roots. Acta Math. **104**, 217–303 (1960).
5. Roots and wreath products. Proc. Cambridge Philos. Soc. **56**, 109–117 (1960).
6. Some remarks on nilpotent groups with roots. Proc. Amer. Math. Soc. **12**, 262–267 (1961).
7. A remark on hyperabelian groups. Arch. Math. (Basel) **12**, 321–323 (1961).
8. A generalization of a theorem of Mal'cev. Arch. Math. (Basel) **12**, 405–408 (1961).
9. Wreath products and finitely presented groups. Math. Z. **75**, 22–28 (1961).
10. A non-hopfian group. Bull. Amer. Math. Soc. **68**, 196–198 (1962).
11. A remark on generalized free products. Proc. Amer. Math. Soc. **13**, 53–54 (1962).
12. Hopficity and abelian groups. Proc. New Mexico Symposium on Abelian Groups, 331–335 (1962).
13. On abelian hopfian groups I. Math. Z. **78**, 53–54 (1962).
14. On generalized free products. Math. Z. **78**, 423–438 (1962).
15. On the residual finiteness of certain generalized free products of nilpotent groups. Trans. Amer. Math. Soc. **106**, 193–209 (1963).
16. Some subgroup theorems for free $\mathfrak{v}$-groups. Trans. Amer. Math. Soc. **108**, 516–525 (1963).
17. On the residual nilpotence of some varietal products. Trans. Amer. Math. **109**, 357–365 (1963).
18. Wreath products and extensions. Math. Z. **81**, 286–299 (1963).
19. Automorphism groups of residually finite groups. J. London Math. Soc. **38**, 117–118 (1963).
20. Groups with one defining relator. J. Austral. Math. Soc. **4**, 385–392 (1964).
21. On a problem of Plotkin concerning locally nilpotent groups. Math. Z. **83**, 25–26 (1964).
22. Finitely presented groups, Proc. Internat. Conf. Theory of Groups Canberra 1965. pp. 37–50 (1967).
23. On free $\mathfrak{D}$-groups Comm. Pure Appl. Math. **18**, 25–30 (1965).
24. Residually finite one-relator groups, Bull. Amer. Math. Soc. **73**, 618–620 (1967).
25. Products of abelian hopfian groups. J. Austral. Math. Soc. **8**, 322–326 (1968).

26. Automorphism groups of nilpotent groups. Amer. Math. J. **91**, 1003–1011 (1969).
27. A non-cyclic one relator group all of whose finite quotients are cyclic. J. Austral. Math. Soc. **10**, 497–498 (1969).
28. Lecture notes on nilpotent groups. Providence: Amer. Math. Soc. 1971.

Baumslag, G., Blackburn, N.
1. Groups with cyclic upper central factors. Proc. London Math. Soc. (3) **10**, 531–544 (1960).

Baumslag, G., Kovács, L. G., Neumann, B. H.
1. On products of normal subgroups. Acta Sci. Math. (Szeged) **26**, 145–147 (1965).

Baumslag, G., Solitar, D.
1. Some two-generator one-relator non-Hopfian groups. Bull. Amer. Math. Soc. **68**, 199–201 (1962).

Baur, H.
1. Ein Kommutativitätskriterium für unendliche auflösbare Gruppen. Arch. Math. (Basel) **11**, 176–182 (1960).

Bender, H.
1. Finite groups having a strongly embedded subgroup. Theory of finite groups, ed. R. Brauer and C. H. Sah,: Benjamin 1969, pp. 21–24.

Beran, L.
1. A note on Chehata's groups. Comment. Math. Univ. Carolinae **7**, 117–121 (1966).

Berlinkov, M. L.
*1. On the lattice of subgroups of a group with finite layers. Uspehi Mat. Nauk. **12**, 267–271 (1957).

Berman, S. D., Lyubimov, V. V.
*1. Groups allowing arbitrary permutation of the factors of their composition series. Uspehi Mat. Nauk. **12**, 181–183 (1957).

Best, E., Taussky, O.
1. A class of groups. Proc. Roy. Irish Acad. Sect. A **47**, 55–62 (1942).

Betten, A.
1. Hinreichende Kriterien für die Hyperzentralität einer Gruppe. Arch. Math. (Basel) **20**, 471–480 (1970).

Birkhoff, G.
1. Transfinite subgroup series. Bull. Amer. Math. Soc. **40**, 847–850 (1934).
2. The structure of abstract algebras. Proc. Cambridge Philos. Soc. **31**, 433–454 (1935).

Blackburn, N.
1. Über das Produkt von zwei zyklischen 2-Gruppen. Math. Z. **68**, 422–427 (1958).
2. On a special class of p-groups. Acta Math. **100**, 45–92 (1958).
3. Nilpotent groups in which the derived group has two generators. J. London Math. Soc. **35**, 33–35 (1960).
4. Some remarks on Černikov p-groups. Illinois J. Math. **6**, 421–433 (1962).
5. Conjugacy in nilpotent groups. Proc. Amer. Math. Soc. **16**, 143–148 (1965).

Bowers, J. F.
1. On composition series of polycyclic groups. J. London Math. Soc. **35**, 433–444 (1960).

Brauer, R.

1. Some applications of the theory of blocks of characters II. J. Algebra **1**, 307–334 (1964).

Brauer, W.

1. Einige Endlichkeitskriterien für Gruppen. J. reine angew. Math. **239/240** 321–332 (1969).

Brenner, J. L.

1. Quelques groupes libres de matrices. C. R. Acad. Sci. Paris **241**, 1689–1691 (1955).
2. The linear homogeneous group III. Ann. of Math. (2) **71**, 210–223 (1960).

Bride, I. M.

1. Second nilpotent *BFC*-groups. J. Austral. Math. Soc. **11**, 9–18 (1970).

Brisley, W., Macdonald, I. D.

1. Two classes of metabelian groups. Math. Z. **112**, 5–12 (1969).

Bruck, R. H.

1. On the restricted Burnside problem. Arch. Math. (Basel) **13**, 179–186 (1962).
2. Engel conditions in groups and related questions, Lecture notes 3rd. Summer Research Inst., Austral. Math. Soc. (1963).

de Bruijn, N. G.

1. Embedding theorems for infinite groups. Nederl. Akad. Wetensch. Proc. Ser. A **60**, 560–569 (1957); Addendum **67**, 594–595 (1964).

Brumberg, N. R.

*1. The connection of wreath products with other operations on groups. Sibirsk. Mat. Z. **4**, 1221–1234 (1963).

*2. On the commutator of two varieties. Mat. Sb. **79**, 37–58 (1969) = Math. USSR-Sb. **8**, 33–51 (1969).

Bryce, R. A.

1. On metabelian groups of prime power exponent. Proc. Roy. Soc. London Ser. A **310**, 393–399 (1969).

Bunt, A. Ya.

*1. On the automorphism group of the generalized wreath product. Latvijas Mat. Ežgodnik Riga **3**, 81–88 (1968).

Burnside, W.

1. The theory of groups of finite order, London: Cambridge Univ. Press 1897, 2nd. ed. 1911.
2. On an unsettled question in the theory of discontinuous groups. Quart. J. Pure Appl. Math. **33**, 230–238 (1901).
3. On soluble groups of linear substitutions. Quart. J. Pure Appl. Math. **33**, 242–244 (1901).
4. On groups in which every two conjugate operations are permutable. Proc. London Math. Soc. **35**, 28–37 (1902).
5. On criteria for the finiteness of the order of a group of linear substitutions. Proc. London Math. Soc. (2) **3**, 435–440. (1905).

Burroughs, J. E., Schafer, J. A.

1. Subgroups of conjugate classes in extensions. Canad. J. Math. **22**, 773–783 (1970).

Busarkin, V. M., Starostin, A. I.

*1. Locally finite groups with a partition. Mat. Sb. **62**, 275–294 (1963).

Calenko, M. S.

*1. Some remarks on infinite simple groups. Sibirsk. Mat. Ž. **4**, 227–231 (1963).

Camm, R.
1. Simple free products. J. London Math. Soc. **28**, 66–76 (1953).

Čan Van Hao.
*1. On semi-simple classes of groups. Sibirsk. Mat. Ž. **3**, 943–949 (1962)
*2. On the minimal radical class containing the class of abelian groups. Dokl. Akad. Nauk. SSSR **149**, 1270–1273 (1963) = Soviet Math. Dokl. **4**, 552–555 (1963).
*3. Nilgroups of finite rank. Sibirsk. Mat. Ž. **5**, 459–464 (1964).

Čarin, V. S.
*1. A remark on the minimal condition for subgroups. Dokl. Akad. Nauk. SSSR **66**, 575–576 (1949).
*2. On complete groups with a radical series of finite length. Dokl. Akad. Nauk. SSSR **66**, 809–811 (1949).
*3. On the theory of locally nilpotent groups. Mat. Sb. **29**, 433–454 (1951) = Amer. Math. Soc. Translations (2) **15**, 33–54 (1960).
*4. On the minimal condition for normal subgroups of locally soluble groups. Mat. Sb. **33**, 27–36 (1953).
*5. On groups of automorphisms of certain classes of soluble groups. Ukrain. Mat. Ž. **5**, 363–369 (1953).
*6. On groups of automorphisms of nilpotent groups. Ukrain. Mat. Ž. **6**, 295–304 (1954).
*7. On the theory of nilpotent groups. Ural. Gos. Univ. Mat. Zap. **19**, 21–25 (1956).
*8. On locally soluble groups of finite rank. Mat. Sb. **41**, 37–48 (1957).
*9. On groups possessing soluble ascending invariant series. Mat. Sb. **41**, 297–316 (1957).
*10. On soluble groups of type A_4. Mat. Sb. **52**, 895–914 (1960).
*11. On soluble groups of type A_3. Mat. Sb. **54**, 489–499 (1961).
*12. A remark on groups possessing an ascending soluble normal series. Ural. Gos. Univ. Mat. Zap. **3**, 50–54 (1962).

Cartier, P.
1. Séminaire Sophus Lie, École Norm. Sup. Paris (1954–1955).

Černikov, S. N.
*1. Extension of a theorem of Frobenius to infinite groups. Mat. Sb. 3, 413–416 (1938).
*2. On a theorem of Frobenius. Mat. Sb. **4**, 531–539 (1938).
*3. On infinite special groups. Mat. Sb. **6**, 199–214 (1939).
*4. Infinite locally soluble groups. Mat. Sb. **7**, 35–64 (1940).
*5. On groups with a Sylow set. Mat. Sb. **8**, 377–394 (1940).
*6. On the theory of infinite special groups. Mat. Sb. **7**, 539–548 (1940).
*7. On the theory of locally soluble groups Mat. Sb. **13**, 317–333 (1943).
*8. On the theory of infinite p-groups. Dokl. Akad. Nauk. SSSR **50**, 71–74 (1945).
*9. On infinite special groups with finite centres. Mat. Sb. **17**, 105–130 (1945).
*10. Complete groups with an ascending central series. Mat. Sb. **18**, 397–422 (1946).
*11. On the theory of finite p-extensions of abelian p-groups. Dokl. Akad. Nauk. SSSR **58**, 1287–1289 (1947).
*12. Infinite groups with finite layers. Mat. Sb. **22**, 101–133 (1948) = Amer. Math. Soc. Translations (1) **56**, 51–102 (1951).

Černikov, S. N.
*13. On the theory of complete groups. Mat. Sb. **22**, 319–348 (1948); Addendum **22**, 455–456 (1948) = Amer. Math. Soc. Translations (1) **56**, 3–49 (1951).
*14. On the theory of special p-groups. Dokl. Akad. Nauk. SSSR **63**, 11–14 (1948).
*15. On the theory of torsion-free groups having ascending central series. Ural. Gos. Univ. Mat. Zap. **7**, 3–21 (1949).
*16. On the theory of locally soluble groups with the minimal condition for subgroups Dokl. Akad. Nauk. SSSR **65**, 21–24 (1949).
*17. On complete groups with an ascending central series. Dokl. Akad. Nauk. SSSR **70**, 965–968 (1950).
*18. On the minimal condition for abelian subgroups. Dokl. Akad. Nauk. SSSR **75**, 345–347 (1950).
*19. Periodic ZA-extensions of complete groups. Mat. Sb. **27**, 117–128 (1950).
*20. On special p-groups Mat. Sb. **27**, 185–200 (1950).
*21. On the centralizer of a complete abelian normal subgroup of an infinite periodic group. Dokl. Akad. Nauk. SSSR **72**, 243–246 (1950).
*22. On locally soluble groups which satisfy the minimal condition for subgroups. Mat. Sb. **28**, 119–129 (1951).
*23. On groups with finite classes of conjugate elements. Dokl. Akad. Nauk. SSSR **114**, 1177–1179 (1957).
*24. On the structure of groups with finite classes of conjugate elements. Dokl. Akad. Nauk. SSSR **115**, 60–63 (1957).
*25. On layer-finite groups. Mat. Sb. **45**, 415–416 (1958).
*26. Finiteness conditions in the general theory of groups. Uspehi Mat. Nauk. **14**, 45–96 (1959) = Amer. Math. Soc. Translations (2) **84**, 1–67 (1969) (with a supplement by the author).
*27. Infinite locally finite groups with finite Sylow subgroups. Mat. Sb. **52**, 647–652 (1960).
*28. Local finiteness conditions for single layer p-groups. Dokl. Akad. Nauk. SSSR **147**, 49–52 (1962) = Soviet Math. Dokl. **3**, 1563–1566 (1963).
29. Endlichkeitsbedingungen in der Gruppentheorie, Math. Forschungsberichte XX, Berlin: VEB Deutscher Verlag der Wissenschaften, 1963; (German translation of [26] with a supplement by the author).
*30. Infinite groups with prescribed properties of their systems of infinite subgroups. Dokl. Akad. Nauk. SSSR **159**, 759–760 (1964) = Soviet Math. Dokl. **5**, 1610–1611 (1964).
*31. Groups with given properties for systems of infinite subgroups. Dokl. Akad. Nauk. SSSR **171**, 806–809 (1966) = Soviet Math. Dokl. **7**, 1565–1568 (1966).
*32. Groups with prescribed properties for systems of infinite subgroups. Ukrain. Mat. Ž. **19**, 111–131 (1967).
*33. The normalizer condition. Mat. Zametki **3**, 45–50 (1968) = Math. Notes **3**, 28–30 (1968).
*34. Periodic groups of automorphisms of extremal groups. Math. Zametki **4**, 91–96 (1968) = Math. Notes **4**, 543–545 1968).
*35. Infinite non-abelian groups with the minimal condition for non-normal abelian subgroups. Dokl. Akad. Nauk. SSSR **184** (1969), 786–789 (1969) = Soviet Math. Dokl. **10**, 172–175 (1969).

*36. An investigation of groups with properties prescribed on their subgroups. Ukrain. Mat. Ž. **21**, 193–209 (1969) = Ukrainian Math. J. **21** (1969).

*37. Infinite non-abelian groups with a minimal condition for non-normal subgroups. Mat. Zametki **6**, 11–18 (1969) = Math. Notes **6**, 465–468 (1969).

Chehata, S.

1. An algebraically simple ordered group. Proc. London Math. Soc. (3) **2**, 183–197 (1952).

Chong-Yun Chao

1. A theorem on nilpotent groups. Proc. Amer. Math. Soc. **19**, 959–960 (1968).

Clapham, C. R. J.

1. Finitely presented groups with word problems of arbitrary degrees of insolubility. Proc. London Math. Soc. (3) **14**, 633–676 (1964).
2. An embedding theorem for finitely generated groups. Proc. London Math. Soc. (3) **17**, 419–430 (1967).

Cleave, J. P.

1. Local properties of systems. J. London Math. Soc. **44**, 121–130 (1969).

Clowes, J. S., Hirsch, K. A.

1. Simple groups of infinite matrices. Math. Z. **58**, 1–3 (1953).

Cohen, D. E.

1. On the laws of a metabelian variety. J. Algebra **5**, 267–273 (1967).

Cohn, P. M.

1. A countably generated group which cannot be covered by finite permutable subsets. J. London Math. Soc. **29**, 248–249 (1954).
2. A non-nilpotent Lie ring satisfying the Engel condition and a non-nilpotent Engel group. Proc. Cambridge Philos. Soc. **51**, 401–405 (1955).
3. A remark on the general product of two infinite cyclic groups. Arch. Math. (Basel) **7**, 94–99 (1956).
4. Groups of order automorphisms of ordered sets. Mathematika **4**, 41–50 (1957).

Collins, D. J.

1. On recognizing Hopf groups. Arch. Math. (Basel) **22**, 235–240 (1969).

Conrad, P.

1. Completions of groups of class 2. Illinois J. Math. **5**, 212–224 (1961).
2. Skew tensor products and groups of class two. Nagoya Math. J. **23**, 15–51 (1963).

Cooper, C. D. H.

1. Power automorphisms of a group. Math. Z. **107**, 335–356 (1968).

Corner, A. L. S.

1. Three examples on hopficity in torsion-free abelian groups. Acta Math. Acad. Sci. Hungar. **16**, 303–310 (1965).

Cossey, J., Whittemore, A.

1. On the Frattini subgroup. Proc. Amer. Math. Soc. **21**, 699–702 (1969).

Coxeter, H. S. M., Moser, W. O. J.

1. Generators and relations for discrete groups. Berlin/Göttingen/Heidelberg: Springer 1957.

Curtis, C., Reiner, I.

1. Representation theory of finite groups and associative algebras, New York: Interscience 1962.

Curzio, M.
1. Sugli automorfismi uniformi nei gruppi a condizione minimale. Riv. Mat. Univ. Parma (2) **1**, 107–122 (1960).
2. Alcuni criteri di finitezza per i gruppi a condizione massimale o minimale. Ricerche Mat. **9**, 248–254 (1960).
3. Sui gruppi di torsione a fattoriali abeliani. Ricerche Mat. **16**, 154–161 (1967).

Curzio, M., Permutti, R.
1. Distributività nel reticolo dei sottogruppi normali di un T-gruppo. Matematiche (Catania) **20**, 46–63 (1965).

Dark, R. S.
1. On subnormal embedding theorems for groups. J. London Math. Soc. **43**, 387–390 (1968).
2. A prime Baer group. Math. Z. **105**, 294–298 (1968).

Dark, R. S., Rhemtulla, A. H.
1. On $\boldsymbol{R}_0$-closed classes and finitely generated groups. Canad. J. Math. **22**, 176–184 (1970).

Dauns, J., Hoffmann, K. H.
1. Nilpotent groups and automorphisms. Acta Sci. Math. (Szeged) **29**, 225–246 (1968).

Dedekind, R.
1. Über Gruppen, deren sämtliche Teiler Normalteiler sind. Math. Ann. **48**, 548–561 (1897).

Dey, I. M. S.
1. Free products of Hopf groups. Math. Z. **85**, 274–284 (1964).
2. Embeddings in non-Hopf groups. J. London Math. Soc. (2) **1**, 745–749 (1969).
3. Free products and residual nilpotency. Bull. Austral. Math. Soc. **1**, 11–13 (1969).

Dey, I. M. S., Neumann, H.
1. The Hopf property of free products. Math. Z. **117**, 325–339 (1970).

Dickson, S. E.
1. On torsion classes of abelian groups. J. Math. Soc. Japan. **17**, 30–35 (1965).

Dietzmann (Dicman), A. P.
*1. On p-groups. Dokl. Akad. Nauk. SSSR **15**, 71–76 (1937).
2. Sur les groupes infinis. C. R. Acad. Sci. Paris **205**, 952–953 (1937).
*3. On the centre of p-groups. Trudy. Sem. Teorii Grupp 30–34 (1938).
*4. Some theorems on infinite groups. Sb. Pamyati Akad. Grave, pp. 63–67 (1940).
5. On an extension of Sylow's theorem. Ann. of Math. (2) **48**, 137–146 (1947).
*6. On Sylow's theorem. Dokl. Akad. Nauk. SSSR **59**, 1235–1236 (1948).
*7. On multigroups whose elements are subsets of a group. Moskov. Gos. Ped. Inst. Učen. Zap. **71**, 71–79 (1953).

Dietzmann, A. P., Kuroš, A. G., Uzkov, A. I.
1. Sylowsche Untergruppen von unendlichen Gruppen. Mat. Sb. **3**, 179–185 (1938).

Dinerstein, N. T.
1. Finiteness conditions in groups with systems of complemented subgroups. Math. Z. **106**, 321–326 (1968).

Dixmier, S.
1. Exposants des quotients des suites centrales descendante et ascendante d'un groupe. C. R. Acad. Sci. Paris **259**, 2751—2753 (1964).

Dixon, J. D.
1. Complete reducibility in infinite groups. Canad. J. Math. **16**, 267—274 (1964).
2. Complements of normal subgroups in infinite groups. Proc. London Math. Soc. (3) **17**, 431—446 (1967).
3. The Fitting subgroup of a linear solvable group. J. Austral. Math. Soc. **7**, 417—424 (1967).
4. The solvable length of a solvable linear group. Math. Z. **107**, 151—158 (1968).

Dixon, J., Poland, J., Rhemtulla, A. H.
1. A generalization of Hamiltonian and nilpotent groups. Math. Z. **112**, 335—339 (1969).

Dlab, V.
1. On cyclic groups. Czechoslovak Math. J. **10**, 244—254 (1960).
2. A note on powers of a group. Acta Sci. Math. (Szeged) **25**, 177—178 (1964).
3. A remark on a paper of Gh. Pic. Czechoslovak Math. J. **17**, 467—468 (1967).
4. On a family of simple ordered groups. J. Austral. Math. Soc. **8**, 591—608 (1968).

Dlab, V., Kořinek, V.
1. The Frattini subgroup of a direct product of groups. Czechoslovak Math. J. **10**, 350—358 (1960).

Duguid, A. M.
1. A class of hyper *FC*-groups. Pacific J. Math. **10**, 117—120 (1960).

Duguid, A. M., Mclain, D. H.
1. *FC*-nilpotent and *FC*-soluble groups. Proc. Cambridge Philos. Soc. **52**, 391—398 (1956).

Dunwoody, M. J.
1. On verbal subgroups of free groups. Arch. Math. (Basel) **16**, 153—157 (1965).

Durbin, J. R.
1. Finite supersolvable wreath products. Proc. Amer. Math. Soc. **17**, 215—218 (1966).
2. Commutativity and n-abelian groups. Math. Z. **98**, 89—92 (1967).
3. Residually central elements in groups. J. Algebra **9**, 408—413 (1968).
4. On normal factor coverings of groups. J. Algebra **12**, 191—194 (1969).

Dyer, J. L.
1. On the residual finiteness of generalized free products. Trans. Amer. Math. Soc. **133**, 131—143 (1968).
2. On the isomorphism problem for polycyclic groups. Math. Z. **112**, 145—153 (1969).

Dyubyuk, P. E.
*1. On subgroups of finite index in infinite groups. Mat. Sb. **10**, 147—150 (1942).

Eidel'kind, D. I.
*1. Finitely generated normal subgroups of polynilpotent groups. Sibirsk. Mat. Ž. **9**, 236—239 (1968) = Siberian Math. J. **9**, 179—181 (1968).

Eidenov, M. I.
*1. Groups without π-torsion. Ural. Gos. Univ. Mat. Zap. **19**, 61–66 (1956).
*2. π-radical groups. Ural. Gos. Univ. Mat. Zap. **3**, 60–68 (1962).

Erdös, J.
1. The theory of groups with finite classes of conjugate elements. Acta Math. Acad. Sci. Hungar. **5**, 45–58 (1954).

Eremin, I. I.
*1. Groups with finite classes of conjugate abelian subgroups. Mat. Sb. **47**, 45–54 (1959).
*2. Groups with finite classes of conjugate infinite subgroups. Perm Gos. Univ. Učen. Zap. **17**, 13–14 (1960).
*3. On central extensions by means of thin layer-finite groups. Izv. Vysš. Učebn. Zaved. Matematika **15**, 93–95 (1960).
*4. Groups with finite classes of conjugate subgroups with a given property. Dokl. Akad. Nauk. SSSR **137**, 772–773 (1961) = Soviet Math. Dokl. **2**, 337–338 (1961).

Fedorov, Yu. G.
*1. On infinite groups every non-trivial subgroup of which has finite index. Uspehi Mat. Nauk. **6**, 187–189 (1951).

Feit, W., Thompson, J. G.
1. Solvability of groups of odd order. Pacific J. Math. **13**, 775–1029 (1963).

Fitting, H.
1. Beiträge zur Theorie der Gruppen endlicher Ordnung. Jahresber. Deutsch. Math. Verein. **48**, 77–141 (1938).

Fluch, W.
1. Über die Nichtlinearität einer gewissen Gruppe. Acta Arith. **10**, 329–332 (1964).
2. Gruppen ohne endlich-dimensionale Darstellungen. Math. Scand. **16**, 164–168 (1965).

Fomin, A. N.
*1. Periodic groups whose maximal abelian subgroups are either normal or normally complemented. Mat. Zametki **3**. 39–44 (1968) = Math. Notes **3**, 25–27 (1968).

Fuchs, L.
1. On groups with finite classes of isomorphic subgroups. Publ. Math. Debrecen **3**, 243–252 (1954).
2. The existence of indecomposable abelian groups of arbitrary power. Acta Math. Acad. Sci. Hungar. **10**, 453–457 (1959).
3. Abelian groups, Oxford: Pergamon 1960.
4. Infinite abelian groups, vol. 1, New York: Academic Press 1970.

Fuchs-Rabinovič, D. I.
1. On the determinators of an operator of the free group. Mat. Sb. **7**, 197–208 (1940).
*2. On the non-simplicity of locally free groups. Mat. Sb. **7**, 327–328 (1940).

Garaščuk, M. S.
*1. On the theory of generalized nilpotent linear groups. Dokl. Akad. Nauk. BSSR **4**, 276–277 (1960).
*2. Locally quasi-nilpotent groups. Vesci Akad. Navuk. BSSR Ser. Fiz.-Mat. Navuk, pp. 5–6 (1967).

Gaschütz, W.
1. Gruppen, in denen das Normalteilersein transitiv ist. J. reine angew. Math. **198**, 87–92 (1957).

Glauberman, G., Krause, E. F., Struik, R. R.
1. Engel congruences in groups of prime power exponent. Canad. J. Math. **18**, 579—588 (1966).

Gluškov, V. M.
*1. On the normalizers of complete subgroups of a complete group. Dokl. Akad. Nauk. SSSR **71**, 421—424 (1950).
*2. On the theory of ZA-groups. Dokl. Akad. Nauk. SSSR **74**, 885—888 (1950).
*3. On locally nilpotent groups without torsion. Dokl. Akad. Nauk. SSSR **80**, 157—160 (1951).
*4. On some questions in the theory of nilpotent and locally nilpotent groups without torsion. Mat. Sb. **30**, 79—104 (1952).
*5. On the central series of infinite groups. Mat. Sb. **31**, 491—496 (1952).

Göbel, R.
1. Kartesisch und residuell abgeschlossene Gruppenklassen. Dissertationes Math. Rozprawy Mat. **63** (1969).
2. Produkte von Gruppenklassen. Arch. Math. (Basel) **20**, 113—125 (1969).

Gol'berg, P. A.
*1. Infinite semi-simple groups. Mat. Sb. **17**, 131—142 (1945).
*2. Sylow π-subgroups of locally normal groups. Mat. Sb. **19**, 451—460 (1946).
*3. Sylow bases of π-separable groups. Dokl. Akad. Nauk. SSSR **64**, 615—618 (1949).
*4. Sylow bases of infinite groups, Mat. Sb. **32**, 465—476 (1953).
*5. On a condition for conjugacy of Sylow π-bases of an arbitrary group. Mat. Sb. **36**, 335—340 (1955).
*6. The S-radical and Sylow bases in infinite groups. Mat. Sb. **50**, 25—42 (1960).

Golod, E. S.
*1. On nil-algebras and residually finite p-groups. Izv. Akad. Nauk. SSSR Ser. Mat. **28**, 273—276 (1964) = Amer. Math. Soc. Translations (2) **48**, 103—106 (1965).
*2. Some problems of Burnside type. Internat. Congress Math. Moscow, pp. 284—298 (1966) = Amer. Math. Soc. Translations (2) **84**, 83—88 (1969).

Golod, E. S., Safarevič, I. R.
*1. On class field towers. Izv. Akad. Nauk. SSSR Ser. Mat. **28**, 261—273 (1964) = Amer. Math. Soc. Translations (2) **48**, 91—103 (1965).

Gorčakov, Yu. M.
*1. Embedding of locally normal groups in a direct product of finite groups. Dokl. Akad. Nauk. SSSR **137**, 26—28 (1961) = Soviet Math. Dokl. **2**, 514—516 (1961).
*2. Locally normal groups. Dokl. Akad. Nauk. SSSR **147**, 537—539 (1962) = Soviet Math. Dokl. **3**, 1654—1656 (1962).
*3. On infinite Frobenius groups. Dokl. Akad. Nauk. SSSR **152**, 787—789 (1963) = Soviet Math. Dokl. **4**, 1397—1399 (1963).
*4. The existence of abelian subgroups of infinite rank in locally soluble groups. Dokl. Akad. Nauk. SSSR **156**, 17—20 (1964) = Soviet Math. Dokl. **5**, 591—594 (1964).
*5. On locally normal groups. Mat. Sb. **67**, 244—254 (1965).
*6. On infinite Frobenius groups. Algebra i Logika **4**, 15—29 (1965).

*7. An example of a G-periodic torsion-free group. Algebra i Logika **6**, 5–7 (1967).
*8. On the central series of the free groups of varieties. Algebra i Logika **6**, 13–24 (1967).
*9. Commutator subgroups. Sibirsk. Mat. Ž. **10**, 1023–1033 (1969) = Siberian Math. J. **10**, 754–761 (1969).

Gorčinskiĭ, Yu. N.
*1. Groups with a finite number of conjugacy classes. Mat. Sb. **31**, 167–182 (1952).
*2. Periodic groups with a finite number of conjugacy classes. Mat. Sb. **31**, 209–216 (1952).

Gregorac, R. T.
1. A note on finitely generated groups. Proc. Amer. Math. Soc. **18**, 756–758 (1967).
2. A note on certain generalized free products. Proc. Amer. Math. Soc. **18**, 754–755 (1967).
3. Residual finiteness of permutational products. J. Austral. Math. Soc. **10**, 423–428 (1969).

de Groot, J.
1. Indecomposable abelian groups. Nederl. Akad. Wetensch. Proc. Ser. A **60**, 137–145 (1957).

Gruenberg, K. W.
1. Two theorems on Engel groups. Proc. Cambridge Philos. Soc. **49**, 377–380 (1953).
2. Residual properties of infinite soluble groups. Proc. London Math. Soc. (3) **7**, 29–62 (1957).
3. The Engel elements of a soluble group. Illinois J. Math. **3**, 151–168 (1959).
4. The upper central series in soluble groups. Illinois J. Math. **5**, 436–466 (1961).
5. The residual nilpotence of certain presentations of finite groups. Arch. Math. (Basel) **13**, 408–417 (1962).
6. The Engel structure of linear groups. J. Algebra **3**, 291–303 (1966).
7. The hypercenter of linear groups. J. Algebra **8**, 34–40 (1968).

Grün, O.
1. Beiträge zur Gruppentheorie I. J. reine angew. Math. **174**, 1–14 (1935).
2. Beiträge zur Gruppentheorie IX. Arch. Math. (Basel) **13**, 49–54 (1962).

Gupta, C. K.
1. On stability groups of certain nilpotent groups. Proc. Internat. Conf. Theory of Groups Canberra **1965**, p. 103 (1967).
2. A bound for the class of certain nilpotent groups. J. Austral. Math. Soc. **5**, 506–511 (1965).
3. On certain soluble groups. Proc. Cambridge Philos. Soc. **66**, 1–4 (1969).

Gupta, C. K., Gupta, N. D.
1. Some groups of prime exponent. J. Combinatorial Theory **5**, 397–407 (1968).

Gupta, C. K., Gupta, N. D., Newman, M. F.
1. Some finite nilpotent p-groups. J. Austral. Math. Soc. **9**, 287–288 (1969).

Gupta, N. D.
1. Some group-laws equivalent to the commutative law. Arch. Math. (Basel) **17**, 97–102 (1966).

2. Groups with Engel-like conditions. Arch. Math. (Basel) **17**, **193–199** (**1966**).
3. On commutation semigroups of a group. J. Austral. Math. Soc. **6**, **36–45** (**1966**).
4. Commutation near-rings of a group. J. Austral. Math. Soc. **7**, **135–140** (**1967**).
5. Metabelian groups in the variety of certain two-variable laws. Proc. Internat. Conf. Theory of Groups Canberra **1965**, pp. **105–109** (**1967**).
6. Polynilpotent groups of prime exponent. Bull. Amer. Math. Soc. **74**, **559–561** (**1968**).
7. Certain locally metanilpotent varieties of groups. Arch. Math. (Basel) **20**, **481–484** (**1969**).
8. The free metabelian group of exponent p^2. Proc. Amer. Math. Soc. **22**, **375–376** (**1969**).

Gupta, N. D., Newman, M. F.
1. On metabelian groups. J. Austral. Math. Soc. **6**, **362–368** (**1966**).
2. Engel congruences in groups of prime-power exponent. Canad. J. Math. **20**, **1321–1323** (**1968**).

Gupta, N. D., Newman, M. F., Tobin, S. J.
1. On metabelian groups of prime power exponent. Proc. Roy. Soc. Ser. A **302**, **237–242** (**1968**).

Gupta, N. D., Rhemtulla, A. H.
1. A note on centre-by-finite-varieties of groups. J. Austral. Math. Soc. **11**, **33–36** (**1970**).

Gupta, N. D., Tobin, S. J.
1. On certain groups with exponent 4. Math. Z. **102**, **216–226** (**1967**).

Gupta, N. D., Weston, K. W.
1. On groups of exponent four. J. Algebra **17**, **59–66** (**1971**).

Guterman, M. M.
1. Normal systems. J. Algebra **4**, **317–320** (**1966**).

Haimo, F.
1. Groups with a certain condition on conjugates. Canad. J. Math. **4**, **309–321** (**1952**).
2. The *FC*-chain of a group. Canad. J. Math. **5** **498–511** (**1953**).
3. Some non-abelian extensions of completely divisible groups. Proc. Amer. Math. Soc. **5**, **25–28** (**1954**).

Hall, M.
1. A topology for free groups and some related groups. Ann. of Math. (2) **52**, **127–139** (**1950**).
2. The theory of groups, New York: MacMillan **1959**.

Hall, P.
1. A contribution to the theory of groups of prime-power order. Proc. London Math. Soc. (2) **36**, **29–95** (**1934**).
2. Verbal and marginal subgroups. J. reine angew. Math. **182**, **156–157** (**1940**).
3. The splitting properties of relatively free groups. Proc. London Math. Soc. (3) **4**, **343–356** (**1954**).
4. Finiteness conditions for soluble groups. Proc. London Math. Soc. (3) **4** **419–436** (**1954**).
5. Finite-by-nilpotent groups. Proc. Cambridge Philos. Soc. **52**, **611–616** (**1956**).

6. Nilpotent groups. Canad. Math. Congress Summer Sem. Univ. Alberta (1957): republished as "The Edmonton Notes on Nilpotent Groups", London: Queen Mary College Mathematics Notes (1969).
7. Some sufficient conditions for a group to be nilpotent. Illinois J. Math. **2**, 787–801 (1958).
8. Periodic FC-groups. J. London Math. Soc. **34** 289–304 (1959).
9. Some constructions for locally finite groups. J. London Math Soc. **34**, 305–319 (1959).
10. On the finiteness of certain soluble groups. Proc. London Math. Soc. (3) **9**, 595–622 (1959).
11. The Frattini subgroups of finitely generated groups. Proc. London Math. Soc. (3) **11**, 327–352 (1961).
12. Wreath powers and characteristically simple groups. Proc. Cambridge Philos. Soc. **58**, 170–184 (1962).
13. On non-strictly simple groups. Proc. Cambridge Philos. Soc. **59**, 531–553 (1963).
14. A note on $\overline{SI}$-groups. J. London Math. Soc. **39**, 338–344 (1964).

Hall, P., Hartley, B.
1. The stability group of a series of subgroups. Proc. London Math. Soc. (3) **16**, 1–39 (1966).

Hall, P., Higman, G.
1. On the p-length of p-soluble groups and reduction theorems for Burnside's problem. Proc. London Math. Soc. (3) **6**, 1–42 (1956).

Hall, P., Kulatilaka, C. R.
1. A property of locally finite groups. J. London Math. Soc. **39**, 235–239 (1964).

Hallett, J. T., Hirsch, K. A.
1. Torsion-free groups having finite automorphism groups I. J. Algebra **2**, 287–298 (1965).
2. Die Konstruktion von Gruppen mit vorgeschriebenen Automorphismengruppen. J. reine angew. Math. **239/240**, 32–46 (1969).

Hartley, B.
1. The order-types of central series. Proc. Cambridge Philos. Soc. **61**, 303–319 (1965).
2. The stability group of a descending invariant series of subgroups. J. Algebra **5**, 133–156 (1967).
3. The residual nilpotence of wreath products. Proc. London Math. Soc. (3) **20**, 365–392 (1970).
4. Wreath products and stability groups. J. London Math. Soc. (2) **2**, 425–428 (1970).

Hartley, B., McDougall, D.
1. Injective modules and soluble groups satisfying the minimal condition for normal subgroups. Bull. Austral. Math. Soc. **4**, 113–135 (1971).

Hartley, B., Stonehewer, S. E.
1. On some questions of P. Hall concerning simple, generalized soluble, groups. J. London Math. Soc. **43**, 739–744 (1968).

Hasse, H.
1. Zahlentheorie, Berlin: Akademie-Verlag **1949**.
2. Vorlesungen über Zahlentheorie, 2.Aufl., Berlin/Göttingen/Heidelberg/New York: Springer **1964**.

Hausen, J.
1. Automorphismengesättigte Klassen abzählbarer Abelschen Gruppen, Studies in abelian groups, Symposium Montpellier pp. **147–181** (1967).
2. The hyporesiduum of the automorphism group of an abelian p-group. Pacific J. Math. **35** 127–139 (1970).
3. Automorphism groups of abelian torsion groups with finite p-ranks. Arch. Math. (Basel) **22**, 128–135 (1971).
4. Abelian torsion groups with artinian primary components and their automorphisms. Fund. Math.
5. Automorphisms of abelian p-groups and hypo-residual finiteness.

Head, T. J.
1. Note on the occurrence of direct factors in a group. Proc. Amer. Math. Soc. **15**, 193–195 (1964).

Heineken, H.
1. Eine Verallgemeinerung des Subnormalteilerbegriffs. Arch. Math. (Basel) **11**, 244–252 (1960).
2. Eine Bemerkung über engelsche Elemente. Arch. Math. (Basel) **11**, 321 (1960).
3. Engelsche Elemente der Länge drei. Illinois J. Math. **5**, 681–707 (1961).
4. Über ein Levisches Nilpotenzkriterium. Arch. Math. (Basel) **12**, 176–178. (1961).
5. Bounds for the nilpotency class of a group. J. London Math. Soc. **37**, 456–458 (1962).
6. Endomorphismenringe und engelsche Elemente. Arch. Math. (Basel) **13**, 29–37 (1962).
7. Liesche Ringe mit Engelbedingung. Math. Ann. **149**, 232–236 (1963).
8. Commutator closed groups. Illinois J. Math. **9**, 242–255 (1965).
9. Linkskommutatorgeschlossene Gruppen. Math. Z. **87**, 37–41 (1965).
10. Gruppen, deren Kommutatorstrukturen gewisse Abschlußeigenschaften haben. Math. Ann. **176**, 96–120 (1968).
11. A class of three-Engel groups. J. Algebra **17**, 341–345 (1971).

Heineken, H., Levin, F.
1. Varieties of groups satisfying one two-variable law. Publ. Math. Debrecen **14**, 211–225 (1967).

Heineken, H., Mohamed, I. J.
1. A group with trivial centre satisfying the normalizer condition. J. Algebra **10**, 368–376 (1968).

Held, D.
1. Closure properties and partial Engel conditions in groups. Illinois J. Math. **8**, 705–712 (1964).
2. Nilpotenz- und Verstreutheitskriterien für artinsche Gruppen. Math. Z. **87**, 49–61 (1965).
3. Gruppen beschränkt Engelscher Automorphismen. Math. Ann. **162**, 1–8 (1965).
4. On abelian subgroups of an infinite 2-group. Acta Sci. Math. (Szeged) **27**, 97–98 (1966).
5. On bounded Engel elements in groups. J. Algebra **3**, 360–365 (1966).
6. Ein Hyperzentralitätskriterium. Arch. Math. (Basel) **18**, 17–22 (1967).

Hering, C.
1. Gruppen mit nichttrivialer Trofimovzahl. Arch. Math. (Basel) **15**, 404–407 (1964).

Higman, G.

1. A finitely related group with an isomorphic proper factor group. J. London Math. Soc. **26**, 59–61 (1951).
2. A finitely generated infinite simple group. J. London Math. Soc. **26**, 61–64 (1951).
3. On a problem of Takahasi. J. London Math. Soc. **28**, 250–252 (1953).
4. On infinite simple permutation groups. Publ. Math. Debrecen **3**, 221–226 (1954).
5. A remark on finitely generated nilpotent groups. Proc. Amer. Math. Soc. **6**, 284–285 (1955).
6. Groups and rings having automorphisms without non-trivial fixed elements. J. London Math. Soc. **32**, 321–334 (1957).
7. Subgroups of finitely presented groups. Proc. Roy. Soc. Ser. A **262**, 455–475 (1961).

Higman, G., Neumann, B. H.

1. On two questions of Itô. J. London Math. Soc. **29**, 84–88 (1954).

Higman, G., Neumann, B. H., Neumann, H.

1. Embedding theorems for groups. J. London Math. Soc. **24**, 247–254 (1949).

Hill, P.

1. On the decomposition of certain infinite nilpotent groups. Math. Z. **113**, 237–248 (1970).

Hirsch, K. A.

1. On infinite soluble groups I. Proc. London Math. Soc. (2) **44**, 53–60 (1938).
2. On infinite soluble groups II. Proc. London Math. Soc. (2) **44**, 336–344 (1938).
3. On infinite soluble groups III. Proc. London Math. Soc. (2) **49**, 184–194 (1946).
4. Eine kennzeichende Eigenschaft nilpotenter Gruppen. Math. Nachr. **4**, 47–49 (1951).
5. On infinite soluble groups IV. J. London Math. Soc. **27**, 81–85 (1952).
6. On infinite soluble groups V. J. London Math. Soc. **29**, 250–251 (1954).
7. Über lokal-nilpotente Gruppen. Math. Z. **63**, 290–294 (1955).

Hirsch, K. A., Zassenhaus, H.

1. Finite automorphism groups of torsion-free groups. J. London Math. Soc. **41**, 545–549 (1966).

Hirshon, R.

1. Some theorems on hopficity. Trans. Amer. Math. Soc. **141**, 229–244 (1969).
2. Cancellation of groups with maximal condition. Proc. Amer. Math. Soc. **24**, 401–403 (1970).
3. On Hopfian groups. Pacific J. Math.

Holland, W. C.

1. The characterization of generalized wreath products. J. Algebra **13**, 152–172 (1969).

Hopkins, C.

1. Finite groups in which conjugate operations are commutative. Amer. J. Math. **51**, 35–41 (1929).

Houang Ki

*1. *S*-complete groups, *SR*-groups and *SD*-groups. Sibirsk. Mat. Ž. **10**, 1427–1430 (1969) = Siberian Math. J. **10**, 1059–1061 (1969).

Houghton, C. H.
1. On the automorphism groups of certain wreath products. Publ. Math. Debrecen **9**, 307–313 (1962).
Hulanicki, A.
1. Note on a paper of de Groot. Nederl. Akad. Wetensch. Proc. Ser. A **61**, 114 (1958).
Hulse, J. A.
1. Automorphism towers of polycyclic groups. J. Algebra **16**, 347–398 (1970).
Humphreys, J. F.
1. Two-generator conditions for polycyclic groups. J. London Math. Soc. (2) **1**, 21–29 (1969).
Huppert, B.
1. Lineare auflösbare Gruppen. Math. Z. **67**, 479–518 (1957).
2. Endliche Gruppen I, Berlin/Heidelberg/New York: Springer 1967.
Hursey, R. J., Rhemtulla, A. H.
1. Ordered groups satisfying the maximal condition locally. Canad. J. Math. **22**, 753–758 (1970).
Itô, N.
1. Note on *S*-groups. Proc. Japan. Acad. **29**, 149–150 (1953).
Ivanjuta, I. D.
*1. Some groups of exponent 4. Dopovīdī Akad. Nauk. Ukraïn. RSR Ser. A **790**, 787–789 (1969).
Iwasawa, K.
1. Einige Sätze über freie Gruppen. Proc. Imp. Acad. Tokyo **19**, 272–274 (1943).
2. On linearly ordered groups. J. Math. Soc. Japan. **1**, 1–9 (1948).
Jabber, M. A.
1. On *S*-groups. Bull. Calcutta Math. Soc. **35**, 111–113 (1943).
Jacobson, N.
1. Structure theory for algebraic algebras of bounded degree. Ann. of Math. (2) **46**, 696–707 (1945).
Jennings, S. A.
1. A note on chain conditions in nilpotent rings and groups. Bull. Amer. Math. Soc. **50**, 759–763 (1944).
2. The group ring of a class of infinite nilpotent groups. Canad. J. Math. **7**, 169–187 (1955).
Kalužnin (Kaloujnine), L. A.
1. Sur quelques propriétés des groupes d'automorphisme d'un groupe abstrait I. C. R. Acad. Sci. Paris **230**, 2067–2069 (1950).
2. Sur quelques propriétés des groupes d'automorphisms d'un groupe abstrait II. C. R. Acad. Sci. Paris **231**, 400–402 (1951).
3. Über gewisse Beziehungen zwischen einer Gruppe und ihrer Automorphismen. Berlin. Math. Tagung pp. 164–172 (1953).
*4. Locally normal groups of higher categories. Algebra i Mat. Logika: Studies in Algebra, Kiev, pp. 62–71 (1966).
*5. The structure of *n*-abelian groups. Mat. Zametki **2**, 455–464 (1967). Math. Notes **2**, 768–772 (1967).
Kalužnin, L. A., Krasner, M.
1. Le produit complet des groupes de permutations et le problème d'extension des groupes. C. R. Acad. Sci. Paris **227**, 806–808 (1948).

2. Produit complet des groupes de permutations et problème d'extension de groupes I. Acta Sci. Math. (Szeged) **13**, 208–230 (1950).
3. Produit complet des groupes de permutations et problème d'extension des groupes II. Acta Sci. Math. (Szeged), **14**, 39–66 (1951).
4. Produit complet des groupes de permutation et problème d'extension de groupes. Acta Sci. Math. (Szeged) **14**, 69–82 (1951).

Kaplansky, I.
1. Infinite abelian groups, Ann Arbor: Univ. Michigan Press 1954.
2. An introduction to differential algebra, Actualités Sci. Industr., Paris: Hermann 1951.

Kappe, L.-C., Kappe, W. P.
1. Potenzen und gruppentheoretische Eigenschaften. Arch. Math. (Basel) **21**, 245–255 (1970).

Kappe, W. P.
1. Die A-Norm einer Gruppe. Illinois J. Math. **5**, 187–197 (1961).
2. Gruppentheoretische Eigenschaften und charakteristische Untergruppen. Arch. Math. (Basel) **13**, 38–48 (1962).
3. Zum Begriff des überauflösbar-eingebetteten Normalteilers. Math. Ann. **149**, 254–257 (1963).
4. On the anticenter of nilpotent groups. Illinois J. Math. **12**, 603–609 (1968).
5. Über gruppentheoretische Eigenschaften, die sich auf t-Produkte übertragen. Acta Sci. Math. (Szeged), **30**, 277–284 (1969).

Kappe, W. P., Parker, D. B.
1. Elements with trivial centralizer in wreath products. Trans. Amer. Math. Soc. **150**, 201–212 (1970).

Karasev, G. A.
*1. The concept of n-nilpotent groups. Sibirsk. Mat. Ž. **7**, 1014–1032 (1966) = Siberian Math. J. **7**, 808–821 (1966).
*2. On n-nilpotent groups. Soobšč. Akad. Nauk. Gruzin. SSR **46**, 309–314 (1967).
*3. The Frattini subgroup of an n-nilpotent group. Sibirsk. Mat. Ž. **8**, 1432–1436 (1967) = Siberian Math. J. **8**, 1082–1085 (1967).

Kargapolov, M. I.
*1. On conjugacy of Sylow p-subgroups of a locally normal group. Uspehi Mat. Nauk. **12**, 297–300 (1957).
*2. On the theory of semi-simple locally normal groups. Naučn. Dokl. Vysš. Fiz. Mat. Nauk. **6**, 3–7 (1958).
*3. On the theory of $\bar{Z}$-groups. Dokl. Akad. Nauk. SSSR **125**, 255–257 (1959).
*4. Some problems in the theory of nilpotent and soluble groups. Dokl. Akad. Nauk. SSSR **127**, 1164–1166 (1959).
*5. Locally finite groups with a normal system with finite factors. Sibirsk. Mat. Ž. **2**, 853–873 (1961).
*6. On the completion of locally nilpotent groups. Sibirsk. Mat. Ž. **3**, 695–700 (1962).
*7. On periodic matrix groups. Sibirsk. Mat. Ž.**3**, 834–838 (1962).
*8. On the π-completion of locally nilpotent groups. Algebra i Logika **1**, 5–13 (1962).
*9. On soluble groups of finite rank. Algebra i Logika **1**, 37–44 (1962).
*10. On generalized soluble groups. Algebra i Logika **2**, 19–28 (1963).
*11. On a problem of O. Yu. Schmidt. Sibirsk. Mat. Ž. **4**, 232–235 (1963).

*12. A remark on the paper "On periodic matrix groups". Sibirsk. Mat. Ž. **4**, 1198–1199 (1963).
*13. Locally finite groups possessing normal systems with finite factors. Sibirsk. Mat. Ž. **2**, 853–873 (1961).
*14. Finitely generated linear groups. Algebra i Logika **6**, 17–20 (1967).
*15. The residual finiteness with respect to conjugacy of supersoluble groups. Algebra i Logika **6**, 63–68 (1967).

Kargapolov, M. I., Merzljakov, Yu. I.
*1. Infinite groups. Algebra Topology Geometry pp. 57–90 (1966). Akad. Nauk. SSSR Inst. Naučn. Informacii, Moscow 1968.

Kargapolov, M. I., Merzljakov, Yu. I., Remeslennikov, V. N.
*1. A method of group adjunction. Perm. Gos. Univ. Učen. Zap. **17**, 9–11 (1960).
*2. Completions of groups. Dokl. Akad. Nauk. SSSR **134**, 518–520 (1960) = Soviet Math. Dokl. **1**, 1099–1101 (1960).

Kargapolov, M. I., Remeslennikov, V. N.
*1. The conjugacy problem for free soluble groups. Algebra i Logika **5**, 15–25 (1966).

Karrass, A., Solitar, D.
1. Some remarks on the infinite symmetric groups. Math. Z. **66**, 64–69 (1956).

Katz, R. A., Magnus, W.
1. Residual properties of free groups. Comm. Pure Appl. Math. **22**, 1–13 (1969).

Kazačkov, B. V.
*1. On a local theorem in the theory of groups. Dokl. Akad. Nauk. SSSR **83**, 525–528 (1952).
*2. The Schur-Zassenhaus Theorem for countable locally finite groups. Mat. Sb. **50**, 499–506 (1960).
*3. Certain conditions for strong group factorizability. Mat. Sb. **78**, 349–354 (1969) = Math. USSR–Sb. **7**, 341–345 (1969).

Kegel, O. H.
1. Lokal endliche Gruppen mit nicht-trivialer Partition. Arch. Math. (Basel) **13**, 10–28 (1962).
2. On the solvability of some factorized linear groups. Illinois J. Math. **9**, 535–547 (1965).
3. Über den Normalisator von subnormalen und erreichbaren Untergruppen. Math. Ann. **163**, 248–258 (1966).
4. Über einfache, lokal endliche Gruppen. Math. Z. **95**, 169–195 (1967).
5. Noethersche 2-Gruppen sind endlich. Monats. Math. **71**, 424–426 (1967).
6. Zur Struktur lokal endlicher Zassenhausgruppen. Arch. Math. (Basel) **18**, 337–348 (1967).
7. Locally finite versus finite simple groups. Symposium, Theory of Finite Groups, Harvard Univ., pp. 247–249 (1969).
8. Lectures on locally finite groups. Math. Inst. Oxford (1969).

Kegel, O. H., Wehrfritz, B. A. F.
1. Strong finiteness conditions in locally finite groups. Math. Z. **117**, 309–324 (1970).

Kemhadze, Š. S.
*1. On the definition of nilpotent groups. Soobšč. Akad. Nauk. Gruzin. SSR **26**, 385–387 (1961) = Amer. Math. Soc. Translations (2) **46**, 162–164 (1965).

*2. Strongly locally nilpotent groups. Trudy I i II Respubl. Konferencii Mat. Vys. Učebn. Zaved. Gruzin. SSR. Izdat. Codna Tbilis. (1964), pp. 75–82.
*3. Quasinilpotent groups. Dokl. Akad. Nauk. SSSR **155**, 1003–1005 (1964) = Soviet Math. Dokl. **5**, 532–535 (1964).
*4. Groups generated by nilpotent and ZA-subgroups. Sibirsk. Mat. Ž. **5**, 827–837 (1964).
*5. On the definition of Baer nil groups. Soobšč. Akad. Nauk. Gruzin. SSR 33, 279–284 (1964).
*6. V- and V'-groups. Sibirsk. Mat. Ž. **8**, 788–797 (1967) = Siberian Math. J. **8**, 595–602 (1967).
*7. Cyclic extensions of ZA-groups. Sakarth. SSR Mecn. Akad. Moambe **50**, 263–266 (1968).

Kikodze, E. B.
*1. On complex commutators of elements of a group. Uspehi Mat. Nauk. **12**, 301–303 (1957).

Kneser, M., Swierszkowski, S.
1. Embedding in groups of countable permutations. Colloq. Math. **7**, 177–179 (1960).

Kogalovskiĭ, S. R.
*1. Structure of characteristic universal classes. Sibirsk. Mat. Ž. **4**, 97–119 (1963).
*2. A theorem of Birkhoff. Uspehi Mat. Nauk. **20**, 206–207 (1965).

Kolchin, E. R.
1. On certain concepts in the theory of algebraic matric groups. Ann. of Math. (2) **49**, 774–789 (1948).

Kontorovič, P. G.
*1. Invariantly covered groups. Mat. Sb. **8**, 423–436 (1940).
*2. On groups with a basis of partition I. Mat. Sb. **12**, 56–70 (1943).
*3. On groups with a basis of partition II. Mat. Sb. **19**, 287–308 (1946).
*4. On the theory of non-commutative torsion-free groups. Dokl. Akad. Nauk. SSSR **59**, 213–216 (1948).
*5. On groups with a basis of partition III. Mat. Sb. **22**, 79–100 (1948).
*6. On groups with a basis of partition IV. Mat. Sb. **26**, 311–320 (1950).
*7. Invariantly covered groups II. Mat. Sb. **28**, 79–88 (1951).
*8. Remarks on the hypercentre of a group. Ural. Gos. Univ. Mat. Zap. **23**, 27–29 (1960).
*9. Three problems on the splitting of groups. Colloq. Math. **17**, 207–208 (1967).

Kopytov, V. M.
*1. On matrix groups. Algebra i Logika **7**, 51–59 (1968) = Algebra and Logic **7**, 162–166 (1968).

Kořinek, V.
1. Les groupes qui ne contiennent pas des sousgroupes caractéristiques propres. Věstn. Kral. Česke Spol. Nauk. 1–20 (1938).
2. Bemerkung über charakteristisch einfache Gruppen. Věstn. Kral. Česke Spol. Nauk. 1–8 (1940).

Kostrikin, A. I.
*1. On the connection between periodic groups and Lie rings. Izv. Akad. Nauk SSSR Ser. Mat. **21**, 289–310 (1957) = Amer. Math. Translations (2) **45**, 165–189 (1965).
*2. Lie rings satisfying the Engel condition. Izv. Akad. Nauk. SSSR Ser.

Mat. **21**, 515—540 (1957) = Amer. Math. Soc. Translations (2) **45**, 191—220 (1965).

*3. On the local nilpotence of Lie rings that satisfy Engel's condition. Dokl. Akad. Nauk. SSSR **118**, 1074—1077 (1958).

*4. The Burnside problem. Izv. Akad. Nauk. SSSR Ser. Mat. **23**, 3—34 (1959).

*5. On Engel properties of groups with identical relation $x^{p^\alpha} = 1$. Dokl. Akad. Nauk. SSSR **135**, 524—426 (1960) = Soviet Math. Dokl. **1**, 1282—1284 (1960).

* Kourov Notebook, Unsolved problems in the theory of groups. Novosibirsk: Izdat Sibirsk. Otdel. Akad. Nauk. SSSR (1965).

Kovács, L. G.

1. Groups with regular automorphisms of order 4. Math. Z. **75**, 277—294 (1961).

Kovács, L. G., Neumann, B. H.

1. An embedding theorem for some countable groups. Acta Sci. Math. (Szeged) **26**, 139—142 (1965).
2. On the existence of Baur soluble groups of arbitrary height. Acta Sci. Math. (Szeged) **26**, 143—144 (1965).

Kovács, L. G., Neumann, B. H., de Vries, H.

1. Some Sylow subgroups. Proc. Roy. Soc. London Ser. A **260**, 304—316 (1961).

Kovács, L. G., Newman, M. F.

1. Direct complementation in groups with operators. Arch. Math. (Basel) **13**, 427—433 (1962).

Krause, E.

1. Groups of exponent 8 satisfy the 14th Engel congruence. Proc. Amer. Math. Soc. **15**, 491—496 (1964).
2. On the collection process. Proc. Amer. Math. Soc. **15**, 497—504 (1964).

Kulatilaka, C. R.

1. Infinite abelian subgroups of some infinite groups. J. London Math. Soc. **39**, 240—244 (1964).

Kurata, Y.

1. A decomposition of normal subgroups in a group. Osaka J. Math. **1**, 201—229 (1964).

Kuroš, A. G.

1. Die Untergruppen der freien Produkte von beliebigen Gruppen. Math. Ann. **109**, 647—660 (1934).
2. Eine Verallgemeinerung des Jordan-Hölderschen Satz. Math. Ann. **111**, 13—18 (1935).

*3. Some recent trends and some outstanding problems in the theory of infinite groups. Uspehi Mat. Nauk. (O. S.) **3**, 5—15 (1937).

*4. Locally free groups. Dokl. Akad. Nauk. SSSR **24**, 99—101 (1939).

*5. Some remarks on the theory of infinite groups. Mat. Sb. **5**, 347—354 (1939).

*6. Theory of groups, Moscow: OGIZ 1944.

*7. Composition systems in infinite groups. Mat. Sb. **16**, 59—72 (1945).

*8. Radicals of rings and algebras. Mat. Sb. 33, 13—26 (1953).

9. The theory of groups, 2nd ed. (2 vols.), New York: Chelsea 1960.
10. Radicaux en theorie des groupes. Bull. Soc. Math. Belg. **14**, 307—310 (1962).

*11. Radicals in the theory of groups. Sibirsk. Mat. Ž. 3, 912–931 (1962).
*12. Lectures on general algebra. Moscow: Izdat. Fiz-Mat. 1962 = General Algebra, New York: Chelsea 1963.
*13. The theory of groups. 3rd augmented ed. (3 vols.), Moscow: Izdat. Nauka 1967.

Kuroš, A. G., Černikov, S. N.
*1. Soluble and nilpotent groups. Uspehi Mat. Nauk. 2, 18–59 (1947) = Amer. Math. Soc. Translations 80 (1953).

Lam, T.-Y.
1. A commutator formula for a pair of subgroups and a theorem of Blackburn. Canad. Math. Bull. 12, 217–219 (1969).

Lazard, M.
1. Sur certaines suites d'éléments dans les groupes libres et leurs extensions. C. R. Acad. Sci. Paris 236, 36–38 (1953).
2. Problèmes d'extension concernant les N-groupes; inversion de la formule de Hausdorff. C. R. Acad. Sci. Paris 237, 1377–1379 (1953).
3. Sur les groupes nilpotents et les anneaux de Lie. Ann. Sci. École Norm. Sup. (3) 71, 101–190 (1954).

Learner, A. L.
1. The embedding of a class of polycyclic groups. Proc. London Math. Soc. (3) 12, 496–510 (1962).
2. Residual properties of polycyclic groups. Illinois J. Math. 8, 536–542 (1964).

Lefschetz, S.
1. Algebraic topology, New York: Amer. Math. Soc. 1942.

Lennox, J. C., Roseblade, J. E.
1. Centrality in finitely generated soluble groups. J. Algebra 16, 399–435 (1970).

Levi, F. W.
1. Über die Untergruppen der freien Gruppen. Math. Z. 37, 90–97 (1933).
2. Groups in which the commutator operation satisfies certain algebraic conditions. J. Indian Math. Soc. 6, 87–97 (1942).

Levi, F. W., van der Waerden, B. L.
1. Über eine besondere Klasse von Gruppen. Abh. Math. Sem. Univ. Hamburg 9, 154–158 (1933).

Levič, E. M.
*1. An example of a simple but not strictly simple group. Latvijas Valsts. Univ. Zinātn. Raksti. 58, 21–26 (1964).
*2. Locally subinvariant elements of a group. Latvijas PSR Zinātnu Akad. Vestis Fiz. Tehn. Zinātnu Sēr. 3, 71–73 (1965).
*3. An example of a locally nilpotent torsion-free group with no accessible elements. Sibirsk. Mat. Ž. 8, 717–719 (1967). = Siberian Math. J. 8, 538–539 (1967).
*4. The representation of soluble groups by matrices over a field of characteristic zero. Dokl. Akad. Nauk. SSSR 188, 520–521 (1969) = Soviet Math. Dokl. 10, 1146–1148 (1969).
*5. A certain problem of P. Hall. Dokl. Akad. Nauk. SSSR 188, 1241–1243 (1969) = Soviet Math. Dokl. 10, 1299–1301 (1969).

Levič, E. M., Tokarenko, A. I.
*1. Note on locally nilpotent torsion-free groups. Sib. Mat. Ž. 11, 1406–1408 (1970) = Siberian Math. J. 11, 1033–1034 (1970).

Levin, F.

1. Solutions of equations over groups. Bull. Amer. Math. Soc. **68**, 603–604 (1962).
2. One variable equations over groups. Arch. Math. (Basel) **15**, 179–188 (1964).
3. On some varieties of soluble groups I. Math. Z. **85**, 369–372 (1964).
4. On some varieties of soluble groups II. Math. Z. **103**, 162–172 (1968).
5. Factor groups of the unimodular group. J. London Math. Soc. **43**, 195–203 (1968).

Lewin, J.

1. A finitely presented group whose group of automorphisms is infinitely generated. J. London Math. Soc. **42**, 610–613 (1967).

Lewin, J., Lewin, T.

1. Semigroup laws in varieties of solvable groups. Proc. Cambridge Philos. Soc. **65**, 1–9 (1969).

Liebeck, H.

1. Concerning nilpotent wreath products. Proc. Cambridge Philos. Soc. **58**, 443–451 (1962).
2. Locally inner and almost inner automorphisms. Arch. Math. (Basel) **15**, 18–27 (1964).

Lihtman, A. I.

*1. On residually nilpotent groups. Sibirsk. Mat. Ž. **6**, 862–866 (1965).

Liman, F. N.

*1. Groups with normal non-cyclic subgroups. Dopovīdī Akad. Nauk. Ukraïn. RSR Ser. A **12**, 1073–1074 (1967).
*2. Non-periodic groups with certain systems of normal subgroups. Algebra i Logika **7**, 70–86 (1968) = Algebra and Logic **7**, 245–254 (1968).
*3. 2-groups with normal non-cyclic subgroups. Mat. Zametki **4**, 75–84 (1968) = Math. Notes **4**, 535–539 (1968).

Livčak, Ya. B.

*1. A locally soluble group which is not an *SN**-group. Dokl. Akad. Nauk. **125**, 266–268 (1959).
*2. On the theory of generalized soluble groups. Sibirsk. Mat. Ž. **1**, 617–622 (1960).

Losey, G.

1. A note on groups of prime power exponent satisfying an Engel congruence. Proc. Amer. Math. Soc. **15**, 209–211 (1964).

Lyndon, R. C.

1. Two notes on nilpotent groups. Proc. Amer. Math. Soc. **3**, 579–583 (1952).

Lyndon, R. C., Ullman, J. L.

1. Pairs of real 2-by-2 matrices that generate free products. Michigan Math. J. **15**, 161–166 (1968).

Macbeath, A. M.

1. Packings, free products and residually finite groups. Proc. Cambridge Philos. Soc. **59**, 555–558 (1963).

McCarthy, D.

1. Infinite groups whose proper quotient groups are finite I. Comm. Pure Appl. Math. **21**, 545–562 (1968).
2. Infinite groups whose proper quotient groups are finite II. Comm. Pure Appl. Math. **23**, 767–789 (1970).

McCleary, S. H.

1. Generalized wreath products viewed as sets with valuation. J. Algebra **16**, 163–182 (1970).

McCool, J.

1. Embedding theorems for countable groups. Canad. J. Math. **22**, 827–835 (1970).
2. Unsolvable problems in groups with solvable word problem. Canad. J. Math. **22**, 836–838 (1970).

McCutcheon, J. J.

1. On certain polycyclic groups. Bull. London Math. Soc. **1**, 179–186 (1969).

Macdonald, I. D.

1. A class of *FC*-groups. J. London Math. Soc. **34**, 73–80 (1959).
2. Some explicit bounds in groups with finite derived groups. Proc. London Math. Soc. (3) **11**, 23–56 (1961).
3. On certain varieties of groups. Math. Z. **76**, 270–282 (1961).
4. On a set of normal subgroups. Proc. Glasgow Math. Assoc. **5**, 137–146 (1962).
5. On central series. Proc. Edinburgh Math. Soc. (2) **13**, 175–178 (1962).
6. On certain varieties of groups II. Math. Z. **78**, 175–178 (1962).
7. On cyclic commutator subgroups. J. London Math. Soc. **38**, 419–422 (1963).
8. Generalizations of a classical theorem about nilpotent groups. Illinois J. Math. **8**, 556–570 (1964).

Macdonald, I. D., Neumann, B. H.

1. A third-Engel 5-group. J. Austral. Math. Soc. **7**, 555–569 (1967).

McDougall, D.

1. Soluble minimax groups with the subnormal intersection property. Math. Z. **114**, 241–244 (1970).
2. Soluble groups with the minimum condition for normal subgroups. Math. Z. **118**, 157–167 (1970).

MacHenry, T.

1. The tensor product and the 2nd nilpotent product of groups. Math. Z. **73**, 134–145 (1960).

McLain, D. H.

1. A characteristically-simple group. Proc. Cambridge Philos. Soc. **50**, 641–642 (1954).
2. A class of locally nilpotent groups. Ph. D. Dissertation, Cambridge Univ. (1956).
3. On locally nilpotent groups. Proc. Cambridge Philos. Soc. **52**, 5–11 (1956).
4. Remarks on the upper central series of a group. Proc. Glasgow Math. Assoc. **3**, 38–44 (1956).
5. Finiteness conditions in locally soluble groups. J. London Math. Soc. **34**, 101–107 (1959).
6. Local theorems in universal algebras. J. London Math. **34**, 177–184 (1959).

MacLane, S.

1. A proof of the subgroup theorem for free products. Mathematika **5**, 13–19 (1958).

Magnus, W.

1. Beziehungen zwischen Gruppen und Idealen in einem speziellen Ring. Math. Ann. **111**, 259–280 (1935).

2. Über Beziehungen zwischen höheren Kommutatoren. J. reine angew. Math. **177**, 105—115 (1937).
3. Residually finite groups. Bull. Amer. Math. Soc. **75**, 305—316 (1969).

Magnus, W., Karrass, A., Solitar, D.
1. Combinatorial group theory, New York: Wiley **1966**.

Mal'cev, A. I.
*1. On the faithful representation of infinite groups by matrices. Mat. Sb. **8**, 405—422 (1940) = Amer. Math. Soc. Translations (2) **45**, 1—18 (1965).
*2. On a general method for obtaining local theorems in group theory. Ivanov. Gos. Ped. Inst. Učen. Zap. **1**, 3—9 (1941).
*3. On groups of finite rank. Mat. Sb. **22**, 351—352 (1948).
*4. Nilpotent torsion-free groups. Izv. Akad. Nauk. SSSR Ser. Mat. **13**, 201—212 (1949).
*5. Generalized nilpotent algebras and their adjoint groups. Mat. Sb. **25**, 347—366 (1949) = Amer. Math. Soc. Translations (2) **69**, 1—21 (1968).
*6. On infinite soluble groups. Dokl. Akad. Nauk. SSSR **67**, 23—25 (1949).
*7. On certain classes of infinite soluble groups. Mat. Sb. **28**, 567—588 (1951) = Amer. Math. Soc. Translations (2) **2**, 1—21 (1956).
*8. Two remarks on nilpotent groups. Mat. Sb. **37**, 567—572 (1955).
*9. Homomorphisms of finite groups. Ivanov. Gos. Ped. Inst. Učen. Zap. **18**, 49—60 (1958).
*10. Model correspondences. Izv. Akad. Nauk. SSSR Ser. Mat. **23**, 313—336 (1959).

Mann, A.
1. Groups with dense normal subgroups. Israel J. Math. **6**, 13—25 (1968).

Medvedeva, R. P.
*1. A generalization of finite groups with the transitivity property for normal subgroups. Sibirsk. Mat. Ž. **6**, 1068—1073 (1965).

Meier-Wunderli, H.
1. Über endliche p-Gruppen, deren Elemente der Gleichung $x^p = 1$ genügen. Comment. Math. Helv. **24**, 18—45 (1950).
2. Metabelsche Gruppen. Comment. Math. Helv. **25**, 1—10 (1951).

Meldrum, J. D. P.
1. On central series of a group. J. Algebra **6**, 281—284 (1967).
2. On nilpotent wreath products. Proc. Cambridge Philos. Soc. **68**, 1—15 (1970).

Mennicke, J. L.
1. Finite factor groups of the unimodular group. Ann. of Math. (2) **81**, 31—37 (1965).

Menogazzo, F.
1. Gruppi nei quali la relazione di quasi-normalità è transitiva. Rend. Sem. Mat. Univ. Padova **40**, 347—361 (1968).
2. Gruppi nei quali la relazione di quasi-normalità è transitiva II. Rend Sem. Mat. Univ. Padova **42**, 389—399 (1969).

Merzljakov, Yu. I.
*1. On the theory of generalized soluble and nilpotent groups. Algebra i Logika **2**, 29—36 (1963).
*2. Locally soluble groups of finite rank. Algebra i Logika **3**, 5—16 (1964).
*3. Central series and commutator series of matrix groups. Algebra i Logika **3**, 49—59 (1964).
*4. Verbal and marginal subgroups of linear groups. Dokl. Akad. Nauk. SSSR **177**, 1008—1011 (1967) = Soviet Math. Dokl. **8**, 1538—1541 (1967).

*5. Matrix representations of automorphisms, extensions and soluble groups. Algebra i Logika **7**, 63–104 (1968) = Algebra and Logic **7**, 169–192 (1968).

*6. Groups which are almost residually finite p-groups. Algebra i Logika **7**, 105–111 (1968) = Algebra and Logic **7**, 62–65 (1968).

Miller, G. A., Moreno, H. C.

1. Non-abelian groups in which every subgroup is abelian. Trans. Amer. Math. Soc. **4**, 398–404 (1903).

Milnor, J.

1. A note on curvature and fundamental group. J. Differential Geometry **2**, 1–7 (1968).
2. Growth of finitely generated solvable groups. J. Differential Geometry **2**, 447–449 (1968).

Mital, J.

1. On residual nilpotence. J. London Math. Soc. (2) **2**, 337–345 (1970).

Mitra, A.

1. A theorem on Engel groups. J. Indian Math. Soc. (N. S.) **25**, 155–161 (1961).

Mohamed, I. J.

1. On a series of subgroups related to groups of automorphisms. Proc. London Math. Soc. (3) **13**, 711–723 (1963).
2. On the class of the stability group of a subgroup chain. J. London Math. Soc. **39** 109–114 (1964).

Moldavanski, D. I.

*1. Certain subgroups of groups with a single defining relation. Sibirsk. Mat. Ž. **8**, 1370–1384 (1967) = Siberian Math. J. **8**, 1039–1048 (1967).

Muhammedžan, H. H.

*1. On the theory of infinite groups with ascending central series. Dokl. Akad. Nauk. SSSR **65**, 269–272 (1949).

*2. On groups with an ascending central series. Mat. Sb. **28**, 185–196 (1951).

*3. On groups possessing an ascending soluble invariant series. Mat. Sb. **39**, 201–218 (1956).

Myagkova, N. N.

*1. On groups of finite rank. Izv. Akad. Nauk. SSSR Ser. Mat. **13**, 495–512 (1949).

Nagrabeckiĭ, V. T.

*1. Invariant coverings of subgroups. Dokl. Akad. Nauk. SSSR **172**, 30–32 (1967) = Soviet Math. Dokl. **8**, 24–26 (1967).

Nelson, E.

1. Finiteness of semigroups of operators in universal algebra. Canad. J. Math. **19**, 764–768 (1967).

Neumann, B. H.

1. Identical relations in groups I. Math. Ann. **114**, 506–525 (1937).
2. Some remarks on infinite groups. J. London Math. Soc. **12**, 120–127 (1937).
3. Groups whose elements have bounded orders. J. London Math. Soc. **12**, 195–198 (1937).
4. Adjunction of elements to groups. J. London Math. Soc. **18**, 4–11 (1943).
5. On a special class of infinite groups. Nieuw Arch. Wisk. **23**, 117–127 (1950).

6. A two-generator group isomorphic to a proper factor group. J. London Math. Soc. **25**, 247–248 (1950).
7. Groups with finite classes of conjugate elements. Proc. London Math. Soc. (3) **1**, 178–187 (1951).
8. On a problem of Hopf. J. London Math. Soc. **28**, 351–353 (1953).
9. Groups covered by permutable subsets. J. London Math. Soc. **29**, 236–248 (1954).
10. An embedding theorem for algebraic systems. Proc. London Math. Soc. (3) **4**, 138–153 (1954).
11. Groups covered by finitely many cosets. Publ. Math. Debrecen **3**, 227–242 (1954).
12. Groups with finite classes of conjugate subgroups. Math. Z. **63**, 76–96 (1955).
13. On a conjecture of Hanna Neumann. Proc. Glasgow Math. Assoc. **3**, 13–17 (1956).
14. Ascending derived series. Compositio Math. **13**, 47–64 (1956).
15. Groups with automorphisms that leave only the neutral element fixed. Arch. Math. (Basel) **7**, 1–5 (1956).
16. Ascending verbal and Frattini series. Math. Z. **69**, 164–172 (1958).
17. Isomorphism of Sylow subgroups of infinite groups. Math. Scand. **6**, 299–307 (1958).
18. Embedding theorems for ordered groups. J. London Math. Soc. **35**, 503–512 (1960).
19. Lectures on topics in the theory of infinite groups. Bombay: Tata Inst. Fundamental Research 1960.
20. Twisted wreath products of groups. Arch. Math. (Basel) **14**, 1–6 (1963).

Neumann, B. H., Neumann, H.

1. Zwei Klassen charakteristischer Untergruppen und ihre Faktorgruppen. Math. Nachr. **4**, 106–125 (1951).
2. Embedding theorems for groups. J. London Math. Soc. **34**, 465–479 (1959).

Neumann, H.

1. Varieties of groups, Berlin/Heidelberg/New York: Springer 1967.

Neumann, P. M.

1. On the structure of standard wreath products of groups. Math. Z. **84**, 343–373 (1964).
2. An improved bound for *BFC p*-groups. J. Austral. Math. Soc. **11**, 19–27 (1970).
3. The inequality of ***SQPS*** and ***QSF*** as operators on classes of groups. Bull. Amer. Math. Soc. **76**, 1067–1069 (1970).

Neuwirth, L.

1. An alternative proof of a theorem of Iwasawa on free groups. Proc. Cambridge Philos. Soc. **57**, 895–896 (1961).

Newell, M. L.

1. A subgroup characterization of soluble Min-by-Max groups. Arch. Math. (Basel) **21**, 128–131 (1970).
2. On normal coverings of groups. Arch. Math. (Basel) **21**, 337–343 (1970).
3. On soluble Min-by-Max groups. Math. Ann. **186**, 282–296 (1970).
4. Finiteness conditions in generalized soluble groups. J. London Math. Soc. (2) **2**, 593–596 (1970).

Newman, M.
1. Pairs of matrices generating discrete free groups and free products. Michigan Math. J. **15**, 155–160 (1968).

Newman, M. F.
1. On a class of metabelian groups. Proc. London Math. Soc. (3) **10**, 354–364 (1960).
2. On a class of nilpotent groups. Proc. London Math. Soc. (3) **10**, 365–375 (1960).
3. On subgroup commutators. Proc. Amer. Math. Soc. **14**, 724–728 (1963).
4. Another non-Hopf group. J. London Math. Soc. **41**, 292 (1966).

Newman, M. F., Wiegold, J.
1. Groups with many nilpotent subgroups. Arch. Math. (Basel) **15**, 241–250 (1964).

Nishigôri, N.
1. On some properties of *FC*-groups. J. Sci. Hiroshima Univ. Ser. A-I Math. **21**, 99–105 (1957/1958).
2. On *FC*-solvable groups. J. Sci. Hiroshima Univ. Ser. A–I **25**, 367–368 (1961).

Nisnevič, V. L.
*1. On groups which are isomorphically representable by matrices over a commutative field. Mat. Sb. **8**, 395–403 (1940).

Northcott, D. G.
1. Ideal theory, London: Cambridge Univ. Press 1953.

Novikov, P. S.
*1. Periodic groups. Dokl. Akad. Nauk. SSSR **127**, 749–752 (1959).

Novikov, P. S., Adjan, S. I.
*1. Infinite periodic groups. Izv. Akad. Nauk. SSSR Ser. Mat. **32**, 212–244, 251–524, 709–731 (1968) = Math. USSR-Izv. **2**, 209–236, 241–479, 665–685 (1968).
*2. Commutative subgroups and the conjugacy problem in free periodic groups of odd order. Izv. Akad. Nauk. SSSR Ser. Mat. **32**, 1176–1190 (1968) = Math. USSR-Izv. **2**, 1131–1144 (1968).
*3. Defining relations and the word problem for free periodic groups of odd order. Izv. Akad. Nauk. SSSR Ser. Mat. **32**, 971–979 (1968) = Math. USSR-Izv. **2**, 935–942 (1968).

Ore, O.
1. Contributions to the theory of groups of finite order. Duke Math. J. **5**, 431–460 (1939).
2. Theory of monomial groups. Trans. Amer. Math. Soc. **51**, 15–64 (1942).

Peluso, A.
1. A residual property of free groups. Comm. Pure Appl. Math. **19**, 435–437 (1966).

Peng, T. A.
1. Engel elements of groups with maximal condition on abelian subgroups. Nanta Math. **1**, 23–28 (1966).
2. Finite groups with pro-normal subgroups. Proc. Amer. Math. Soc. **20**, 232–234 (1969).
3. Finite soluble groups with an Engel condition. J. Algebra **11**, 319–330 (1969).
4. On groups with nilpotent derived group. Arch. Math. (Basel) **22**, 251–253 (1969).

Phillips, R. E.
1. On direct products of generalized solvable groups. Bull. Amer. Math. Soc. **73**, **973–975** (**1967**).
2. f-systems in infinite groups. Arch. Math. (Basel) **20**, **345–355** (**1969**).

Phillips, R. E., Combrink, C. R.
1. A note on subsolvable groups. Math. Z. **92**, **349–352** (**1966**).

Phillips, R. E., Hickin, K. K.
1. On ascending series of subgroups in infinite groups. J. Algebra **16**, **153–162** (**1970**).

Phillips, R. E., Robinson, D. J. S., Roseblade, J. E.
1. Maximal subgroups and chief factors of certain generalized soluble groups. Pacific J. Math. **37**, **475–480** (**1971**).

Pic, G.
1. On a theorem of B. H. Neumann. Studia Univ. Babeş-Bolyai Ser. Math.-Phys. **1**, **21–26** (**1960**).
2. Une propriété des groupes FC-nilpotents. Com. Acad. RPR **12**, **969–972** (**1962**).

Pickel, P. F.
1. Finitely generated nilpotent groups with isomorphic finite quotients. Bull. Amer. Math. Soc. **77**, **216–219** (**1971**).

Pilgrim, D.
1. Engel conditions on groups. Proc. Iowa Acad. Sci. **71**, **377–383** (**1964**).

Platonov, V. P.
*1. Periodic subgroups of algebraic groups. Dokl. Akad. Nauk. SSSR **153**, **270–272** (**1963**) = Soviet Math. Dokl. **4**, **1653–1656** (**1963**).
*2. The Frattini subgroup of linear groups and finite approximability. Dokl. Akad. Nauk. SSSR **171**, **798–801** (**1966**) = Soviet Math. Dokl. **7**, **1557–1560** (**1966**).
*3. Linear groups with identical relations. Dokl. Akad. Nauk. BSSR **11**, **581–583** (**1967**).
*4. Some remarks on linear groups. Mat. Zametki **4**, **635–638** (**1968**) = Math. Notes **4**, **873–874** (**1968**).
*5. A certain problem for finitely generated linear groups. Dokl. Akad. Nauk. BSSR **12**, **492–494** (**1968**).
*6. On a problem of Mal′cev. Mat. Sb. **79**, **621–624** (**1969**) = Math. USSR-Sb. **8**, **599–602** (**1969**).

Plotkin, B. I.
*1. On the theory of non-commutative torsion-free groups. Dokl. Akad. Nauk. SSSR **73**, **655–657** (**1950**).
*2. On the theory of locally nilpotent groups. Dokl. Akad. Nauk. SSSR **76**, **639–641** (**1951**).
*3. On the theory of soluble torsion-free groups. Dokl. Akad. Nauk. SSSR **84**, **665–668** (**1952**).
*4. On nilgroups. Dokl. Akad. Nauk. SSSR **94**, **999–1001** (**1954**).
*5. On the nil-radical of a group. Dokl. Akad. Nauk. SSSR **98**, **341–343** (**1965**).
*6. On some criteria of locally nilpotent groups. Uspehi Mat. Nauk. **9**, **181–186** (**1954**) = Amer. Math. Soc. Transl. (2) **17**, **1–7** (**1961**).
*7. Radical groups. Mat. Sb. **37**, **507–526** (**1955**) = Amer. Math. Soc. Translations (2) **17**, **9–28** (**1961**).
*8. On the theory of soluble groups with finiteness conditions. Dokl. Akad. Nauk. SSSR **100**, **417–420** (**1955**).

*9. On the theory of torsion-free soluble groups. Mat. Sb. **36**, 31–38 (1955).

*10. On groups with finiteness conditions for abelian subgroups. Dokl. Akad. Nauk. SSSR **107**, 648–651 (1956).

*11. Radical and semi-simple groups. Trudy Moskov. Mat. Obšč. **6**, 299–336 (1957).

*12. Generalized soluble and nilpotent groups. Uspehi Mat. Nauk. 13, 89–172 (1958) = Amer. Math. Soc. Translations (2) **17**, 29–115 (1961).

*13. On certain classes of infinite groups. Uspehi Mat. Nauk. **13**, 189–192 (1958).

*14. Radicals and nil-elements in groups. Izv. Vysš Učebn. Zaved. Matematika **1**, 130–135 (1958).

*15. On a radical of automorphism groups of a group with maximal condition. Dokl. Akad. Nauk. SSSR **130**, 977–980 (1960) = Soviet Math. Dokl. **1**, 117–121 (1960).

*16. Radical groups whose radical has an ascending central series. Ural. Gos. Univ. Mat. Zap. **23**, 40–43 (1960).

*17. Algebraic elements in abstract groups. Ural. Gos. Univ. Mat. Zap. **23**, 44–48 (1960).

*18. Some properties of automorphisms of nilpotent groups. Dokl. Akad. Nauk. SSSR **137**, 1303–1306 (1961) = Soviet Math. Dokl. **2**, 471–474 (1961).

*19. Radicals in group pairs. Dokl. Akad. Nauk. SSSR **140**, 1019–1022 (1961) = Soviet Math. Dokl. **2**, 1312–1315 (1961).

*20. Locally stable groups of automorphisms. Sibirsk. Mat. Ž. **2**, 101–114 (1961).

*21. Normal subgroups which bound a group. Mat Sb. **53**, 343–352 (1961).

*22. Radicals connected with group representations. Dokl. Akad. Nauk. SSSR **144**, 52–55 (1962) = Soviet Math. Dokl. **3**, 668–671 (1962).

*23. On certain radicals of automorphism groups. Uspehi Mat. Nauk. **17**, 165–171 (1962).

24. Äußere lokale Eigenschaften von Automorphismengruppen. Arch. Math. (Basel) **13**, 401–407 (1962).

*25. Infinite-dimensional linear groups. Dokl. Akad. Nauk. SSSR **153**, 42–45 (1963) = Soviet Math. Dokl. **4**, 1617–1620 (1963).

*26. Generalized stable and generalized nilpotent groups of automorphisms. Sibirsk. Mat. Ž. **4**, 1389–1403 (1963).

*27. Some remarks on group pairs. Ural. Gos. Univ. Mat. Zap. **4**, 63–69 (1963).

*28. Groups of automorphisms of algebraic systems, Moscow: Izdat. Nauk. 1966.

*29. Semi-simple groups. Sibirsk. Mat. Ž. **9**, 623–631 (1968) = Siberian Math. J. **9**, 468–473 (1968).

*30. On a semi-group of radical classes. Sibirsk. Mat. Ž. **10**, 1091–1108 (1969) = Siberian Math. J. **10**, 805–817 (1969).

Plotkin, B. I., Kemhadze, Š. S.

*1. A scheme for constructing radicals in groups. Sibirsk. Mat. Ž. **6**, 1197–1201 (1965).

Plotkin, B. I., Vilyacer, V. G.

*1. On the theory of locally stable groups of automorphisms. Dokl. Akad. Nauk. SSSR **134**, 529–532 (1960) = Soviet Math. Dokl. **1**, 1108–1111 (1960).

Polovickiĭ, Ya. D.
*1. Layer-extremal groups. Dokl. Akad. Nauk. SSSR **134**, 533–535 (1960) = Soviet Math. Dokl. **1**, 1112–1113 (1960).
*2. On locally extremal groups and groups with the π-minimal condition. Dokl. Akad. Nauk. SSSR **138**, 1022–1024 (1961) = Soviet Math. Dokl. **2**, 780–782 (1961).
*3. Layer-extremal groups. Mat. Sb. **56**, 95–106 (1962).
*4. Locally extremal and layer-extremal groups. Mat. Sb. **58**, 685–694 (1962).
*5. Groups with the π-minimal condition on subgroups. Sibirsk. Mat. Ž. **3**, 582–590 (1962).
*6. A condition for abelian groups. Perm. Gos. Univ. Učen. Zap. **22**, 41–42 (1962).
*7. Groups with extremal classes of conjugate elements. Sibirsk. Mat. Ž. **5**, 891–895 (1964).
*8. Periodic automorphism groups of extremal groups. Perm. Gos. Univ. Učen. Zap. **131**, 58–62 (1966).

Poss, S.
1. A residual property of free groups. Comm. Pure Appl. Math. **23**, 749–765 (1970).

Rae, A., Roseblade, J. E.
1. Automorphism towers of extremal groups. Math. Z. **117**, 70–75 (1970).

Ree, R.
1. On ordered finitely generated soluble groups. Trans. Roy. Soc. Canada Sect. III (3) **48**, 39–42 (1954).

Reidemeister, K.
1. Einführung in die kombinatorische Topologie, Braunschweig: Vieweg 1952.

Remak, R.
1. Über minimale invariante Untergruppen in der Theorie der endlichen Gruppen. J. reine angew. Math. **162**, 1–16 (1930).

Remeslennikov, V. N.
*1. Conjugate subgroups in nilpotent groups. Algebra i Logika **6**, 61–76 (1967).
*2. Finite approximability of metabelian groups. Algebra i Logika **7**, 106–113 (1968) = Algebra and Logic **7**, 268–272 (1968).
*3. Representations of finitely generated metabelian groups by matrices. Algebra i Logika **8**, 72–75 (1969) = Algebra and Logic **8**, 39–40 (1969).
*4. Conjugacy in polycyclic groups. Algebra i Logika **8**, 712–725 (1969) = Algebra and Logic **8**, 404–411 (1969).

Rhemtulla, A. H.
1. A minimality property of polycyclic groups. J. London Math. Soc. **42**, 456–462 (1967).
2. A problem of bounded expressibility in free products. Proc. Cambridge Philos. Soc. **64**, 573–584 (1968).
3. A property of groups with no central factors. Canad. Math. Bull. **12**, 467–470 (1969).
4. Commutators of certain finitely generated soluble groups. Canad. J. Math. **21**, 1160–1164 (1969).

Ridley, J. N.
1. The free centre-by-metabelian group of rank two. Proc. London Math. Soc. (3) **20**, 321–347 (1970).

Riles, J. B.

1. The near Frattini subgroups of infinite groups. J. Algebra **12**, 155–171 (1969).

Rips, I. A.

*1. Two propositions about Baer groups. Dokl. Akad. Nauk. SSSR **186**, 264–267 (1969) = Soviet Math. Dokl. **10**, 589–592 (1969).

Robertson, E. F.

1. On certain subgroups of $GL(R)$. J. Algebra **15**, 293–300 (1970).

Robinson, D. J. S.

1. Groups in which normality is a transitive relation. Proc. Cambridge Philos. Soc. **60**, 21–38 (1964).
2. Joins of subnormal subgroups. Illinois J. Math. **9**, 144–168 (1965).
3. On finitely generated soluble groups. Proc. London Math. Soc. (3) **15**, 508–516 (1965).
4. On the theory of subnormal subgroups. Math. Z. **89**, 30–51 (1965).
5. Wreath products and indices of subnormality. Proc. London Math. Soc. (3) **17**, 257–270 (1967).
6. On soluble minimax groups. Math. Z. **101**, 13–40 (1967).
7. Residual properties of some classes ot infinite soluble groups. Proc. London Math. Soc. (3) **18**, 495–520 (1968).
8. A note on finite groups in which normality is transitive. Proc. Amer. Math. Soc. **19**, 933–937 (1968).
9. Finiteness conditions for subnormal and ascendant abelian subgroups. J. Algebra **10**, 333–359 (1968).
10. A property of the lower central series of a group. Math. Z. **107**, 225–231 (1968).
11. Infinite soluble and nilpotent groups. London: Queen Mary College Mathematics Notes (1968).
12. A note on groups of finite rank. Compositio Math. **31**, 240–246 (1969).
13. Groups which are minimal with respect to normality being intransitive. Pacific J. Math. **31**, 777–785 (1969).
14. On the theory of groups with extremal layers. J. Algebra **14**, 182–193 (1970).
15. A theorem on finitely generated hyperabelian groups. Invent. Math. **10**, 38–43 (1970).
16. Intersections of primary powers of a group. Math. Z. **124**, 119–132 (1972).

Romalis, G. M., Sesekin, N. F.

*1. Metahamiltonian groups. Ural. Gos. Univ. Mat. Zap. **5**, 101–106 (1966).

Romanovskiĭ, N. S.

*1. On the residual finiteness of free products with respect to occurrence. Izv. Akad. Nauk. SSSR Ser. Mat. **33**, 1324–1329 (1969) = Math. USSR-Izv. **3**, 1245–1249 (1969).

Rosati, L. A.

1. Sui gruppi ogni sottogruppo ciclico dei quali e caratteristico. Boll. Un. Mat. Ital. (3) **11**, 544–552 (1956).
2. Sui gruppi a fattoriali abeliani. Matematiche (Catania) **13**, 138–147 (1958).

Roseblade, J. E.

1. On certain classes of locally soluble groups. Proc. Cambridge Philos. Soc. **58**, 185–195 (1962).
2. The automorphism group of McLain's characteristically simple group. Math. Z. **82**, 267–282 (1963).

3. On certain subnormal coalition classes. J. Algebra **1**, 132–138 (1964).
4. The permutability of orthogonal subnormal subgroups. Math. Z. **90**, 365–372 (1965).
5. A note on subnormal coalition classes. Math. Z. **90**, 373–375 (1965).
6. On groups in which every subgroup is subnormal. J. Algebra **2**, 402–412 (1965).
7. A note on disjoint subnormal subgroups. Bull. London Math. Soc. **1**, 65–69 (1969).
8. The derived series of a join of subnormal subgroups. Math. Z. **117**, 57–69 (1970).

Roseblade, J. E., Stonehewer, S. E.
1. Subjunctive and locally coalescent classes of groups. J. Algebra **8**, 423–435 (1968).

Rosenberg, A.
1. The structure of the infinite general linear group. Ann. of Math. (2) **68**, 278–294 (1958).

Rosenlicht, M.
1. On a result of Baer. Proc. Amer. Math. Soc. **13**, 99–101 (1962).

Rotman, J. J.
1. The theory of groups, an introduction. Boston: Allyn & Bacon 1965.

Rozenfel'd, H. M.
*1. Some classes of infinite groups with prescribed indices. Perm. Gos. Univ. Učen. Zap. **22**, 58–64 (1962).
*2. Locally normal groups with a condition on the indices of invariant subgroups. Perm. Gos. Univ. Učen. Zap. **103**, 73–76 (1963).

Šain, B. M.
*1. On the Birkhoff-Kogalovskiĭ theorem. Uspehi Mat. Nauk. **20**, 173–174 (1965).

Sanov, I. N.
*1. On systems of relations in periodic groups with prime power periods. Izv. Akad. Nauk. SSSR Ser. Mat. **15**, 477–502 (1951).

Saşiada, E.
1. Construction of directly indecomposable abelian groups of a power higher than that of the continuum. Bull. Acad. Polon. Sci. Sér. Sci. Math. Astronom. Phys. **5**, 701–703 (1957), **7**, 23–26 (1959).

Schenkman, E.
1. A generalization of the central elements of a group. Pacific J. Math. **3**, 501–504 (1953).
2. Two theorems on finitely generated groups. Proc. Amer. Math. Soc. **5**, 497–498 (1954).
3. On the structure of the automorphism group of a nilpotent group. Portugal. Math. **13**, 129–135 (1954).
4. A splitting theorem and the principal ideal theorem for some infinitely generated groups. Proc. Amer. Math. Soc. **7**, 870–873 (1956).
5. The similarity between the properties of ideals in commutative rings and the properties of normal subgroups of groups. Proc. Amer. Math. Soc. **9**, 375–381 (1958).
6. On the norm of a group, Illinois J. Math. **4**, 150–152 (1960).
7. Group theory, Princeton: Van Nostrand 1965.
8. A replacement theorem for nilpotent groups with maximum condition. Illinois J. Math. **12**, 307–311 (1968).

Schenkman, E.
9. The general product of two finitely generated abelian groups. Proc. Amer. Math. Soc. **21**, 202—204 (1969).
10. Some criteria for nilpotency in groups and Lie algebras. Proc. Amer. Math. Soc. **21**, 714—718 (1969).

Schiefelbusch, L.
1. The Trofimov number of some infinite groups with finiteness conditions. Arch. Math. (Basel) **18**, 122—127 (1967).

Schlette, A.
1. Artinian, almost abelian groups and their groups of automorphisms. Pacific J. Math. **29**, 403—425 (1969).

Schmidt, O. J. (Šmidt, O. Yu.)
*1. Abstract theory of groups. Kiev 1916, 2nd. ed. Moscow 1933.
*2. On groups every proper subgroup of which is special. Mat. Sb. (O. S.) **31**, 366—372 (1924).
3. Über unendliche Gruppen mit endlicher Kette. Math. Z. **29**, 34—41 (1928).
*4. On infinite special groups. Mat. Sb. **8**, 363—375 (1940).
*5. Infinite soluble groups. Mat. Sb. **17**, 145—162 (1945).
*6. Selected works, Moscow: Matematika 1959.
7. Die locale Endlichkeit einer Klasse unendlicher periodischer Gruppen. Math. Forschungsberichte XX, Berlin: VEB Deutscher Verlag der Wissenschaften 1963, pp. 79—81.

Schoenwaelder, U.
1. Centralizers of abelian, normal subgroups of hypercyclic groups. Pacific J. Math. **31**, 197—208 (1969).

Schreier, J., Ulam, S.
1. Sur le groupe des permutations de la suite des nombres naturels. C. R. Acad. Sci. Paris **197**, 737—738 (1933).

Schreier, O.
1. Die Untergruppen der freien Gruppen. Abh. Math. Sem. Univ. Hamburg **5**, 161—183 (1927).

Schur, I.
1. Neuer Beweis eines Satzes über endliche Gruppen. Sitzber. Akad. Wiss. Berlin 1013—1019 (1902).
2. Über die Darstellungen der endlichen Gruppen durch gebrochene lineare Substitutionen. J. reine angew. Math. **127**, 20—50 (1904).
3. Untersuchungen über die Darstellung der endlichen Gruppen durch gebrochene lineare Substitutionen. J. reine angew. Math. **132**, 85—137 (1907).
4. Über Gruppen periodischer linearer Substitutionen. Sitzber. Preuss. Akad. Wiss. 619—627 (1911).

Scott, W. R.
1. Groups and cardinal numbers. Amer. J. Math. **74**, 187—197 (1952).
2. On infinite groups. Pacific J. Math. **5**, 589—598 (1955).
3. On a result of B. H. Neumann. Math. Z. **66**, 240 (1956).
4. Solvable factorizable groups II. Illinois J. Math. **4**, 652—655 (1960).
5. Group theory, Englewood Cliffs: Prentice Hall 1964.

Scott, W. R., Sonneborn, L. M.
1. Supersolvable wreath products. Math. Z. **92**, 154—163 (1966).

Scruton, T.
1. Bounds for the class of nilpotent wreath products. Proc. Cambridge Philos. Soc. **62**, 165—169 (1966).

Ščukin, K. K.
*1. Prime groups. Kišinev. Gos. Univ. Učen. Zap. **39**, 209–218 (1959).
*2. Prime semi-maximal normal subgroups. Kišinev. Gos. Univ. Učen. Zap. **54**, 3–7 (1960).
*3. Locally soluble radicals of some classes of groups. Kišinev. Gos. Univ. Učen. Zap. **54**, 95–100 (1960).
*4. An RI^*-soluble radical of groups. Dokl. Akad. Nauk. SSSR **132**, 541–544 (1960) = Soviet Math. Dokl. **1**, 615–618 (1960).
*5. The RI^*-soluble radical of groups. Mat. Sb. **52**, 1021–1031 (1960).
*6. On the theory of radicals in groups. Dokl. Akad. Nauk. SSSR **142**, 1047–1049 (1962) = Soviet Math. Dokl. **3**, 260–263 (1962).
*7. Remarks on prime groups. Kišinev. Gos. Univ. Učen. Zap. **50**, 9–11 (1962).
*8. On the theory of radicals in groups. Sibirsk. Mat. Ž. **3**, 932–942 (1962).

Seksenbaev, K.
*1. On the theory of polycyclic groups. Algebra i Logika **4**, 79–83 (1965).
*2. On the anticentre of bundles of groups. Izv. Akad. Nauk. Kazah. Ser. Fiz.-Mat., 20–24 (1966).
*3. The finite approximability of a finite extension of a nilpotent group with respect to conjugacy. Algebra i Logika **6**, 29–31 (1967).

Selberg, A.
1. On discontinuous groups in higher dimensional symmetric spaces. Internat. Colloq. Function Theory, Tata Institute of Fundamental Research, Bombay, pp. 147–164 (1960).

Sesekin, N. F.
*1. On the theory of special groups without torsion. Dokl. Akad. Nauk. SSSR **70**, 185–188 (1950).
*2. On the theory of torsion-free locally nilpotent groups. Dokl. Akad. Nauk. SSSR **84**, 225–228 (1952).
*3. On locally nilpotent groups without torsion. Mat. Sb. **32**, 407–442 (1953).
*4. On the classification of finitely generated torsion-free metabelian groups. Ural. Gos. Univ. Mat. Zap. **19**, 27–41 (1956).
*5. On the product of complete abelian permutable groups. Ural. Gos. Univ. Mat. Zap. **3**, 45–49 (1962).
*6. Products of finitely-connected abelian groups. Sibirsk. Mat. Ž. **9**, 1427–1430 (1968) = Siberian Math. J. **9**, 1070–1072 (1968).

Sesekin, N. F., Širokovskaya, O. S.
*1. A class of bigraded groups. Mat. Sb. **46**, 133–142 (1958).

Sesekin, N. F., Starostin, A. I.
*1. On a class of periodic groups. Uspehi Mat. Nauk. **9**, 225–228 (1954).

Ševcov, G. S.
*1. Semi-direct products of locally cyclic groups. Perm. Gos. Univ. Učen. Zap. **131**, 76–80 (1966).
*2. Locally generalized Schmidt groups. Sibirsk. Mat. Ž. **9**, 1431–1434 (1968) = Siberian Math. J. **9**, 1073–1075 (1968).
*3. Semi-direct products of rational groups. Izv. Vysš. Učebn. Zaved. Matematika **1**, 184–201 (1958).

Shepperd, J. A. H., Wiegold, J.
1. Transitive permutation groups and groups with finite derived groups. Math. Z. **81**, 279–285 (1963).

Shores, T. S.

1. A note on products of normal subgroups. Canad. Math. Bull. **12**, 21–24 (1969).
2. On $\bar{Z}$-hypercentral normal subgroups. Arch. Math. (Basel) **21**, 344–348 (1970).

Simon, H.

1. Noethersche Gruppen mit endlicher Hyperzentrumsfaktorgruppe. Illinois J. Math. **8**, 231–240 (1964).
2. Noethersche Gruppen mit nilpotenten Normalteilern von endlichem Index. Illinois J. Math. **8**, 241–247 (1964).
3. Eine Verallgemeinerung eines Satzes von Mal'cev und Baer. Arch. Math. (Basel), **17**, 289–291 (1966).
4. Artinian and noetherian hypercentral groups. Proc. Amer. Math. Soc. **17**, 1407–1409 (1966).
5. Noethersche Gruppen mit überauflösbar eingebetteten Normalteilern von endlichem Index. Arch. Math. (Basel) **18**, 8–14 (1967).
6. Extensions of torsion free groups by torsion groups. Proc. Amer. Math. Soc. **23**, 433–438 (1969).

Širšov, A. I.

*1. On certain near-Engel groups. Algebra i Logika **2**, 5–18 (1963).

Skolem, T.

1. On the existence of a multiplicative basis for an arbitrary algebraic field. Norske Vid. Selsk. Forh. (Trondheim) **20**, 4–7 (1947).

Slotterbeck, O. A.

1. Finite factor coverings of groups. J. Algebra **17**, 67–73 (1971).

Šmel'kin, A. L.

*1. Nilpotent products of torsion-free nilpotent groups. Sibirsk. Mat. Ž. **3**, 625–640 (1962).
*2. A property of semi-simple classes of groups. Sibirsk. Mat. Ž. **3**, 950–951 (1962).
*3. Free polynilpotent groups. Dokl. Akad. Nauk. SSSR **151**, 73–75 (1963) = Soviet Math. Dokl. **4**, 950–953 (1963).
*4. Free polynilpotent groups. Izv. Akad. Nauk. SSSR Ser. Mat. **28**, 91–122 (1964).
*5. On soluble products of groups. Sibirsk. Mat. Ž. **6**, 212–220 (1965).
*6. Wreath products and varieties of groups. Izv. Akad. Nauk. SSSR Ser. Mat. **29**, 149–170 (1965).
*7. On complete nilpotent groups. Algebra i Logika **6**, 111–114 (1967).
*8. Polycyclic groups. Sibirsk. Mat. Ž. **9**, 234–235 (1968) = Siberian Math. J. **9**, 178 (1968).
*9. On free products of groups. Mat. Sb. **79**, 616–620 (1969) = Math. USSR-Sb. **8**, 593–597 (1969).

Smirnov, D. M.

*1. On the theory of locally nilpotent groups. Dokl. Akad. Nauk. SSSR **76**, 643–646 (1951).
*2. On automorphisms of soluble groups. Dokl. Akad. Nauk. SSSR **84**, 891–894 (1952).
*3. Infra-invariant subgroups. Ivanov. Gos. Ped. Inst. Učen. Zap. **4**, 92–96 (1953).
*4. On groups of automorphisms of soluble groups. Mat. Sb. **32**, 365–384 (1953).
*5. Groups with upper central series. Mat. Sb. **33**, 471–484 (1953).

*6. Some classes of infinite groups. Ivanov. Gos. Ped. Inst. Učen. Zap. **5**, 57–60 (1954).
*7. Two classes of soluble groups of finite rank. Ivanov. Gos. Ped. Inst. Učen. Zap. **18**, 67–74 (1958).
*8. On the theory of finitely approximable groups. Ukrain. Mat. Ž. **15**, 453–457 (1963).
*9. On generalized soluble groups and their group rings. Mat. Sb. **67**, 366–383 (1965).

Smirnov, D. M., Taĭclin, M. A.
*1. On finitely approximable abelian multi-operator groups. Uspehi Mat. Nauk. **17**, 137–142 (1962).

Sokolov, V. G.
*1. Frattini subgroups. Algebra i Logika **7**, 85–93 (1968) = Algebra and Logic **7**, 122–126 (1968).

Solian, A.
1. Über die *n*-Vollständigkeit in Gruppen. Rev. Roumaine Math. Pures Appl. **1**, 5–22 (1956).

Specht, W.
1. Gruppentheorie, Berlin/Göttingen/Heidelberg: Springer 1956.
2. Beiträge zur Gruppentheorie I. Lokalendliche Gruppen. Math. Nachr. **18**, 39–56 (1958).

Stanley, T. E.
1. Generalizations of the classes of nilpotent and hypercentral groups. Math. Z. **118**, 180–190 (1970).

Starostin, A. I.
*1. Periodic locally soluble groups with a complete partition. Izv. Vysš. Učebn. Zaved. Matematika **15**, 168–177 (1960).

Starostin, A. I., Eidenov, M. I.
*1. Sylow bases of infinite groups. Sibirsk. Mat. Ž. **3**, 273–279 (1962).

Stebe, P. F.
1. On free products of isomorphic free groups with a single finitely generated amalgamated subgroup. J. Algebra **11**, 359–362 (1969).
2. A residual property of certain groups. Proc. Amer. Math. Soc. **26**, 37–42 (1970).

Steenrod, N. E.
1. Universal homology groups. Amer. J. Math. **58**, 661–701 (1936).

Stewart, A. G. R.
1. On the class of certain nilpotent groups. Proc. Roy. Soc. Ser. A. **292**, 374–379 (1966).
2. On centre-extended-by-metabelian groups. Math. Ann. **185**, 285–302 (1970).

Stewart, I. N.
1. The minimal condition for subideals of Lie algebras. Math. Z. **111**, 301–310 (1969).
2. An algebraic treatment of Mal'cev's theorems concerning nilpotent Lie groups and their Lie algebras. Compositio Math. **22**, 289–312 (1970).
3. Infinite-dimensional Lie algebras in the spirit of infinite group theory. Compositio Math. **22**, 313–331 (1970).

Stonehewer, S. E.
1. Abnormal subgroups of a class of periodic locally soluble groups. Proc. London Math. Soc. (3) **14**, 520–536 (1964).
2. Locally soluble *FC*-groups. Arch. Math. (Basel) **16**, 158–177 (1965).

3. Formations and a class of locally soluble groups. Proc. Cambridge Philos. Soc. **62**, 613–635 (1966).
4. Some finiteness conditions in locally soluble groups. J. London Math. Soc. **43**, 689–694 (1968).
5. Group algebras of some torsion-free groups. J. Algebra **13**, 143–147 (1969).
6. The join of finitely many subnormal subgroups. Bull. London Math. Soc. **2**, 77–82 (1970).

Stöppler, S.
1. *k*-Auflösbare Gruppen. Rend. Sem. Mat. Univ. Padova **43**, 141–175 (1970).

Stroud, P. W.
1. On a property of verbal and marginal subgroups. Proc. Cambridge Philos. Soc. **61**, 41–48 (1965).

Struik, R. R.
1. Notes on a paper by Sanov. Proc. Amer. Math. Soc. **8**, 638–641 (1957).
2. Notes on a paper by Sanov II. Proc. Amer. Math. Soc. **12**, 758–763 (1961).

Strunkov, S. P.
*1. Subgroups of periodic groups. Dokl. Akad. Nauk. SSSR **170**, 279–281 (1966) = Soviet Math. Dokl. **7**, 1201–1203 (1966).
*2. On the problem of O. Yu. Schmidt. Sibirsk. Mat. Ž. **7**, 476–479 (1966) = Siberian Math. J. **7**, 388–390 (1966).
*3. Normalizers and abelian subgroups of certain classes of groups. Izv. Akad. Nauk. SSSR Ser. Mat. **31**, 657–670 (1967) = Math. USSR-Izv. **1**, 639–650 (1967).

Sulinski, A., Anderson, R., Divinsky, N.
1. Lower radical properties for associative and alternative rings. J. London Math. Soc. **41**, 417–424 (1966).

Šunkov, V. P.
*1. On groups decomposable into a uniform product of their *p*-subgroups. Dokl. Akad. Nauk. SSSR **154**, 542–544 (1964) = Soviet Math. Dokl. **5**, 147–149 (1964).
*2. On the theory of generalized soluble groups. Dokl. Akad. Nauk. SSSR **160** 1279–1282 (1965) = Soviet Math. Dokl. **6**, 326–329 (1965).
*3. Abstract characterization of a simple projective group of type *PGL* (2, *K*) field *K* of characteristic $r \neq 0$ or 2. Dokl. Akad. Nauk. SSSR **163**, over a 837–840 (1965) = Soviet Math. Dokl. **6**, 1043–1047 (1965).
*4. A contribution to the theory of locally finite groups. Dokl. Akad. Nauk. SSSR **168**, 1272–1274 (1966) = Soviet Math. Dokl. **7**, 841–843 (1966).
*5. On the theory of periodic groups. Dokl. Akad. Nauk. SSSR **175**, 1236–1237 (1967) = Soviet Math. Dokl. **8**, 1011–1012 (1967).
*6. A locally finite group with extremal Sylow *p*-subgroups for some prime *p*. Sibirsk. Mat. Ž. **8**, 213–229 (1967) = Siberian Math. J. **8**, 161–171 (1967).
*7. A certain generalization of Frobenius' theorem to periodic groups. Algebra i Logika **6**, 113–124 (1967).
*8. Minimality problem for subgroups in locally finite groups. Dokl. Akad. Nauk. SSSR **181**, 294–295 (1968) = Soviet Math. Dokl. **9**, 840–842 (1968).
*9. On a periodic group with an almost regular involution. Algebra i Logika **7**, 113–121 (1968) = Algebra and Logic **7**, 66–69 (1968).
*10. On the minimality problem for locally finite groups. Algebra i Logika **9**, 220–248 (1970) = Algebra and Logic **9**, 137–151 (1970).

Suprunenko, D. A.
*1. Soluble and nilpotent linear groups. Minsk (1958) = Amer. Math. Soc. Translations of Mathematical Monographs **9** (1963).
*2. Locally nilpotent subgroups of infinite symmetric groups. Dokl. Akad. Nauk. SSSR **167**, 302—304 (1967) = Soviet Math. Dokl. **7**, 392—394 (1966).
Suprunenko, D. A., Garaščuk, M. S.
*1. Linear nil groups. Dokl. Akad. Nauk. BSSR **4**, 407—408 (1960)
*2. Linear groups with Engel's condition. Dokl. Akad. Nauk. BSSR **6**, 277—280 (1962).
*3. Linear groups with a category. Dokl. Akad. Nauk. BSSR **6**, 411—414 (1962).
Suprunenko, D. A., Medvedeva, R. P.
*1. Irreducible nilpotent linear groups over the field of rational numbers. Dokl. Akad. Nauk. BSSR **2**, 363—364 (1958).
Suprunenko, D. A., Tyškevič, R. I.
*1. Reducible locally nilpotent linear groups. Izv. Akad. Nauk. SSSR Ser. Mat. **24**, 787—806 (1960).
Swan, R. G.
1. Representations of polycyclic groups. Proc. Amer. Math. Soc. **18**, 573—574 (1967).
Szász, F.
1. Groups in which every non-trivial power is cyclic. Magyar Tud. Akad. Fiz. Oszt. Közl. **5**, 491—492 (1955).
2. Über Gruppen, deren sämtliche nicht-triviale Potenzen zyklische Untergruppen sind. Acta Math. Sci. (Szeged) **17**, 83—84 (1956).
3. A characterization of the cyclic troups. Rev. Roumaine Math. Pures Appl. **1**, 13—16 (1956).
4. On cyclic groups. Fund. Math. **43**, 238—240 (1956).
5. Reduktion eines gruppentheoretischen Problems von O. J. Schmidt. Bull. Acad. Polon. Sci. Ser. Sci. Math. Astronom. Phys. **7**, 369—372 (1959).
6. Bemerkung zu meiner Arbeit "Über Gruppen, deren sämtliche nicht-triviale Potenzen zyklische Untergruppen sind". Acta Sci. Math. (Szeged) **23**, 64—66 (1962).
Szekeres, G.
1. Metabelian groups with two generators. Proc. Internat. Conf. Theory of Groups Canberra **1965**, pp. 323—346 (1967).
Szép, J., Itô, N.
1. Über die Factorisation von Gruppen. Acta Sci. Math. (Szeged) **16**, 229—231 (1955).
Taiclin, M. A.
*1. Finite approximability of Ω-groups. Sibirsk. Mat. Ž. **3**, 95—102 (1962).
Takahasi, M.
1. Note on chain conditions in free groups. Osaka Math. J. **3**, 221—225 (1951).
Thompson, J. G.
1. Finite groups with fixed-point-free automorphisms of prime order. Proc. Nat. Acad. Sci. USA **45**, 578—581 (1959).
Timošenko, E. I.
*1. Conjugacy in free metabelian groups. Algebra i Logika **6**, 89—94 (1967).
Tobin, S.
1. On a theorem of Baer and Higman. Canad. J. Math. **8**, 263—270 (1956).
2. Simple bounds for Burnside p-groups. Proc. Amer. Math. Soc. **11**, 704—706 (1960).

Tokarenko, A. I.

*1. On linear groups over rings. Sibirsk. Mat. Ž. **9**, 951—959 (1968) = Siberian Math. J. **9**, 708—713 (1968).

*2. Radicals and stability in linear groups over commutative rings. Sibirsk. Mat. Ž. **9**, 165—176 (1968) = Siberian Math. J. **9**, 125—132 (1968).

*3. Locally nilpotent groups over commutative rings. Proc. Riga Sem. Algebra Latv. Gos. Univ. Riga, pp. 267—279 (1969).

*4. A remark on finitely generated linear groups. Proc. Riga Sem. Algebra Latv. Gos. Univ. Riga, pp. 280—281 (1969).

Tomkinson, M. J.

1. Local conjugacy classes. Math. Z. **108**, 202—212 (1969).

2. Local conjugacy classes II. Arch. Math. (Basel) **20**, 567—571 (1969).

3. Formations of locally soluble *FC*-groups. Proc. London Math. Soc. (3) **19**, 675—708 (1969).

4. F-injectors of locally soluble *FC*-groups. Glasgow Math. J. **10**, 130—136 (1969).

Tovbin, A. V.

*1. On the existence of a centre in infinite groups. Dokl. Akad. Nauk. SSSR **31**, 198 (1941).

Trofimov, P. I.

*1. A study of the influence of the greatest common divisor of the orders of the classes of conjugate non-normal Sylow subgroups of a finite group and its properties. Sibirsk. Mat. Ž. **4**, 236—239 (1963).

Turner-Smith, R. F.

1. Marginal subgroup properties for outer commutator words. Proc. London Math. Soc. (3) **14**, 321—341 (1964).

2. Finiteness conditions for verbal subgroups. J. London Math. Soc. **41**, 166—176 (1966).

Tyškevič, R. F.

*1. A generalization of some theorems on finite groups. Dokl. Akad. Nauk. BSSR **6**, 471—474 (1962).

Ušakov, V. I.

*1. *P*-radical groups. Izv. Vysš. Učebn. Zaved. Matematika **19**, 233—238 (1960).

Valiev, M. K.

*1. A theorem of G. Higman. Algebra i Logika **7**, 9—22 (1968) = Algebra and Logic **7**, 135—143 (1968).

Vil'jams, N. N.

*1. Metadedekind and metahamiltonian groups. Mat. Sb. **76**, 634—654 (1968) = Math. USSR-Sb. **5**, 599—616 (1968).

Vilyacer, V. G.

*1. On the theory of locally nilpotent groups. Uspehi Mat. Nauk. **13**, 163—168 (1958).

*2. Stable groups of automorphisms. Dokl. Akad. Nauk. SSSR **131**, 728—730 (1960) = Soviet Math. Dokl. **1**, 313—315 (1960).

*3. Some examples of groups of automorphisms. Dokl. Akad. Nauk. SSSR **139**, 1283—1288 (1961) = Soviet Math. Dokl. **2**, 1069—1072 (1961).

Vol'vačev, R. T.

*1. Periodic nilpotent linear groups over the field of rational numbers. Mat. Sb. **70**, 368—379 (1966) = Amer. Math. Soc. Translations (2) **66**, 99—110 (1968).

de Vries, H., de Miranda, A. B.

1. Groups with a small number of automorphisms. Math. Z. **68**, 450–464 (1958).

Wang, H.-C.

1. Discrete subgroups of soluble Lie groups I. Ann. of Math. (2) **64**, 1–19 (1956).

Wedderburn, J. H. M.

1. Note on algebras. Ann. of Math. (2) **38**, 854–856 (1937).

Wehrfritz, B. A. F.

1. Conjugacy theorems in locally finite groups. J. London Math. Soc. **42**, 679–686 (1967).
2. Conjugacy theorems in locally finite groups II. Arch. Math. (Basel) **18**, 470–473 (1967).
3. A note on periodic locally soluble groups. Arch. Math. (Basel) **18**, 577–579 (1967).
4. Frattini subgroups in finitely generated linear groups. J. London Math. Soc. **43**, 619–622 (1968).
5. Locally nilpotent linear groups. J. London Math. Soc. **43**, 667–674 (1968).
6. Transfer theorems for periodic linear groups. Proc. London Math. Soc. (3) **19**, 143–163 (1969).
7. A residual property of free metabelian groups. Arch. Math. (Basel) **20**, 248–250 (1969).
8. Sylow subgroups of locally finite groups with Min-p. J. London Math. Soc. (2) **1**, 421–427 (1969).
9. Infinite linear groups, London: Queen Mary College Mathematics Notes 1969.
10. Groups of automorphisms of soluble groups. Proc. London. Math. (3) **20**, 101–122 (1970).
11. Supersoluble and locally supersoluble linear groups. J. Algebra **17**, 41–58 (1971).
12. On locally finite groups with Min-p. J. London Math. Soc. (2) **3**, 121–128 (1971).

Weidig, I.

1. Gruppen mit abgeschwächter Normalteilertransitivität. Rend. Sem. Mat. Univ. Padova **36**, 185–215 (1966).

Weston, K. W.

1. ZA-groups which satisfy the mth Engel condition. Illinois J. Math. **8**, 458–472 (1964).
2. The lower central series of metabelian Engel groups. Notices Amer. Math. Soc. **12**, 81 (1965).
3. ZA-groups with a cyclic terminal upper central factor. Portugal. Math. **26**, 185–191 (1967).

Whittemore, A.

1. On the Frattini subgroup. Trans. Amer. Math. Soc. **141**, 323–333 (1969).

Wiegold, J.

1. Groups with boundedly finite classes of conjugate elements. Proc. Roy. Soc. London Ser. A **238**, 389–401 (1957).
2. Nilpotent products of groups with amalgamations. Publ. Math. Debrecen **6**, 131–168 (1959).
3. On direct factors in groups. J. London Math. Soc. **35**, 310–320 (1960).
4. Embedding group amalgams in wreath products. Math. Z. **80**, 148–153 (1962).

5. Adjunction of elements to nilpotent groups. J. London Math. Soc. **38**, 17–26 (1963).
6. Multiplicators and groups with finite central factor-groups. Math. Z. **89**, 345–347 (1965).
7. Periodic series. Mat. Časopis Sloven. Akad. Vied. **18**, 81–82 (1968).
8. Commutator subgroups of finite p-groups. J. Austral. Math. Soc. **104**, 480, (1969).

Wielandt, H.
1. Eine Kennzeichnung der direkten Produkte von p-Gruppen. Math. Z. **41**, 281–282 (1936).
2. Eine Verallgemeinerung der invarianten Untergruppen. Math. Z. **45**, 209–244 (1939).
3. Vertauschbare nachinvariante Untergruppen. Abh. Math. Sem. Univ. Hamburg **21**, 55–62 (1957).
4. Über den Normalisator der subnormalen Untergruppen. Math. Z. **69**, 463–465 (1958).
5. Unendliche Permutationsgruppen. Lecture Notes. Tübingen 1959/1960.

Wilson, J. S.
1. Some properties of groups inherited by normal subgroups of finite index. Math. Z. **114**, 19–21 (1970).
2. Groups satisfying the maximal condition for normal subgroups. Math. Z. **118**, 107–114 (1970).
3. Groups with every proper quotient finite. Proc. Cambridge Philos. Soc. **69**, 373–392 (1971).

Winter, D. J.
1. Representations of locally finite groups. Bull. Amer. Math. Soc. **74**, 145–148 (1968).

Witt, E.
1. Treue Darstellung Liescher Ringe. J. reine angew. Math. **177**, 152–160 (1937).
2. Über die Kommutatorgruppe kompakter Gruppen. Rend. Mat. e Appl. (5) **14**, 125–129 (1954).

Wittmann, E.
1. Paare von Gruppenklassen mit der Holomorph-Eigenschaft. Arch. Math. (Basel) **21**, 23–30 (1970).

Wolf, J. A.
1. Growth of finitely generated soluble groups and curvature of Riemannian manifolds. J. Differential Geometry **2**, 421–446 (1968).

Woodhouse, D.
1. Some finiteness properties of certain Baer-soluble groups. Quart. J. Math. Oxford Ser. **19**, 57–65 (1968).

Wos, L. T.
1. On commutative prime power subgroups of the norm. Illinois J. Math. **2**, 271–284 (1958).

Wright, C. R. B.
1. On the nilpotency class of a group of exponent four. Pacific J. Math. **11**, 387–394 (1968).

Zacher, G.
1. Caratterizzazione dei t-gruppi finiti risolubili. Ricerche Mat. **1**, 287–294 (1952).

2. I gruppi risolubili finiti in cui i sottogruppi di composizione coincidono con i sottogruppi quasi-normali. Atti Accad. Naz. Lincei Rend. Cl. Sci. Fis. Mat. Natur. (8) **37**, 150—154 (1964).

Zaičev, D. I.

*1. Groups with the condition of commutator minimality. Sibirsk. Mat. Ž. **6**, 1014—1020 (1965).

*2. Stably soluble and stably nilpotent groups. Dokl. Akad. Nauk. SSSR **176**, 509—511 (1967) = Soviet Math. Dokl. **8**, 1122—1125 (1967).

*3. Stably nilpotent groups. Mat. Zametki **2**, 337—346 (1967).

*4. Groups satisfying a weak minimum condition. Dokl. Akad. Nauk. SSSR **178**, 780—782 (1968) = Soviet Math. Dokl. **9**, 194—197 (1968).

*5. On solvable groups of finite rank. Dokl. Akad. Nauk. SSSR **181**, 13—14 (1968) = Soviet Math. Dokl. **9**, 783—785 (1968).

*6. Groups which satisfy a weak minimality condition. Ukrain. Mat. Ž. **20**, 472—482 (1968) = Ukrainian Math. J. **20** (1968).

*7. The existence of stably nilpotent subgroups in locally nilpotent groups. Mat. Zametki **4**, 361—368 (1968) = Math. Notes **4**, 708—712 (1968).

*8. On groups which satisfy a weak minimality condition. Mat. Sb. **78**, 323—331 (1969) = Math. USSR-Sb. **7**, 315—322 (1969).

*9. Stably soluble groups. Izv. Akad. Nauk. SSSR Ser. Mat. 33, 765—780 (1969) = Math. USSR-Izv. **3**, 723—736 (1969).

Zalesskiĭ, A. E.

*1. Locally finite groups with the minimal condition for centralizers. Izv. Akad. Nauk. BSSR Ser. Fiz.-Mat. **3**, 127—129 (1965).

*2. Sylow p-subgroups of the general linear group over a division ring. Izv. Akad. Nauk. SSSR Ser. Mat. **31**, 1149—1158 (1967) = Math. USSR-Izv. **1**, 1099—1108 (1967).

*3. The structure of certain classes of matrix groups over a division ring. Sibirsk. Mat. Ž. **8**, 1284—1298 (1967) = Siberian Math. J. **8**, 978—988 (1967).

*4. On the structure of finitely generated modules over the group algebra of a polycyclic group. Dokl. Akad. Nauk. BSSR **14**, 977—980 (1970).

Zappa, G.

1. Sui gruppi di Hirsch supersolubili I. Rend. Sem. Mat. Univ. Padova **12**, 1—11 (1941).

2. Sui gruppi di Hirsch supersolubili II. Rend. Sem. Mat. Univ. Padova **12**, 62—80 (1941).

3. Sugli automorfismi uniformi nei gruppi di Hirsch. Ricerche Mat. **7**, 3—13 (1958).

Zassenhaus, H.

1. Beweis eines Satzes über diskrete Gruppen. Abh. Math. Sem. Univ. Hamburg **12**, 289—312 (1938).

2. Über Lie'sche Ringe mit Primzahlcharakteristik. Abh. Math. Sem. Univ. Hamburg **13**, 1—100 (1940).

3. The theory of groups. 2nd ed. New York: Chelsea 1958.

4. On linear Noetherian groups. J. Number Theory **1**, 70—89 (1969).

Zel'manzon, M. E.

*1. Groups in which all subgroups are cyclic. Uspehi Mat. Nauk. **16**, 109—113 (1961).

*2. On groups all of whose subgroups are metabelian. Ukrain. Mat. Ž. **17**, 67—73 (1965).

Zorn, M.

1. Nilpotency of finite groups. Bull. Amer. Math. Soc. **42**, 485—486 (1936).

2. On a theorem of Engel. Bull. Amer. Math. Soc. **43**, 401—404 (1937).

Author Index

Subject Index

DG 721/2/72

Ergebnisse der Mathematik und ihrer Grenzgebiete

1. Bachmann: Transfinite Zahlen.
2. Miranda: Partial Differential Equations of Elliptic Type.
4. Samuel: Méthodes d'Algèbre Abstraite en Géométrie Algébrique.
5. Dieudonné: La Géométrie dés Groupes Classiques.
7. Ostmann: Additive Zahlentheorie. 1. Teil: Allgemeine Untersuchungen.
8. Wittich: Neuere Untersuchungen über eindeutige analytische Funktionen.
11. Ostmann: Additive Zahlentheorie. 2. Teil: Spezielle Zahlenmengen.
13. Segre: Some Properties of Differentiable Varieties and Transformations.
15. Zeller/Beckmann: Theorie der Limitierungsverfahren.
16. Cesari: Asymptotic Behavior and Stability Problems in Ordinary Differential Equations.
17. Severi: Il teorema di Riemann-Roch per curve-superficie e varietà questioni collegate.
18. Jenkins: Univalent Functions and Conformal Mapping.
19. Boas/Buck: Polynomial Expansions of Analytic Functions.
20. Bruck: A Survey of Binary Systems.
23. Bergmann: Integral Operators in the Theory of Linear Partial Differential Equations.
25. Sikorski: Boolean Algebras.
26. Künzi: Quasikonforme Abbildungen.
27. Schatten: Norm Ideals of Completely Continuous Operators.
28. Noshiro: Cluster Sets.
29. Jacobs: Neuere Methoden und Ergebnisse der Ergodentheorie.
30. Beckenbach/Bellmann: Inequalities.
31. Wolfowitz: Coding Theorems of Information Theory.
32. Constantinescu/Cornea: Ideale Ränder Riemannscher Flächen.
33. Conner/Floyd: Differentiable Periodic Maps.
34. Mumford: Geometric Invariant Theory.
35. Gabriel/Zisman: Calculus of Fractions and Homotopy Theory.
36. Putnam: Commutation Properties of Hilbert Space Operators and Related Topics.
37. Neumann: Varieties of Groups.
38. Boas: Integrability Theorems for Trigonometric Transforms.
39. Sz.-Nagy: Spektraldarstellung linearer Transformationen des Hilbertschen Raumes.
40. Seligman: Modular Lie Algebras.
41. Deuring: Algebren.
42. Schütte: Vollständige Systeme modaler und intuitionistischer Logik.
43. Smullyan: First-Order Logic.
44. Dembowski: Finite Geometries.

45. Linnik: Ergodic Properties of Algebraic Fields.
46. Krull: Idealtheorie.
47. Nachbin: Topology on Spaces of Holomorphic Mappings.
48. A. Ionescu Tulcea/C. Ionescu Tulcea: Topics in the Theory of Lifting.
49. Hayes/Pauc: Derivation and Martingales.
50. Kahane: Séries de Fourier Absolument Convergents.
51. Behnke/Thullen: Theorie der Funktionen mehrerer komplexer Veränderlichen.
52. Wilf: Finite Sections of Some Classical Inequalities.
53. Ramis: Sous-ensembles analytiques d'une variété banachique complexe.
54. Busemann: Recent Synthetic Differential Geometry.
55. Walter: Differential and Integral Inequalities.
56. Monna: Analyse non-archimédienne.
57. Alfsen: Compact Convex Sets and Boundary Integrals.
58. Greco/Salmon: Topics in $\mathfrak{m}$-Adic Topologies.
59. López de Medrano: Involutions on Manifolds.
60. Sakai: C*-Algebras and W*-Algebras.
61. Zariski: Algebraic Surfaces.
62. Robinson: Finiteness Conditions and Generalized Soluble Groups, Part 1.
63. Robinson: Finiteness Conditions and Generalized Soluble Groups, Part 2.
64. Hakim: Topos annelés et schémas relatifs.
65. Browder: Surgery on Simply-Connected Manifolds.
66. Pietsch: Nuclear Locally Convex Spaces.
67. Dellacherie: Capacités et processus stochastiques.
68. Raghunathan: Discrete Subgroups of Lie Groups.